Geothermal Energy in the Western United States

Sheldon L. Bierman
David F. Stover
Paul A. Nelson
William J. Lamont

The Praeger Special Studies program—utilizing the most modern and efficient book production techniques and a selective worldwide distribution network—makes available to the academic, government, and business communities significant, timely research in U.S. and international economic, social, and political development.

Geothermal Energy in the Western United States

Innovation versus Monopoly

Praeger Publishers New York London

Library of Congress Cataloging in Publication Data

Main entry under title:

Geothermal energy in the Western United States.

Includes bibliographical references and index.
1. Geothermal engineering. 2. Geothermal resources—The West. I. Bierman, Sheldon L.
TJ280.7.G42 621.4 77-16078

PRAEGER SPECIAL STUDIES
200 Park Avenue, New York, N.Y., 10017, U.S.A.

Published in the United States of America in 1978
by Praeger Publishers,
A Division of Holt, Rinehart and Winston, CBS, Inc.

89 038 987654321

Printed in the United States of America

ACKNOWLEDGMENTS

The authors wish to express their appreciation of information and assistance furnished by Messrs. Henry Banta, John Galloway, Phineas Indritz, Joseph Lerner, and Walter Measday. Valuable assistance was also rendered by Whitfield A. Russell. Our thanks also go to Tania R. Barton, who extended the hospitality of the Postal Rate Commission Library. Many employees of government agencies and industrial firms supplied us with needed information; their contributions are noted in the text. This book is based in part on work done under National Science Foundation Grant No. APR 75-18321-A02, to the law firm of Pearce and Brand. That study was edited by C. Jack Pearce and reflected assistance by Wallace E. Brand; they have not, however, reviewed the present volume. Any opinions, findings, conclusions, or recommendations expressed in this volume are those of the authors and do not necessarily reflect the views of NSF. The preparation of the present volume was assisted by a grant from the Energy Research and Development Administration, and again, the views herein expressed are those of the authors and not necessarily those of ERDA. We wish to acknowledge the helpful assistance of the Bureau of Industrial Development at Michigan Technological University.

While one of the authors is a member of the staff of the Postal Rate Commission, it should be emphasized that the views expressed in this volume are his own and do not purport to reflect the opinions, or carry the endorsement, of the Postal Rate Commission, any of its members, or any other member of its staff.

While acknowledging the valuable assistance received from many quarters, the authors of course accept responsibility for any errors that may remain.

CONTENTS

Chapter		Page

LIST OF TABLES AND FIGURES

LIST OF ABBREVIATIONS AND ACRONYMS

A.B.A.	American Bar Association
a.c.	alternating current
AEC	Atomic Energy Commission
AIF	Atomic Industrial Forum
AMPS	Associated Mountain Power Systems
APPA	Arizona Power Pooling Association
APS	Arizona Public Service Company
API	American Petroleum Institute
ARB	Air Resources Board
ARCO	Atlantic Richfield Company
BLM	Bureau of Land Management
BP	British Petroleum
BPA	Bonneville Power Administration
Btu	British thermal unit
CALPP	California Power Pool
CAP	Central Arizona Project
CD	certificate of deposit
CEP	Canadian Entitlement Power
CEQA	California Environmental Quality Act
CFP	Compagnie Française des Petroles
CPI	P. T. Caltex Pacific Indonesia
CRSP	Colorado River Storage Project
CVP	Central Valley Project
d.c.	direct current
DISC	Domestic International Sales Corporations
DWR	Department of Water Resources
EA	Environmental Assessment
EIR	Environmental Impact Report
EIS	Environmental Impact Statement

ERDA	Energy Research and Development Administration
FAC	Federal Advisory Committee
FEA	Federal Energy Administration
FPC	Federal Power Commission
FTC	Federal Trade Commission
GAO	General Accounting Office
GKI	Geothermal Kinetics, Inc.
GRI	Geothermal Resources International
HAS	Humble (now Exxon), ARCO, and SOCAL
HIRI	Hawaiian Independent Refining, Inc.
IBLA	Interior Board of Land Appeals
ICC	Interstate Commerce Commission
ICPA	Intermountain Consumers Power Association
I.D.	Irrigation District
IDC	intangible drilling cost
IEA	International Energy Agreement
IEC	Intercontinental Energy Corporation
INEL	Idaho National Engineering Laboratory
IRC	Internal Revenue Code
KGRA	Known Geothermal Resources Area
kv	kilovolt
kw	kilowatt
kwh	kilowatt-hour
LADWP	Los Angeles Department of Water and Power
LBL	Lawrence Berkeley Laboratory
LL&E	Louisiana Land & Exploration Co.
LMU	Logical Mining Unit
LNG	Liquefied Natural Gas
MDC	Maximum Dependable Capacity
MER	Maximum Efficient Recovery Rate

MPR	Maximum Production Rate
mw	megawatt
NCPA	Northern California Power Agency
NEPA	National Environmental Policy Act
NMPP	New Mexico Power Pool
NPC	National Petroleum Council
NPC	Nevada Power Company
NRC	Nuclear Regulatory Commission
NSF	National Science Foundation
NWPP	Northwest Power Pool
OCS	Outer Continental Shelf
OPEC	Organization of Petroleum Exporting Countries
PAD	Petroleum Administration for Defense
PDC	Project's Dependable Capacity Rating
PEC	Pacific Energy Corporation
PGE	Portland General Electric
PG&E	Pacific Gas and Electric
PGT	Pacific Gas Transmission Company
PLATO	Pennzoil Louisiana and Texas, Inc.
PNCA	Pacific Northwest Coordinating Agreement
POGO	Pennzoil Offshore Gas Operators, Inc.
PP&L	Pacific Power and Light Company
PSCC	Public Service Company of Colorado
PUC	Public Utility Commission
PUD	Public Utility District
R&D	Research and Development
REA	Rural Electrification Administration
REC	Rural Electric Cooperative
RMPP	Rocky Mountain Power Pool
SCE	Southern California Edison

SCMG	Southern California Municipal Group
SDAPUD	Shasta Dam Area Public Utility District
SDG&E	San Diego Gas and Electric Company
SEC	Securities and Exchange Commission
SMUD	Sacramento Municipal Utility District
SNG	Synthetic Natural Gas
SOCAL	Standard Oil of California
SOHIO	Standard Oil of Ohio
SRP	Salt River Project
TAPS	Trans-Alaska Pipeline System
TG&E	Tucson Gas & Electric Company
THUMS	Texaco, Humble (Exxon), Union, Mobil, and Shell
TMPC	The Montana Power Company
TVA	Tennessee Valley Authority
UP&L	Utah Power and Light Company
USBM	United States Bureau of Mines
USBR	United States Bureau of Reclamation
USGS	United States Geological Survey
WSCC	Western Systems Coordinating Council
WPPSS	Washington Public Power Supply System

Geothermal Energy in the Western United States

1
INTRODUCTION

This is a book about the energy industry in the American West, but it is not a simple survey of that industry. Instead, it has a special focus: the relationship between the western energy industry and an emerging (or potentially emerging) energy source. Geothermal energy, it appears, could have a significant place in the western energy industry. To understand the nature and scope of its potential role, a study has been made of the structure of the electrical raw fuels sectors of that industry as they are today. This book details what has been found about the way the western energy industry functions today, in the infancy of geothermal development, as well as about the way further exploitation of geothermal resources is governed by the roles and relationships of raw fuel companies, electric utilities, and the government in the West.

Geothermal energy provides a particularly striking example of how industry structure affects the development of a new energy source. Because geothermal development draws upon, and potentially affects, both the petroleum and the electric power industries, it is heavily influenced by the structure of both those industries and the relationships between them.

Geothermal energy is derived from steam and hot water found within the earth.* The product of a series of wells in a geothermal field is brought together to provide energy to drive electric generators. While some is used for heating, the bulk of geothermal energy must be utilized in generating electricity. Geothermal resources occur in only a limited number of sites, and the steam or hot water

---

*In the future, molten rock may take its place as a geothermal source.

produced from them cannot be transported for long distances, except in the form of electricity. Thus the primary geothermal energy must be converted to electric power and carried to consuming markets over high voltage transmission lines. Access to these transmission grids is thus a necessity for any entity hoping to develop geothermal resources.

The development of geothermal resources entails the expenditure of considerable time and effort. A reserve of adequate size and quality must be found and sufficient wells drilled to support an electric generating unit before product revenues can be realized. Environmental protection requirements must be met. Frequently land tracts must be leased from the federal government. These matters add time and expense. The resulting cash flow lag is a significant barrier to entry into the field.

The overall capability of geothermal resources to support cost-competitive generation is quite limited, compared to coal or uranium. Still, geothermal generation could be developed in sufficient quantity to displace substantial absolute volumes of other fuels in specific areas. (Over 15 years, with vigorous development, about as much capacity might be developed as is used by the largest single electric utility in California today.)

Since efficient geothermal energy units will be smaller than those required for coal or nuclear projects, availability of these small units promises special benefit to the smaller utility systems. Smaller systems--frequently owned by governmental units or cooperatives--often require additional capacity, and seek to control it independently. Unless they use a newer technology such as geothermal energy to build their own units, they must obtain new base-load power from coal or nuclear units. New nuclear or coal capacity for small systems will require joint enterprises or arrangements with large privately owned utilities which compete, sometimes acrimoniously, with these small systems.

Actual plans for geothermal generation appear to run far behind the potential for development of such generation on a basis cost-competitive with other types of generation in such applications. Instead, the infant geothermal industry's activity is primarily directed at locating and leasing tracts of land that appear to have promising resources. Exploratory drilling has been done in a limited number of cases. Electricity production occurs, in the United States, only in the Geysers area of California.

Firms involved in this industry range from small enterprises acquiring tracts and selling technological services such as geophysical testing, to medium-sized and large industrial firms that often have extensive holdings in other fuels. Certain utilities also participate in various aspects of the industry. Smaller firms wishing to participate in exploratory activity typically enter into joint enterprises with these utilities and industrial firms to obtain financing. Otherwise, these small firms are generally limited to speculation in land.

This situation leaves the development pace to the utilities and petroleum firms. The petroleum firms seek a price for their geothermal product tied to the price of crude oil. The utilities do not wish to embark upon a course of innovative power generation without some fuel cost benefits and so they seek a lower purchase price not tied to oil prices. Furthermore, larger utilities, confronted with the problems of allocating their limited financial and professional resources to meet rising demands on huge systems, seem to be inclined to focus on units of 500 to 1,000 megawatts (mw) rather than on development of a geothermal resource with 50 to 150mw units.

This study of the geothermal industry has been carried out along the lines of an orthodox antitrust inquiry. First, an understanding was obtained of the steps involved in developing geothermal resources and the magnitude of the resource involved. Then the participant firms and government agencies and their roles were identified, and with this background relevant sales and supply markets, as well. These markets were examined to see if conditions in them were competitive and development was progressing, and what the factors were that make for more or for less competition and development. Where factors retarding competition and development were found, an attempt was made to identify their causes and find available remedies.

CHAPTER

2

GEOTHERMAL ENERGY AND ITS USE

Geothermal energy is derived from the internal heat of the earth. This energy is usable at those points where it occurs in concentration in an extractable form (such as hot water or steam) within sufficient proximity to the surface.* Unfortunately, it cannot be transported in the form of heat over distances to market; rather, it must be converted to electricity and shipped by wire if it is to be distributed for use.†

Geothermal resources vary in their temperature and other physical characteristics such as steam purity. The hotter a stream of geothermal fluid is, and the lower its content of brines and substances other than water, the easier it is to use for power generation.

As Figure 2.1 shows, useful geothermal fluids (hydrothermal convection systems) become sources as their quality improves. Only a few geothermal systems have sufficient quality to permit ready development with current technology. Access to these resources is essential for utility generation. This is particularly the

---

*Concentrations of extractable heat at depths less than three kilometers (about 9,900 feet) may be considered economic resources. With the present technology and economic limits imposed by the opportunity cost of substitute fuels, drilling can be done profitably only to a depth of between 9,000 and 10,000 feet. At greater well depths, drilling costs increase at an exponential rate, and problems of the cooling of rising fluid in wells commencing to flow increase.

†Localized use of geothermal energy for space and water heating occurs in the West. However, the geographic coincidence of such demand and supply is limited.

case for enterprises requiring low risks, such as small firms with strained capital budgets. The developers of these resources will likely have financial and technical advantages over later entrants.

FIGURE 2.1

Temperature Distribution of Geothermal Spring Systems

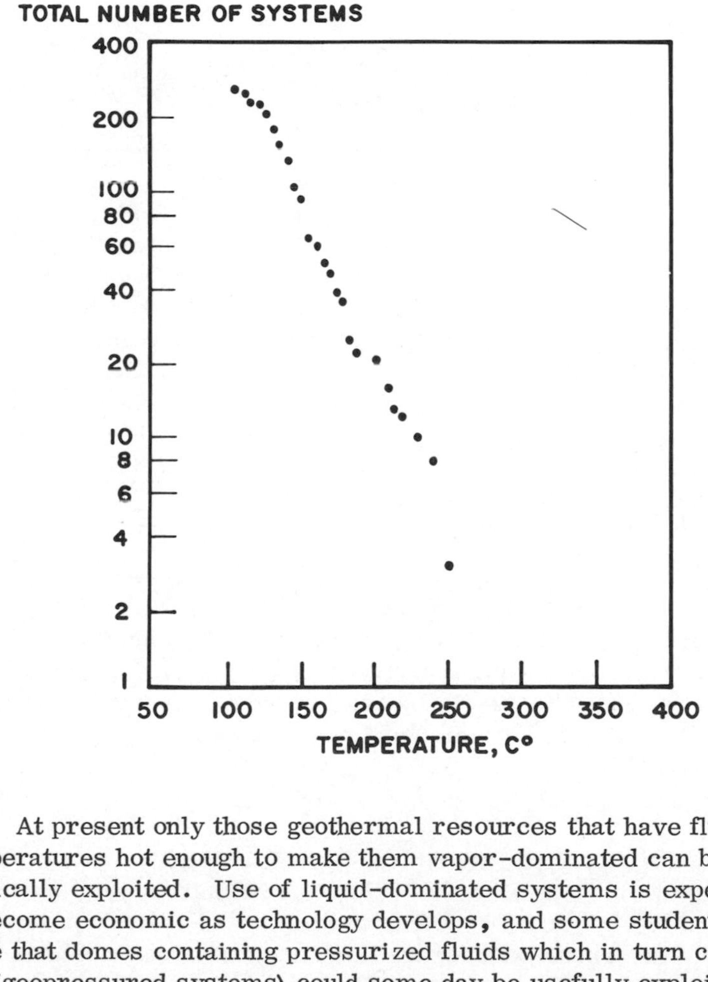

At present only those geothermal resources that have fluid temperatures hot enough to make them vapor-dominated can be economically exploited. Use of liquid-dominated systems is expected to become economic as technology develops, and some students believe that domes containing pressurized fluids which in turn contain gas (geopressured systems) could some day be usefully exploited.

The general magnitude of the vapor-dominated resource is equivalent in heat energy to about 17 billion barrels of oil. Geothermal resources are located in the western states, particularly in Idaho, Nevada, Oregon, and California. California is the predominant location for very hot water systems (temperature over

150°C). Large lower temperature resources occur in Idaho and Oregon. The large resource found in Wyoming at Yellowstone National Park is not available for commercial extraction.

Building on data from the U.S. Geological Survey (USGS), Dr. LaMori developed the following table estimating the total number of hydrothermal convection systems in the United States if the distribution of such resources follows that reported to date:

TABLE 2.1

Hydrothermal Convection Systems

	USGS	If 2/3 Discovered	If 1/2 Discovered	If 20 Percent Discovered
Electric Utilization				
T 210°C	14	18	36	90
T 150°C-210°C	52	82	164	410
Nonelectric utilization				
T 100°C-150°C	196	325	650	1,625
T 50°C-100°C	--	1,400	2,800	7,000
Number at 15°C	--	5,000	10,000	25,000

As regards the size of the geothermal resource, LaMori concludes that there are not likely to exist thousands of systems capable of electric production using presently available technology, that is, greater than 150°C; that it appears that only a few hundred systems exist; and that a conservative estimate of about 100 systems cannot be ruled out.

THE POTENTIAL CONTRIBUTION OF GEOTHERMAL ENERGY

To determine whether the large accumulations of geothermal energy in the western states are an economically extractable energy reserve requires looking at the cost of harnessing this energy relative to the costs of other types of electric power generation. Costs for employing geothermally driven power generation can be confidently estimated only with respect to vapor-dominated and very hot water systems. Figure 2.2 presents such an estimate for the costs

FIGURE 2.2

Geothermal Supply

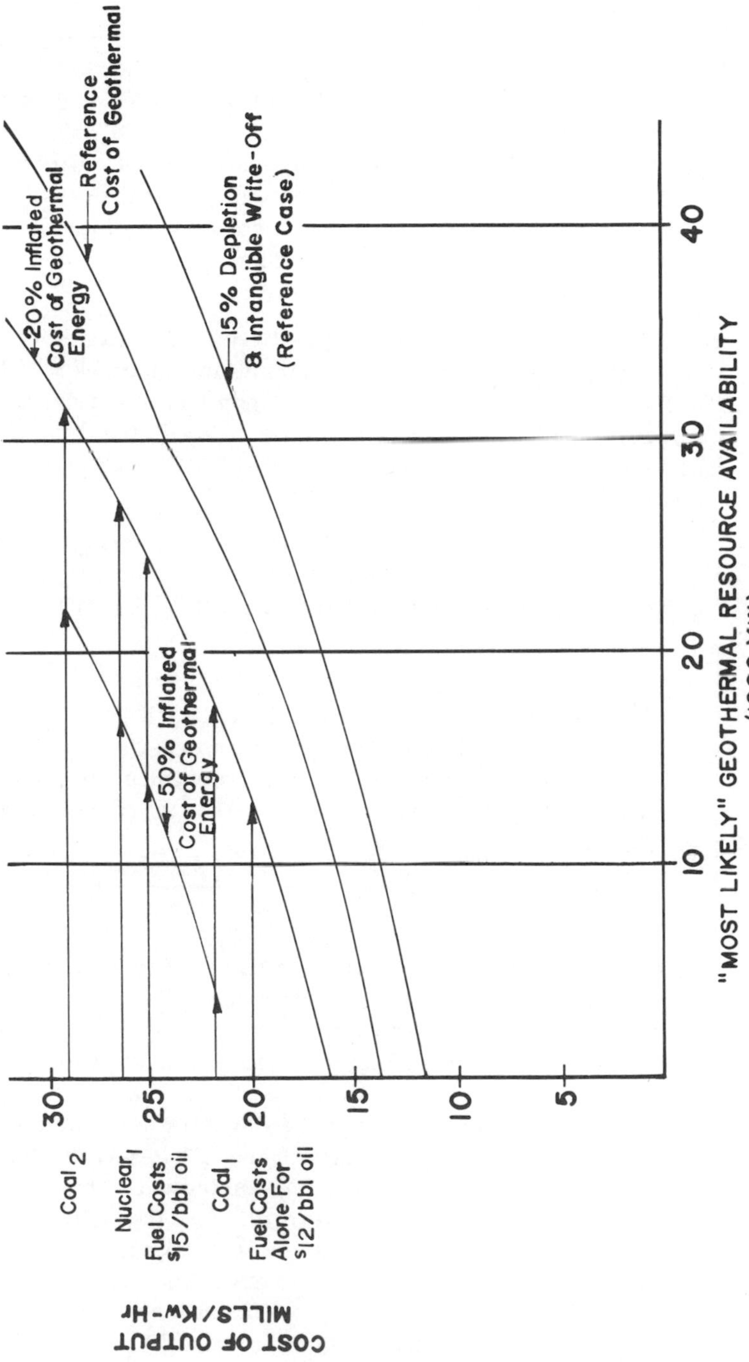

1. $Coal_1$

The busbar price (22.5 mills) of coal-fired generation with capital costs of coal-fired generation of 15.2 mills/kwh ($600 per kw), operating and maintenance costs of 3.3 mills and fuel costs of 4 mills. No provision is made for transmission. Thus, this cost is comparable to busbar costs of geothermal generation at sites distant from load centers.

2. $Coal_2$

The busbar price of coal-fired generation having capital costs of 15.2 mills, operation and maintenance costs of 3.3 mills, fuel costs of 4 mills, and transmission related costs of 7.5 mills--29.5 mills total. This roughly represents the competitive price of coal from east of California.

3. $Nuclear_1$

The 27.3 mill busbar price represents a current unit cost of $650 per kilowatt (68 percent load factor), operating and maintenance costs of 3.3 mills, and fuel costs of 5.8 mills.

4. Residual Fuel Oil

Fuel oil costs for $15 and $12 per barrel residual fuel oil are included (at 10,000 Btu per kwh heat rates) to show the cost at which geothermal could be supplied as energy sans capacity. Actual oil use costs should be raised several mills to reflect operating charges. This would show $12 per barrel oil at about the cost of coal.

of generation erected at one point in time. The supply curve reflects the variation in quality among geothermal sites by showing increasing costs as increasingly poorer sites are employed; the total volume of good quality sites is reflected in the total amount of generation, shown as mw of installed capacity, on the curve's horizontal axis.

The supply curve is derived from work done for the Energy Research and Development Administration (ERDA) at the Battelle Memorial Laboratory in Richland, Washington. This work, while disputed in part by some observers who feel that it is projecting geothermal costs that are too low, is quite detailed. An attempt has been made here to make a very conservative correction for underestimates of cost--an endemic problem with new technologies--by adding additional cost curves showing costs 50 percent higher than those estimated.

This curve assumes higher costs not offset by increased tax deductions or tax benefits. When this curve's costs for geothermal energy are compared with the costs of nuclear generation or coal-fired generation, or fuel oil generation, it appears that a substantial amount of geothermal generation could be developed at less cost than competing energy sources.

The costs shown in Figure 2.2 are as of December 1974 and should be increased by 20 percent to account for inflation through December 1976.

Costs shown reflect the following items:

Use of the sum-of-the-years digits depreciation over 30 years for power plants and use of depreciation periods for classes of wells. These vary with well type and typically are 10 years. Firms have adopted a practice of using 10 to 30 years for their cost estimates.

Provision for well replacement. *

Well capacity at 20 percent above the design capacity required to run the turbines.

Fixed charge rates of 17 percent for plant costs; and rates of return of approximately 20 percent for field investment, based on a 15 percent return on equity and an 8 percent average cost of debt applied to a capital structure that is 42 percent debt (as per 20 large oil companies).

Well spacing estimated at 20 acres.

Financial and tax assumptions are:

	Reservoir (percent)	Powerplant (percent)
Capitalization		
Debt	42	59
Equity	58	41
Bond interest rate	8	8
Return on equity	15	12
Federal income tax rate	48	48
Property tax rate	2.5	2.5
Royalty payment	10	--
Revenue tax rate	--	4

---

*Use of accelerated depreciation by a utility on its tax return may not be paralleled in its rate cases where straight-line depreciation may be used. The tax saving resulting from accelerated depreciation is not required to be passed through to consumers in some states. This normalization, in which tax expenses are charged rate payers as if straight-line depreciation had been used on utility tax returns, provides interest-free capital.

In this study it was assumed that the plant would be down for maintenance and repair 20 percent of the year. Thus, the plant availability is 80 percent.

Investment tax credits are not reflected. These are estimated at 2 to 3 percent of costs for a 10 percent credit. No contingency fund is included. The curves treat all units of capacity as being simultaneously installed.

In actual fact, tax costs may be reduced by changes in the Internal Revenue Code and by employing forms of business enterprise discussed hereinafter. The returns on investment required to bring forth capital will vary with the types of enterprises engaged in geothermal development and their alternative investment opportunities.

The time required for development of a series of wells in a geothermal field and a generating facility appears to be on the order of five to seven years. This span is comparable to those for competing fuels: It is about equal to the lead time for coal-fluid generation, and is several years less than the lead time for a nuclear unit. The risks of inflation may be expected to weigh more heavily on more capital-intensive nuclear and coal projects, while the earlier stage of geothermal technology may cause greater risks (and opportunities for cost cutting) to be attributed to that form of energy.

As indicated, a considerable amount of geothermal capacity (over 12,000 mw) would be developed at busbar cost below that for fuel alone (not counting any special charges for oil-burning equipment). This generation could be developed at dry steam and hot water sources of high temperature and lower salinity.

The uppermost curve reflects costs of geothermal development increased by 50 percent to account for inflation and cost overruns. At current prices for residual fuel oil on the West Coast ($15+ per barrel), the total cost of geothermal generation from many "dry steam" prospects could be recovered at less than the energy cost for oil.

The costs shown are those estimated as the actual costs of producing power from geothermal resources. They should not be confused with the price that would have to be charged at the generating plant outlet if geothermal energy were sold to utilities at prices based on the cost of alternative fuels, that is, based on something other than actual cost. Under that pricing mechanism, higher busbar costs would result and few geothermal units, in all likelihood, would be built.

This comparison indicates that 5,000 to 15,000 mw might be obtained from geothermal sources at a cost below that expected for nuclear or fossil-fueled base-load generating units. Additionally, development of geothermal energy would result in avoidance of other external costs associated with pollution and with the use of foreign-derived fuel that might be incurred with other fuels.

For the future, costs for nuclear units are expected to rise to around $1,200 per kw (33+ mills per kwh). Nuclear-unit busbar costs will exceed 40 mills. Nuclear fuel availability is a matter of concern, as are fuel costs which are rising rapidly.[1]

The delivered costs of electric energy include transmission expenses in addition to the plant busbar cost. At high load factors, transmission costs may be expected to be between $5 and $10 per kw per year (or 0.5 to 1 mill per kwh) for distances of a few hundred miles, using publicly owned transmission systems, and about 20 percent more on private transmission grids. The addition of one or two mills to the plant outlet cost to cover transmission and line losses would permit base-load geothermal energy to be transmitted 400-800+ miles at Bonneville Power Administration (BPA) transmission rates, and 320-600+ miles at rates 20 percent higher.

It can be seen that with the use of such wheeling rates over intrastate distances in California, the cost of geothermal power from dry steam or hot low-salinity sources is very competitive with coal or nuclear power, for those interested in units of 50-100 mw. Accordingly, the market for geothermal energy in California appears to be California utilities. In the other regions of the West, such as Utah, local systems would be principal users of locally generated geothermal energy.

Geothermal generating units must be located near either load centers or high voltage lines. Because even large geothermal fields may be expected to be developed and produced gradually, it is unlikely that very high voltage lines will be built for substantial distances to serve such development; the line capacity cost is too great. This factor constitutes a limitation of geothermal energy development.

Geothermal generation, provided at cost (including a reasonable return), would encounter cost competition from oil-, coal-, and nuclear-fueled generation at the points indicated in Figure 2.2 as intercepts of the geothermal cost curves and the costs shown respectively for generation of electricity from other energy sources. Thus, if geothermal generation has a cost of 20 to 50 percent over the December 1974 reference cost, between approximately 14,000 and 17,000 mw of geothermal capacity could be developed at a cost less than that for nuclear units. This is the equivalent of about 12 to 14 nuclear generating units.

It should of course be noted that the geothermal costs shown are estimates for one point in time. They do not reflect "learning curves" or temporary cost shifts. The costs of coal and nuclear generation are not represented as supply curves, but rather are single point estimates of busbar costs. It is unclear what the supply curve for nuclear power is, in view of cost of money and fueling uncertainties. Nuclear power plants are very capital-intensive, and their costs directly track construction lead times. Furthermore, they have been subject to siting problems on the West Coast, as well as severe cost inflation. The supply curve for coal in the range dealt with here could be expected to be rather flat in the West with its large strip-mineable resources. Several factors indicate that coal and nuclear costs picked for comparison are likely to be high relative to the estimated geothermal costs. Coal costs at 25 to 31 mills and nuclear plant costs of 35 mills per kwh permit the costs of high temperature, low salinity geothermal resources to escalate by 50 percent and still be attractive on a per mw basis for locations near load centers. While coal is clearly the price-competitive fuel at present, a break in cartel pricing could make residual fuel oil very attractive. Oil-fired capacity costs only 13-14 mills.*

One thousand mw of installed geothermal capacity operating at a 70 percent capacity factor could produce as much energy as 10.22 million barrels of residual fuel oil per year (6,132 million kwh). In 1974, West Coast utilities received 68,032,000 barrels of residual fuel oil, while mountain state utilities received 7,239,800 barrels.[2] Thus, California and mountain state oil consumption is seven to eight times the expectable equivalent fuel output from 1,000 mw of geothermal services.

GEOTHERMAL POWER PROJECT COSTS

Two different cycles may be used for geothermal power generation, depending on the heat energy per unit of volume in the geothermal fluid. Figure 2.3 illustrates schematically the cycle in which the fluid (or a portion segregated from the rest of the fluid) is a steam having sufficient heat energy and volume to drive a turbine directly. Where the geothermal fluid has less heat energy per unit of fluid, it must be used to vaporize a second (binary) fluid, having a lower boiling point. The vaporized binary fluid then drives the turbine. Figure 2.4 describes this binary process. Geothermal units are necessarily planned to run on a continuous basis because any cycling of operations requires a substantial amount of start-up time due to problems associated with cooling of wells. They are what are known as base-load units; their output is not varied to follow short-term demand changes.

---

*Oil fuel costs are now 20 to 25 mills per kwh, assuming a heat rate of 10,000 Btu (British thermal units) per kwh. Busbar costs of fuel can be computed by dividing fuel prices by the heat content (Btu) in a barrel of oil and then multiplying by the heat rate.

FIGURE 2.3

Flashed Binary Unit Scheme

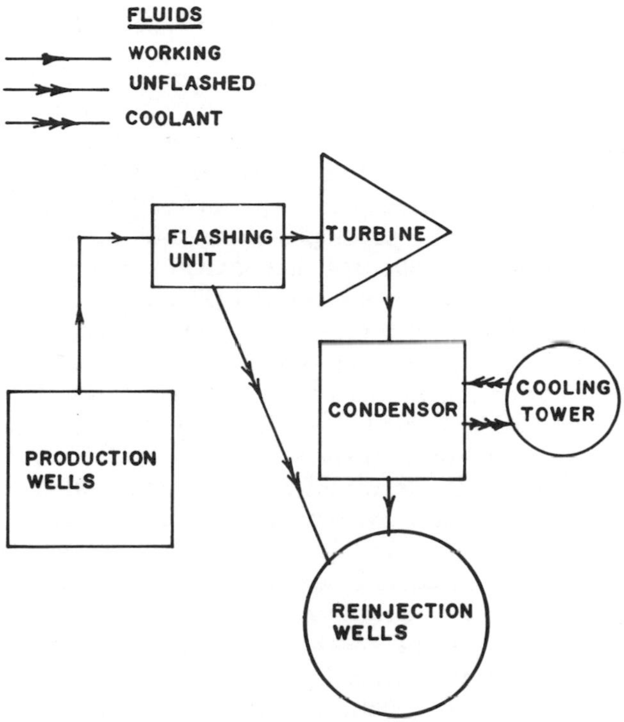

FIGURE 2.4

Binary Cycle for Geothermal Exploitation

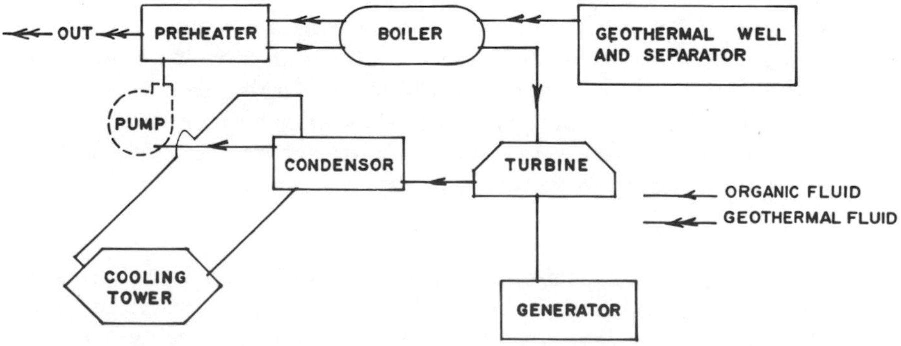

To the extent that alternate base-load power sources are not available when load growth projections indicate they are needed, geothermal units could have a special attractiveness, if sufficient geothermal capacity could be licensed and developed quickly enough and if prerequisite interconnection and coordination agreements were entered into.

Timing is a crucial element in the costs of geothermal development as it is in other capital-intensive energy projects. The possibility of delay of geothermal development lowers the attractiveness of geothermal energy for smaller utility systems that cannot readily absorb slippage of scheduled new units. In addition, the devastating results of poor timing in a development program are only worsened when losses and interest charges cannot be shared with the tax collector--as is the case for public power systems.*

If a geothermal field developed for a 55mw unit and containing 12 wells (costing $500,000 each) is delayed from production for one year, the cost to an energy enterprise that can earn 20 percent on its money would be $1.2 million--subject, of course, to partial recoupment through rising fuel prices.

The location of a geothermal resource involves the use of geophysical art and science to evaluate a range of prospects. This evaluation process for narrowing a list of prospects employs a series of tests to pick the prospects that appear to warrant drilling for geothermal development. Tracts warranting drilling, a matter as to which professional opinions may differ, must then be assembled. When drilling rights are secured around a potential site, exploratory holes are sunk. If their results appear promising, further evaluative drilling takes place. If the results are favorable a series of development wells with attendant fluid transmission piping are installed and tested. This complex of wells is coupled to a generating unit to convert the geothermal heat into transportable electrical energy.

The state of geophysical knowledge is such that some wells drilled may be expected not to produce. The risk of encountering such "cold" holes appears to be less than that incurred in wildcat exploration for petroleum. This lower risk apparently reflects the present early stage of geothermal exploration in which sites with surface explorations remain unexplored. As this easy layer of sites is tested, the successive layers of sites may be more risky to explore. If technology does not develop or perform as projected, risks may shift. A further risk is the marginal well; here the determination

---

*The most recent geothermal units in the Geysers have gone into operation far behind their original schedule. Delays in scheduled operation of new units are endemic in the power industry.

as to whether a well is or is not economically exploitable may well depend on uncertainties, or differing judgments, as to technology. Drillers seek to manage their risks either by exploring a number of prospects, in the case of larger firms, or by attempting to employ capital invested by others, a method favored by some smaller firms, or by use of both methods.

The capital costs of an exploration program rise sharply when the drilling phase is entered into, and again rise substantially when postexploratory development of a field begins.[3] During the successive stages of development, lags between cash outlays and receipts become more expensive to bear.

Another illustration of the costs of a complete exploration and development program is set forth in Figure 2.5 which shows planned expenditures for the city of Burbank's $16 million program to develop a 50mw geothermally driven generating unit.

FIGURE 2.5

Planned Expenditures for City of Burbank
50mw Geothermal Generating Unit

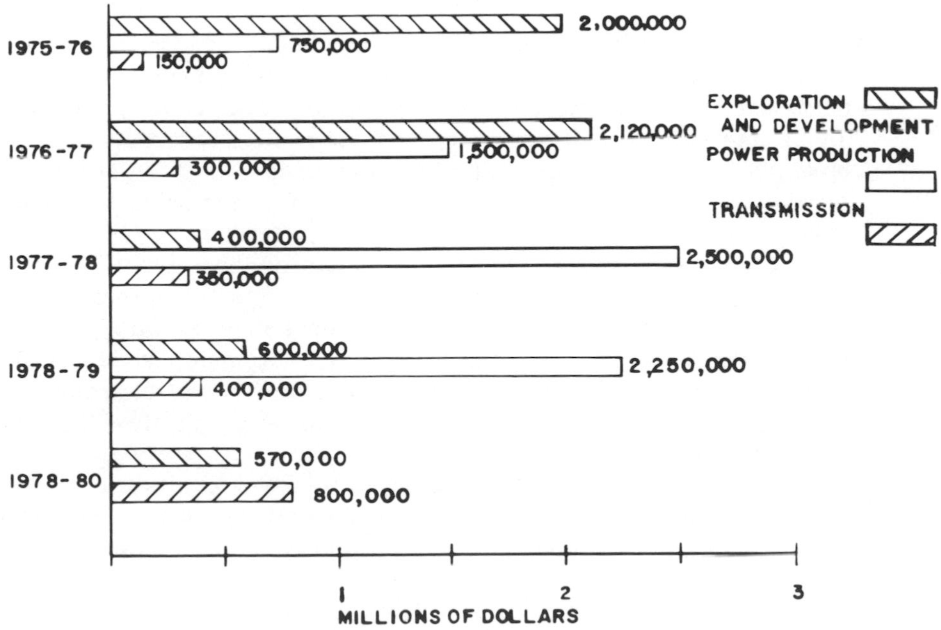

While actual drilling has been infrequent to date, enough data exist to indicate that the two or three successful wells needed to indicate the presence of a geothermal field are generally estimated to cost several million dollars. For a wildcat area, where the drilling success ratio is far below that of the Geysers, the Imperial Valley, and some other areas,* employees of a major company have stated that a $16 million drilling program would be required, while estimates by an independent firm were in the $10 to $15 million range. The drilling program cost forecast, which ranges from a maximum negative cashflow of $16 million down to $11 million, assumes a 25 percent rate of return; $10 million would be required at a 15 percent rate of return. Rates of return of 20 to 25 percent were stated by several petroleum company personnel to be realistic for geothermal. With 350°F water, a price of 30 mills would be required for a 15 to 20 percent return on investment; 20 mills required for a 15 to 20 percent return with water of 400°F.

Lease bonus amounts are determined by predicting program producing risk and then adding increments of bonus sums until the anticipated rate of return is reduced to the lowest level deemed acceptable. Projected costs are lower for smaller programs and those less oriented toward wildcatting or perhaps otherwise anticipating a higher success rate. Government personnel estimate costs ranging from $4.5 to $8 million to find and develop a field for electric generation.

Other geothermal enterprises have stated that lesser sums are required, asserting, for instance, that the drilling costs of smaller firms are substantially below those of larger enterprises. Several small enterprises seek to develop $2 to $3 million programs to drill up to six wells. They are looking for two or three successes which could confirm a field and serve as the basis for raising development funds.

Costs of drilling geothermal wells exceed those normal for shallow onshore oil and gas exploration. Wells in the Geysers run from $600,000 to $800,000, although one cost about $1.3 million. Use of directional drilling from drilling islands might add $200,000 to well costs.[4] These higher drilling costs may retard the rate of drilling, consequently limiting the rate at which geophysical theories are tested.

Moreover, data obtained during drilling are kept confidential. Limited drilling and the failure to disclose drilling logs are likely to result in slower development of field exploration techniques. Slower

---

*For instance, two of the three wells drilled by Magma and Union Oil at Brady Hot Springs, Nevada, appear to be commercial.

development of these techniques increases and maintains levels of exploratory risk.[5] Drilling information may be shared by parties to a joint venture or to a plan for "unitized" development.

Developers can, however, shift part of the cost of geothermal development to taxpayers. The tax devices used to accomplish this, as shown below, encourage the use of joint enterprises and appear to be more favorable to established enterprises than to would-be entrants.

With the possible exception of outlays for dry steam wells, investment is capitalized and subsequently amortized for income tax purposes. (Expenses incurred for research and for leases subsequently abandoned are sometimes expenses rather than amortized.) Geothermal investment can receive benefits of investment tax credits and accelerated depreciation, and rental and interest expenses of third-party financing are deductible. Costs shown heretofore reflect financing by private firms. Municipal enterprises could, subject to problems discussed in the barriers section of this study, reduce their costs through use of tax-exempt financings, or by joint venturing. Geothermal developers generally cannot take tax deductions for depletion (for petroleum companies 22.5 percent of gross revenues up to 50 percent of net revenue) or for intangible drilling expenses (expenses incurred for labor, fuel, repairs, hauling, and supplies used in drilling and completing a well--a substantial portion of total drilling costs).[6]

For dry steam geothermal systems, intangible drilling cost (IDC) deductions are allowed and have been estimated to reduce total production cost figures by 1.2 mills (from 14.5 mills to 13.3 mills per kwh), providing tax reductions are not offset by higher royalty or financing charges. Dry steam developments may also use a 22.5 percent depletion allowance. For hot water sources, requiring a greater fluid flow and number of wells to support a unit, cost estimates could be reduced by as much as 2 mills if IDC deductions were available and by up to 3 mills if a 15 percent depletion allowance were granted (28.9 mills without tax deduction, and 23.9 mills with both deductions [December 1974 costs]).

Complete passing along of tax savings to reduce costs at the busbar would result in no additional capital formation via retained earnings by field discoverers or producers. If, as may be expected, the wildcat explorers are different from the field developers, percentage depletion will confer little benefit on explorers in tax savings that would result prior to production.

Investors in oil and gas drilling funds have, in the past, obtained a tax shelter through the practice of using nonrecourse loans. Such loans have been made to a drilling program, and the funds have been used for intangible drilling costs. These costs are then deductible on the tax returns of the program's partners. Use of such loans has been resisted by the Internal Revenue Service (IRS) and, under recent tax legislation,[7] its use by individuals may be ended in regard to future leases. If the incentives are offered to limited partners in joint ventures, geothermal development will still be faced with a time-lag problem that may make oil and gas drilling more inviting.

Smaller firms' net income would change if costs were reduced and tax benefits were not just passed along. This change in net income could occur only if, and to the extent that, product prices did not follow marginal costs of production.* In a competitive market, reduced costs would be expected to be mirrored in lowered prices. Absent competitive conditions in the geothermal market, other factors, such as opportunity costs reducing investment, or increases in rents for scarce items, negate effects of a depletion allowance. Tax exempt entities, if depletion and intangible drilling costs deductions are permitted, will find it necessary to carry out geothermal work through joint ventures. Unless they can share costs with the taxpayers in this manner, they cannot be competitive. Municipal entities may appear to have competitive advantages--chiefly their low-cost financing ability. But this advantage cannot be brought into play until drilling costs have been successfully borne. At that earlier stage, entities for advantageous deductions available for drilling expenses are in a more favorable position.

While there is evidence that the availability of intangible drilling costs deductions would reduce the exploratory risk of geothermal venture investors, it is far from clear that without such incentives energy could not be produced and sold at the better sites which will be first developed at a return on investment adequate to attract development capital without a subsidy from the taxpayers.

As depletion would be available only upon successful production, it would only increase capital available if additional money were invested in anticipation of receiving this deduction; or if depreciation deductions were not passed along to investors as dividends or to buyers in lower prices, but rather were retained by the firm and plowed back into geothermal work; or if properties that are explored by firm A would command a better price when sold to firm B for development in consequence of the availability of a depletion allowance.

The cash-flow lag and the overall levels of outlay for geothermal energy development constitute barriers to entry to smaller enterprises. The tax treatment given geothermal projects favors those developing the more promising vapor-dominated systems, and those with other income to shelter from taxes. It is not as favorable to

---

*Taxes are just another cost.

entrant firms seeking limited partnership financing as is tax treatment for petroleum.

FIGURE 2.6

The Effect of Plant Size and Power Cycle on Generating Costs for Electricity

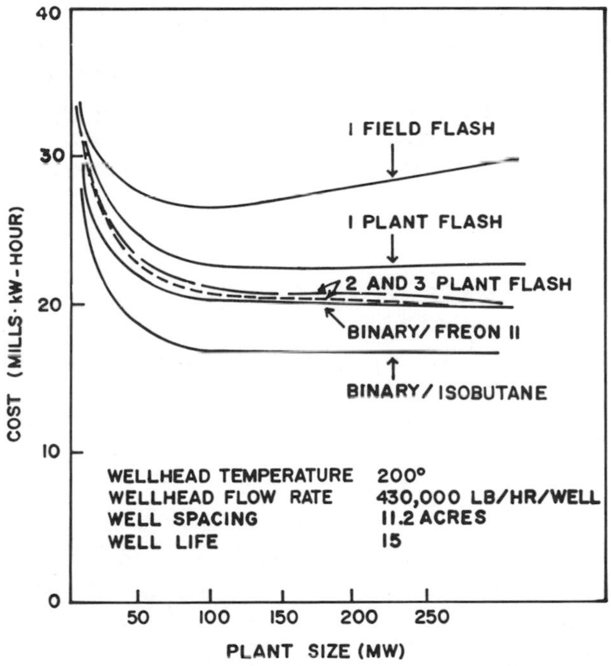

Source: Clarence Bloomster, Economic Analysis of Geothermal Energy Costs (Richland, Wash.: Battelle Pacific Northwest Laboratories, November 1975).

Geothermally driven generating units are limited in their economic size by the large size of the low pressure turbines they employ. Figure 2.6 shows the costs of generating capacity, as affected by size and type of power unit, for exploitation of a low salinity source, and illustrates two important points regarding costs. First, the unit economies of scale cease to increase at about 55 mw. Second, binary cycle units are cheapest per unit of output. However, binary power cycle costs are particularly sensitive to the rate and efficiency at which heat is transferred from well-derived fluid to the material driving the turbine. Poor heat transfer coefficients* result in higher costs, as larger heat transfer areas (heat exchangers) must be constructed. Heat exchanger costs vary directly with well fluid temperature, unit costs of exchangers, and heat transfer coefficients.

As Figure 2.6 shows, costs of units above 55mw size are uniform. The dominant factors determining costs of power are the wellhead temperature, wellflow rate, and the cost of wells--that is, the heat delivered per hour related to the cost of installing the field plant to produce this heat. For a given temperature, there is a strong relationship between flow rates and energy costs. Once economies associated with larger units are exhausted, it makes more sense to replicate units. This replication improves system reliability, confirming the effects of the unavailability of a unit.

Geothermal units thus are small when compared with 800mw coal-fuel units or 1,200mw nuclear units--the economic system for such energy sources.

FORECASTS OF GEOTHERMAL USE

Definite plans for utility employment of geothermal energy are quite limited. The Western States Coordinating Council, reporting generating resource projects for all states from Montana and Colorado west, forecast 1980 and 1989 capacity as follows:

	Megawatts	
	1980	1989
Hydro	51,194	55,332
Thermal	70,608	99,414
Miscellaneous	477	1,247
Total	122,279	155,993

The projected 1980-84 increase in thermal capacity is 28,806 mw; that for geothermal is 770 mw.[8]

When the projected growth of 28,806 mw is compared with the mw of geothermal capacity that would be economical if priced at cost, it is seen that geothermal power could play a meaningful role in the market for new western electric generating capacity, that is, the market for generation to go on line in the early and mid-1980s. For this to occur, there must be a combination of intent to accomplish this task with a reduction in barriers discussed below.

---

*Rate of heat transfer per hour per unit area of heat exchanger per unit temperature difference between fluids (Btu/hr/ft^2/^{o}F).

Longer range projections for supply (1985-94) forecast total additions of 91,000 mw:

	Percent
Hydro	9.2
Nuclear	62.1
Fossil	26.1
Other	2.6*

These data show that utility plans for future generation envision reliance on nuclear energy and fossil fuel.

The utility planning choice among geothermal, nuclear, and coal may make use of, but does not necessarily depend upon, estimates of supply cost. An engineering analysis may show, for instance, that under certain factual conditions the supply cost of geothermal energy would be lower than that of either coal or nuclear fuel. However, utility planners may feel constrained to add an element of cost, perhaps not quantified, or quantified only roughly to the geothermal estimate, in order to recognize their perception of the risk in exploiting a novel energy source.

Moreover, an engineering analysis will reflect supply cost, and not necessarily price. It is price with which utility planners must be concerned, at least so long as they are purchasing primary energy rather than developing it for themselves. Therefore, their perception of the differences between ideal competitive conditions and actual market conditions may lead them to hypothesize a corresponding difference between the supply cost estimate (based on engineering analysis) and the estimated price (based on, inter alia, supply cost, energy market concentration, and the apparent pricing policies of those who control geothermal and other primary energy sources).

It is interesting to note the substantial decline in government projections for geothermal utilization. In October 1975, ERDA estimated that 6,000 mw of geothermal electric capacity would be available by 1985, given a successful federal program implementation.[9] This was considerably more ambitious, or optimistic, than the electric utility group's expectations.

Considering development of hydrothermal and geopressured resources in the absence of a federal program, ERDA estimated that by 1985 1,500 mw will be on line, and perhaps by the year 2000

---

*2.6 percent of 91,000 megawatts is 2,366 megawatts. These 2,366 mw include geothermal, fuel cells, and so forth.

the geopressured resource may be added to the line of resource options. ERDA explains that the bulk of the 1,500 mw is planned expansion in the Geysers, with limited additional development of liquid-dominated hydrothermal resources in southern California and scattered small-scale, nonelectric applications.[10]

By January 1977, ERDA's forecast of geothermal capacity available by 1985 had fallen to 3,000-4,000 mw. Table 2.2 presents this revised forecast.

TABLE 2.2

Intended Commercial Geothermal Utilization Potential, Given Successful Federal Program Implementation

Utilization	1985	2000	2020
Electric capacity (mw)	3,000-4,000	20,000-40,000	70,000-140,000
Electric applications			
Equivalent fossil fuel energy (quads per year)	0.2-0.3	1.5-3.0	5-10
Nonelectric applications (quads per year)	0.1	1	8
Total energy (quads per year)	0.3-0.4	2.5-4.0	13-18

Source: ERDA, Program Approval Document, Fiscal Year 1977, Geothermal Energy Development (Washington, D.C.: U.S. Government Printing Office), January 17, 1977.

These ERDA forecasts suggest that no geothermal capacity is expected to be developed in the Raft River area in the near future, while development in the Imperial Valley of California is anticipated. However, discussions with electric utility personnel and a review of scheduled load additions indicate that no commercial units have been scheduled for the Imperial Valley. Plans announced by California municipal systems may result in additional capacity, especially in the Geysers.* Some flashed steam capacity (50-100 mw) may be installed in Utah. The scheduling of binary cycle units, necessary for

*The Northern California Power Agency (NCPA) seeks to develop 165 mw ultimately, with 66 mw by 1979-80. The city of Burbank seeks to develop several hundred mw and develop geothermal power, preferably in the Geysers.

development in such hot water areas, will probably await the results of ERDA research.*

Municipal utility enterprises active in geothermal development--such as the city of Burbank, California; Bountiful, Utah; and NCPA--need to acquire a supplementary source of power. Their load growth continues without an offsetting increase in the available federal power. Similarly, the Sacramento Municipal Utility District is seeking base-load power to fill the supply void left by indefinite postponement of a nuclear unit. All these systems are engaged in geothermal programs. Bountiful, Burbank, NCPA, and the city of Santa Clara are actively leasing and have entered into joint enterprises to explore for geothermal energy. Because of concern about a capacity shortage on the West Coast about 1980, capacity that might be available in the interim is specially appealing. Unfortunately geothermal lead times are such as to make little relief likely by 1980. Thus it is unlikely that a geothermal program could be started in time to take over a substantial portion of the California State Department of Water Resources' present power supply arrangements (800 mw) which lapse in 1983.

The smaller economical size of geothermal units, relative to nuclear or fossil-fueled base-load generating units, makes geothermal generation appealing to small power systems seeking to control their own generation and wishing to reduce wholesale purchases from larger utilities. This small size detracts from their attractiveness to large utilities.

Smaller utilities are reluctant to venture into what are for them the unknown waters of geothermal energy unless they can anticipate obtaining energy at a price below that of petroleum. These smaller systems appear to believe that they must develop their own geothermal resources, alone or in a joint enterprise with another enterprise (probably another small system), to obtain energy at prices other than those tied to prices for alternate fuels.

The plans of the larger systems must be firmly developed years in advance of fruition. Among larger utilities, Pacific Gas and Electric Company's (PG&E) plans for the Geysers are the only

---

*ERDA-sponsored work with San Diego Gas and Electric (SDG&E) in the Imperial Valley on development of heat exchangers is going to take several years, and will then require testing in a 10-50mw unit. The rate at which other systems can be expected to amend their ten-year planning cycle to add a novel type of generation is limited. This is especially so in view of the dissuasive effect that federal power marketing programs and loan guarantee proposals may be expected to have on investment by public bodies.

definite project for geothermal development. These plans as reported to the Federal Power Commission (FPC) (1975) do not envision geothermal development elsewhere. Plans submitted to ERDA for research work by SDG&E do foresee limited pilot development of 10 to 50 mw over a decade. Utah Power & Light also may install some generation. For other utilities, plans for geothermal generation are not firm, although some of them are or have been engaged in exploratory work. Work on one demonstration plant has begun. All of these larger systems' geothermal plans are an outgrowth of uncertainties regarding their ability to construct nuclear units in California, and their ability to construct coal-fired stations in the mountain states and build transmission lines.

Industrial concerns interested in self-generation of power are a potential alternative market for geothermal energy. Not a few such firms have engaged in geothermal programs. Several, such as Dow Chemical and AMAX, are so engaged in a major way.

The number of firms generating their own electricity has decreased over time, although, as late as 1968, self-generation provided 17 percent of industry electric power. The economies of scale and resulting lower rates achieved in the larger generating units built by utilities, coupled with managerial preferences and anticompetitive utility practices in regard to providing back-up and other coordination services, usually make power purchases cheaper than self-generation, although the rapid rise in power costs may change this situation.

The following reasons were listed in a recent report by the Dow Chemical Company:[11]

Corporate income tax rates that favor expense over capital investment and foster shorter-term thinking.

Risk differences requiring industrial firms to seek higher returns than do utilities.

Decrease in central station power rates, leading to substitution of power for steam and decreasing the industrial steam base.

Unavailability of labor skilled in handling coal-fired boilers. (Gas- and oil-fired package boilers--requiring little supervision--are generally not suitable for power generation.)

Shortened work weeks in many industries, leading to operating problems and decreased usage of capital equipment.

Widening of the gap in size--and hence economy of operation--between central station and industrial generating equipment.

A management attitude that "We're not in the power business"--or one that attempts to impose on power operations the policies regarding operation, maintenance, spares, and so forth, which govern the principal business of the firm. (Power plant maintenance generally comes infrequently, but is expensive when it is needed. Managers without a power background may attempt to keep production up by arbitrary economies, which succeed temporarily because power plants normally have substantial built-in reserves. This policy may result in sudden and serious breakdowns, and a number of firms have abruptly switched to purchased power after such a shutdown. The process is seldom reversed.)

Utility practices that have long discouraged the generation of power by any other type of entity; for example, rate schedules favoring the large industrial user (whether or not cost-justified), and heavy demand charges (levied even if no power is used), making it uneconomic to use the utility to back up industrially owned generation.

Dow reports that industry continues to generate steam for lack of a good alternative, but that the above factors have led to a change from self-generation to reliance on utilities for power, and a resultant neglect of the economic potential of by-product power generation.

GEOTHERMAL LANDS

An examination of the ownership of rights to find and produce geothermal steam reveals some important things about the structure, functioning, and future prospects of the geothermal industry. Among the land-related factors relevant to this inquiry are the pattern of ownership of rights to explore and develop; the way in which owners of these rights make use of them; and the effect of land-acquisition mechanisms now in use (chiefly government and private leasing) on ownership and development patterns and industry structure. Among the relevant questions are: Do geothermal rights tend to fall into the hands of firms owning other types of primary energy resources? Are geothermal prospects being stockpiled, acquired for speculation, or obtained for prompt development? How do geothermal leasing practices influence the type of firm that can successfully assemble a tract? Do lease terms encourage or discourage rapid development?

In reading the description of governmental and private leasing that follows, it should be borne in mind that there is a significant disproportion between lease acquisitions and actual drilling.* While

---

*In 1976 about 1.4 million acres were federally leased for geothermal exploration and potential development. In the same year, five new wells were drilled.

acquisition of prospects is sometimes accompanied by geological and geophysical evaluation (done by the acquiring firms, or for them by independent contractors), drilling and development need not accompany acquisition, and often do not. This fact suggests that the acquisition of rights is the principal business of many firms in the geothermal industry.

The right to enter and test the possibilities of a trace of land is usually obtained through a lease. Prospective acreage is leased from federal and state governments and from private parties.* Leases on privately held lands are acquired through negotiation, while leases on government lands are acquired under competitive bidding procedures or by filing. Geothermal rights are held by a variety of enterprises: large and small petroleum companies, specialized geothermal firms, private utilities and their subsidiaries, municipal and Rural Electrification Administration (REA) cooperative utilities, and industrial concerns.

Federal Leases

Leasing of geothermal tracts in the federal domain is of interest both because a substantial portion of geothermal resources is believed to be located there (or on adjacent acreage) and because of the availability of information on these extensive and widespread lands. The federal lands appearing to have some promise or interest as geothermal resource tracts are classified as Known Geothermal Resources Areas (KGRA) and are leased by competitive bidding; other federal lands may be leased upon application.

Lease sales did not commence until January 1974, some three years after the enactment of the Geothermal Steam Act of 1970.[12] Since then, only half the units offered or reoffered have been bid on, and in many sales bidding has been limited. Sales by competitive bidding consist of auctions in which would-be lessees submit sealed bids for "bonus" amounts of cash they offer to pay the government for a lease. The bids are accompanied by deposits.

Tables 2.3 and 2.4 show that the federal leases issued by competitive bidding have gone primarily to petroleum companies, with the balance to geothermal enterprises. Petroleum company lessees range from an independent, Anschutz, to large firms such as Shell Oil, Standard Oil of California, and Phillips Petroleum.

---

*Also, some prospective tracts held by firms involved in the geothermal industry were acquired primarily for other purposes, for example, by a land company or a utility. Federal lands may be entered for "casual work" by non-lessees upon filing of a notice with the Interior Department.

TABLE 2.3

Federal Competitive Leases Issued
as of June 30, 1975

State	Party	Number of Leases	Acreage
Arizona	None		
California	Geothermal Resources International	2[a]	3,710
	Magma	3	5,065
	Occidental	2	382
	Republic GI	4	8,478
	Shell	2	3,874
	Signal[f]	2[b]	876
	Thermogenics	1	175
	Union	5[c]	3,448
	Total	21	26,008
Colorado	Anschutz	1	916
	Phillips	2	4,120
	Total	3	5,036
Idaho	None		
Montana	None		
Nevada	Calvert	2	5,123
	Chevron	5	11,591
	Geothermal Resources	1	1,772
	Getty	2	4,940
	Natomas	2	3,980
	Southern Union	1	2,402
	Sun Oil	1	2,161
	Total	14	31,969
New Mexico	Anadarko	9	18,477
Oregon	Republic GI	1	1,347
Utah	American Geothermal	3	7,579
	Getty	1	1,920
	Phillips	10[d]	18,912[e]
	Union	7	14,390
	Total	21	42,801
Washington	None		
Wyoming	None		

[a]Leases granted under grandfather rights: Chevron, 1 lease, 1,815 acres; and Getty, 1 lease, 1,895 acres.

[b]One lease granted under grandfather rights for Natomas, 626 acres.

[c]One lease granted under grandfather rights for Signal, 737 acres.

[d]One grandfather lease granted for 40 acres from A. L. McDonald and William L. McDonald.

[e]Another document provided by the Bureau of Land Management, Utah State Office, indicates this figure should be 17,721 acres.

[f]Aminoil acquired former Signal Oil and Gas properties from Burmah Oil and Gas.

Source: Authors' survey of state offices of U.S. Bureau of Land Management.

TABLE 2.4

Distribution of Federal Geothermal Leases, by Company

KGRA	Owner	Bonus (dollars)	Acreage
Geysers	Shell Oil	4,500,000	3,874
	Signal Oil (Aminoil)	2,130,600	876
	Occidental Petroleum	447,004	382
	Union Oil	220,933	3,448
	Thermogenics	22,050	175
Mono-Long Valley	Republic Geothermal	515,767	1,773
	Getty Oil	98,592	1,895
	Chevron Oil	18,459	1,815
East Mesa	Republic Geothermal	650,106	6,705
	Magma Power	11,398	5,065
Vale	Republic Geothermal	13,831	1,347
Roosevelt Hot Springs	Phillips Petroleum	798,856	18,872
	Union Oil	51,993	2,560
	Getty Oil	24,000	1,920
	A. L. McDonald et al.	2,335	40
Brady-Hazen	Natomas Oil Company	88,912	5,074
	Southern Union Production	15,108	2,402
	Geothermal Resources, International	14,000	1,772
Beowawe	Chevron Oil	595,652	6,890
	Getty Oil	75,602	4,940
Hot Springs Points	Chevron Oil	240,893	4,701
All KGRA	Occidental Petroleum	447,004	382
	Union Oil Company	272,926	6,008
	Getty Oil	198,194	8,775
	Natomas Oil	88,912	5,074
	Thermogenics	22,050	175
	Southern Union Production	15,108	2,402
	Geothermal Resources, International	14,000	1,772
	Magma Power	11,398	5,065
	A. L. McDonald et al.	2,385	40

Source: Bureau of Land Management, Upland Minerals Division, Branch of Mineral Leasing, 1976.

The average price per acre in federal lease sales has been $53.71 per acre* with specific prices ranging widely. At the sales where prices exceeded $50 an acre, with one exception the high bidders have all been petroleum companies that are significant producers and suppliers of fuel to West Coast utilities from source areas such as Indonesia. The one such sale in which the highest bidder was not a petroleum company drew few bids. Table 2.5 shows the price bid per acre. The prospect of bidding against major petroleum companies has dismayed a number of smaller public and private enterprises which have concluded that they had better seek noncompetitive leases. The prospect of adding a heavy bonus to the high start-up costs of a geothermal project is a significant deterrent.

TABLE 2.5

Federal Geothermal Leases, Competitive Bids

Company	Aggregate Bonus (dollars)	Acres	Price Bid per Acre (dollars)
Phillips Petroleum	798,856	18,872	42.33
Chevron	855,004	13,406	63.78
Republic Geothermal	1,179,704	9,825	120.07
Getty Oil	188,194	8,775	22.59
Union Oil	272,926	6,008	45.43
Natomas	88,912	5,074	17.52
Magma Power	11,398	5,065	2.25
Shell Oil	4,500,000	3,874	1,161.59
So. Union Production	15,108	2,402	6.29
Geothermal Resources Int'l.	14,000	1,772	7.90
Burmah Oil (Aminoil)	2,130,600	876	2,432.19
Occidental Petroleum	447,004	382	1,170.16
Thermogenics	22,050	175	126.00
A. L. McDonald et al.	2,385	40	59.63

Source: Bureau of Competition and Economics, "Report to the Federal Trade Commission on Federal Energy Land Policy: Efficiency, Revenue, and Competition" (October 1975), reprinted by Senate Interior Committee, 94th Cong., 2 Sess. as Serial No. 94-28 (92-118), 1976.

---

*For the 282,787 acres leased by competitive sale as of June 30, 1976.

Small enterprises have greater representation in the ownership of noncompetitive leases, as Table 2.6 shows.

While 939 noncompetitive applications have been filed in California (560 on lands administered by BLM and 379 on Forest Service land), 387 await action and only five leases had been issued as of July 31, 1976: one to Eason Oil Co., two to Mobil Oil Co., and two to Southern Union Products Co. The Bureau of Land Management (BLM) has been slow in granting leases upon noncompetitive applications, and a large backlog of such applications has developed.

State Leases

The land holdings of many western states encompass substantial geothermal prospects. However, while many of these states have made provision for geothermal leases, no state lands have been leased for geothermal activity in Arizona, Montana, Nevada, Washington, or Wyoming. In some states, lack of information as to the possible extent and value of geothermal resources has deterred leasing.

In California, only a portion of the more than one-half million acres of state-owned lands within potential geothermal areas has been explored for development. Under the current framework for leasing California state lands (established by the Geothermal Resources Act of 1967, California Public Resources Code [Division Six]), the State Lands Commission may issue short-term geothermal prospecting permits on a first-come, first-served basis, or it may issue long-term leases, either preferentially, under certain conditions, or by competitive bid. Applications may not be for less than 640 acres nor for more than 2,560 acres for onshore lands. In the case of submerged lands, the maximum application allowed is for 5,760 acres. No permittee may have an interest in more than 25,600 acres.

California prospecting permits give permittees an exclusive right to explore the land for three years with a possible two-year extension. Upon discovery of a geothermal resource under a prospecting permit, the permittee has a preferential right to a geothermal lease for a period of 20 years and for so long thereafter as geothermal resources are produced up to a lease term of 99 years.

The State Lands Division of California is currently experimenting with a program of leasing geothermal lands through bidding on a net royalty basis. In such bidding, the state is offered a share of the net profits from a leased tract. Such leasing methods, it is hoped, will permit bidding by entities such as public bodies wishing to develop geothermal energy but unable or reluctant to increase

TABLE 2.6

Federal Noncompetitive Leases Issued
as of June 30, 1975

State	Party	Number of Leases	Acreage
Arizona	Phillips	4	6,507.77
California	--	--	--
Colorado	--	--	--
Idaho	Nancy Anschutz	2	3,763.63
	ARCO	6	13,545.02
	Bill Maddox	4	7,749.01
	Sue Rodgers	5	8,521.72
	Steam Corp.	2	3,685.16
	Sun Oil	5	6,306.22
	Union	1	640.00
	Total	25	44,210.76
Maryland	--	--	--
Montana	--	--	--
Nevada	Al-Aquitaine	8	19,068.76
	Amer. Thermal	5	8,346.02
	Anadarko	8	20,416.05
	Anschutz Corp.	6	13,882.15
	Burmah	2	4,471.00
	Calvert Geothermal[a]	1	635.00
	Chevron	2	3,510.25
	Lowell Harrison	2/3	1,067.46
	Geothermal Resources	4	8,097.14
	Francis Grinnin et al.	1	1,595.60
	Richard Hoefle	2	1,920.00
	Peter Hummel	1	1,440.00
	Douglas Hunt	5	8,418.22
	Lamar Hunt	2	2,580.15
	Nelson Hunt	1	480.00
	Magma	4	9,229.52
	Mobil	5	11,591.43
	Pacific Geothermal	1	640.00
	James Palmer	2/3	1,067.46
	Phillips	2	2,840.00
	Marcellene Sands	1	640.00
	Caroline Schoellkopf	6	11,133.56
	Southern Union	5	11,055.00
	Sun Oil	10	11,043.11
	Thermex	6	14,228.44

(continued)

TABLE 2.6 (continued)

State	Party	Number of Leases	Acreage
Nevada (cont.)	Edward Towne	2-2/3	5,561.56
	Union Oil of California	5	10,705.07
	Total	87	185,662.94[b]
New Mexico	Jack Grimm	5	9,568.61
	Lamar Hunt	6	15,353.40
	Nelson Hunt	8	18,618.90
	Caroline Hunt	3	6,365.56
	Norma Hunt	7	15,419.20
	W. H. Hunt	8	20,058.20
	Thomas Hunt	6	15,169.40
	Nancy Hunt	4	7,040.00
	Total	47	107,593.27[c]
Oregon	Franklin W. Baumgartner	4	4,505.00
	Getty Oil	1	640.00
	Pacific Energy Corp.	1	640.00
	Total	6	5,785.00[d]
Utah	James Becker		5,961.27
	Chevron		6,175.70
	Earth Sciences		1,040.40
	Milton Fisher		5,401.26
	Geothermal Exploration		5,754.52
	Lamar Hunt		12,195.95
	Nancy Hunt		7,801.98
	Nelson Hunt		6,339.76
	Norma Hunt		640.00
	Thomas Hunt		1,900.01
	Malcolm Justice		2,182.24
	Gary Seltzer		1,284.07
	Steam Corp.		3,416.50
	Thermex		3,737.65
	Union Oil		3,806.06
	Trevar Windsor		3,868.72
	Total		71,506.09[e]
Washington	--	--	--
Wyoming	--	--	--

[a]Acquired by Sun Oil Company.

[b]Four noncompetitive leases have been assigned by Chevron to Geothermal Resources International totaling 6,244.92 acres. Another BLM source counts 103 leases with 194,163 acres.

[c]Another BLM source gives 104,210 acres.

[d]Another BLM source gives 2 leases of 1,268 acres.

[e]Another BLM source counts 40 leases totaling 62,331 acres.

Source: Bureau of Land Management, U.S. Department of the Interior.

front-end risk investment, or by smaller independent firms now effectively barred by the capital requirements of bonus bids. One such lease sale has occurred. At that sale, the city of Santa Clara was high bidder on 135 acres in the Geysers area.

Table 2.7 summarizes grants of prospecting permits and leases.

TABLE 2.7

California State Geothermal Prospecting Permits and Leases

Permittee or Lessee	Number of Permits	Acreage
Current prospecting permits (as of November 24, 1975)		
American Thermal	4	8,898
Getty	2	9,857
Getty-Mono	3	14,738
Gulf	4	14,918
Total	13	48,411
Current leases (as of June 27, 1975)		
Imperial	4*	535
Union	2	3,988
Total	6	4,523

*Inactive; produced only hot brine.
Source: California State Lands Division.

In Colorado, geothermal leases are let only upon application. The 1974 Colorado Geothermal Steam Act is part of the state's water laws; permission is required from the State Engineer before wells may be dug. Regulation is by the Oil and Gas Conservation Commission, subject to the State Engineer's appraisal of the seniority of water uses under Colorado prior appropriation doctrines. Table 2.8 summarizes Colorado leases.

In Utah, state lands are leased upon application, with competitive bidding required only for tracts on which leases have terminated or lapsed. Table 2.9 summarizes grants of leases in Utah.

Under Utah law geothermal production would occur under the state water laws, and accordingly applications for water rights appropriations must be made to the Division of Water Rights.* Interference with prior appropriations will become a problem only if communication occurs between surface and geothermal waters. This matter is under investigation by the State Engineer, and all applications are pending.

TABLE 2.8

Colorado State Geothermal Leases as of June 1975

Lessee	Number of Leases	Acreage
Antares Oil	1	5,760
Austral Oil	1	640
M. L. Gillespie	3	6,640
Gulf Oil	2	1,920
Mapco[a]	2	68,351
Petro-Lewis[b]	1	11,764
C. A. Underwood	16	47,437
Total	26	142,512

[a]The only lessee reported to have drilled, and its well has been plugged.

[b]Leasing and exploring in the Mount Princeton area under a letter of intent from Public Service Company of Colorado to acquire steam.

Source: Geothermal Energy 3 (June 1975).

Utah Power & Light Company (UP&L) has blanketed the state with applications for approximately 100,000 wells on ten-acre spacing by filing on all likely prospects. UP&L would have a priority of appropriation and protection against offset drilling. Table 2.10 summarizes these applications, all of which are pending.

The State of New Mexico owns 11 percent of the state's land area. Leasing tracts are selected by following trend maps and applications to state and federal governments, and are leased by competitive bidding with a royalty of 10 percent of gross revenue or 8

*Lessees may not hold the entire beneficial interest in a tract.

TABLE 2.9

Geothermal Leases for Utah State Lands

Company	Number of Leases	Acreage
Thermal Power Co. of Utah	15	16,027
American Oil Shale Corp.	4	5,567
Resource Leasing Corp.	2	1,279
Chevron Oil Co.	2	2,584
Phillips Petr. Co.	18	23,562
Sonja V. McCormick	14	11,242
Steam Corp. of America	4	2,528
Wm. A. Stevenson	4	2,166
J. W. Covello	6	5,454
R. E. Puckett	36	53,009
Chas. L. Golding	7	6,240
Gerald C. Harrison	2	2,824
Davon Inc.	3	3,050
Malcolm F. Justice Jr.	4	2,784
Milan S. Papulak	1	240
Donald F. X. Finn	4	4,453
James A. Murphy, Jr.	1	2,167
Roy Barnes	3	791
L. Doral Christensen	1	80
W. O. Darley	1	600
Worley Valley Oil Operation	1	160
Chris A. Marks	1	361
Intercontinental Energy	1	364
Eliz. A. Justice	2	960
Malcolm V. Justice Jr.	1	120
Calvin F. Beckstrom	5	4,887
Total	157	158,594

Note: All let since January 1973.

Source: Conversation with Val Finlayson, Utah Power and Light, and Morgan, Office of the State Engineer.

percent of net revenue.[13] State leases on June 27, 1974, embraced about 90,881 acres. Leases have primary terms of five years with the right to renew for succeeding five-year terms if geothermal resources are, or are capable of being, produced or utilized in commercial quantities from the leased land or from lands utilized with the lease. Annual rentals are one dollar per acre, and delay rents after the discovery of commercial quantities of geothermal resources are two dollars per acre. The lessor reserves the right to prescribe a development program and to approve or disapprove development programs submitted by lessee. A $5,000 bond is required prior to operations on a lease tract.[14] Table 2.11 summarizes New Mexico lease information.

TABLE 2.10

Utah Water Rights Applications

Company	Number	Quantity (second feet)	Percent of Total
UP&L and Thermal Power Co.	15	41,632	19
UP&L and Geo-Drilling Co.	11	87,552	40
UP&L and Geothermal Kinetics Systems Corp.	10	64,400	29
UP&L subtotal			88
Chevron Oil Co.	6	18,484	8
Union Oil Co.	1	2,000	1
Phillips Petroleum	1	1,680	1
Frank J. Allan	3	2,176	1
J. K. Letts	2	1,536	1
A. L. McDonald	3	75	--
Wm. L. McDonald	3	75	--
Utah State Land Board	1	25	--
Total		219,635	

Source: Utah State Engineers Office.

Idaho's geothermal resources are believed to be the hot water type;[15] and drilling for them is regulated by the State Department of Water Resources.[16] Table 2.12 is a summary of the very considerable leasing of state lands by the Idaho Department of Public Lands. Leases on private lands are reported to be held by Gulf Oil, Anschutz, and Phillips Petroleum. Union Oil, Sun Oil, and Magma have also indicated varying degrees of interest.

TABLE 2.11

New Mexico State Geothermal Leases
as of June 27, 1975

Lessee	Number of Leases	Acreage
Antwell, Alan J.	4	1,948
AMAX[a]	29	11,771
Burmah (Aminoil)	65	25,853
Calvert Geothermal[b]	8	5,423
The Cherokee	7	4,433
Chevron[a]	14	6,695
Deuterium	17	7,210
Fogelson, E. E.	9	4,080
Folmar, Cecil J.	5	3,109
Grimm, Jack	1	640
Gulf Oil	16	6,471
Hodges, Leland	11	3,865
Kelly, John[c]	5	2,624
Southern Union	18	3,869
Thermal Expl.	3	2,890
Total	212	90,881

[a]AMAX is controlled (20 percent equity ownership) by Standard Oil of California (SOCAL). Chevron is a subsidiary of SOCAL.

[b]Sun Oil Company.

[c]Former Assistant Secretary (Mineral Resources), U.S. Department of the Interior.

Source: Minerals Division, New Mexico Land Office.

Drilling has been undertaken by the Anschutz Corporation (in the Grandview-Bruno area), by Gulf Oil, and by Geothermal Kinetics Inc. (in a venture with UP&L).*

The Raft River Rural Electric Cooperative (REC) has leased 100,000 acres that include some state lands. With financial support from ERDA, and to a lesser extent from the Northwest Public Power Council, it has drilled two wells, currently being tested. Raft River is seeking further ERDA support to drill two more wells, and to

*UP&L is reported to have a geothermal lease on Fort Hall Indian Reservation in Idaho.

TABLE 2.12

Idaho State Geothermal Leases
as of March 1, 1975

Lessee	Number of Leases	Acreage
Anschutz Corp.	164	55,213
Nancy P. Anschutz	165	75,787
Malcolm Mossmann	72	25,561
Chevron Oil	19	3,113
Gen Oil	23	14,271
Don Gould	4	1,980
Gulf Oil	174	68,853
F. Joe Kanta	4	1,491
Phillips Petroleum	10	5,089
Raft River REC	12	8,235
Warren Sorrells	12	7,692
Sun Oil	2	880
Total	661	268,111

Source: Idaho Department of Water Resources.

TABLE 2.13

Oregon State Geothermal Leases
as of December 1, 1975

Lessee	Number of Leases	Acreage
AMAX[a]	1	1,280
Chevron[a]	1	2,800
Intercontinental Energy	2	1,920
Max Millis[b]	1	2,240
Total	5	8,240

[a]Only one drilling; some exploratory drilling in Harvey County stopped several months before.

[b]Controlled by Standard Oil of California.

Note: Of 458 geothermal lease applications in Oregon, Sun Oil Co. has 73 and Chevron Oil Co. has 48.

Source: Geothermal Energy 2 (November 18, 1974).

build a 10mw pilot plant which it hopes to have operational by 1978. This binary-flashing unit, if successful, would be followed by a larger commercial unit.[17]

Oregon has a number of geothermal resource areas. Among the oil and geothermal firms exploring in this state is a joint venture of Weyerhaeuser, Pacific Power and Light (PP&L), and Portland General Electric (PGE), all holding leases of private lands in Klamath County.[18] Table 2.13 lists firms holding leases on Oregon State lands in Lake, Harvey, and Malheur Counties; Weyerhaeuser and Hydro Search had applications pending in Klamath County. By the end of 1975, only three wells had been drilled in Oregon, one by Gulf, one by Magma, and one by San Juan Oil (MAPCO, Inc.), none successful.

GEOTHERMAL EXPLORATION AND DRILLING

While extensive acreage has been subjected to geothermal leases, drilling has gone forward on a more limited basis. In 1975 and 1976 drilling was limited to respectively 45 and 51 wells in the continental western states, six of them by the federal government.* Very little drilling occurred in new areas; all California drilling except for one well was in the Imperial Valley and at the Geysers.† Table 2.14 lists wells drilled by region and Table 2.15 lists wells by operators.

In 1975, 24 wells were drilled in the Geysers area and 18 in 1976. Outside of California, no wells were drilled in Colorado, Arizona, or New Mexico in 1976, and only 15 elsewhere: six in Nevada, one in Oregon, six in Utah, and two in Idaho, including one by the Idaho National Laboratory (INEL). In 1975, more wells were drilled in California, two wells were drilled in New Mexico, six in Utah, two in Idaho (by INEL), four in Nevada, and one in Oregon.

It is interesting to note that the Union Oil Company, which drilled over one-third of the wells in the 1975-76 period, is a crude-short company that has traditionally had an interest in nontraditional energy sources such as oil shale. The role of small specialized

---

*The Idaho National Engineering Laboratory (INEL), working in the Raft River area, Idaho. ERDA funded this work and also a well in Hawaii.

†The Geysers drilling included field extension efforts. Of the 14 privately drilled wells outside of California, Magma drilled four in Nevada, and Phillips Petroleum six at Roosevelt Hot Springs.

TABLE 2.14

Geothermal Drilling in the Western United States, 1975-76, by Region

State	Operator	Number of Wells	Number of Observation Wells
California			
Geysers Region	Union	24	--
	Aminoil	14	--
	Shell	6	--
	Pacific Energy	4	--
	McCullock	1	--
	Chevron	1	--
	AMAX	--	3
	Magma	1	--
Imperial Valley	Republic	10	--
	Union	--	3
	Chevron	--	4
	Magma	2	--
Long Valley	Republic	1	--
Idaho	Idaho National	--	--
	Engineering Lab	4	2
Nevada	Magma	4	--
	Chevron	1	--
	Union	--	3
	Phillips	2	--
New Mexico	Union	2	--
Oregon	San Juan (Mapco)	1	--
	Thermal Power	1	--
Utah	Phillips	7	--
	Union	3	--
	McCullock	2	--
	Thermal	2	--

Source: Based on Witham and Reed, unpublished paper for Area Geothermal Supervisor's Office, Conservation Division, U.S. Geological Survey, May 1976; Smith and Matlick, "Summary of 1975 Geothermal Drilling: Western United States," Geothermal Energy 4 (June 1976).

firms and independents is particularly pronounced outside of the Geysers. The dominance of major and very large oil firms in acquiring prime prospects is evidenced in drilling at the very promising site in south central Utah (Beaver and Millard counties). There seven of the ten wells drilled in 1975-76 were by Phillips, with one by Union.

TABLE 2.15

Geothermal Drilling in the Western United States, 1975-76, by Operator

Operator	Number of Wells Drilled 1975	1976
Union	15	19(6)
Aminoil	8	6
Republic Geothermal	4	7
McCullock	--	3
Shell	3	3
Chevron	(5)	2(1)
Magma	4	3
Phillips	5	3
Thermal Power	--	3
INEL	--	2(2)
Pacific Energy	3	1
AMAX	--	(3)
San Juan (Mapco)	1	52(12)

Note: Figures in parentheses indicate observation holes.

Source: Based on Withan and Reed, unpublished paper for Area Geothermal Supervisor's Office, Conservation Division, U.S. Geological Survey, May 1976; Smith and Matlick, "Summary of 1975 Geothermal Drilling: Western United States," Geothermal Energy 4 (June 1976).

In total, as of August 1975, nine geothermal wells had been drilled on federal lands, two were in progress, and applications were pending for six more. In 1976, five wells were drilled. Table 2.16 lists those companies which had drilled or received permits to drill geothermal wells on federal lands as of August 28, 1975.

TABLE 2.16

Exploration and Drilling on Federal Lands
as of August 1975

State	Party	Notice of Intent	Application to Drill	Wells Drilled
California	Burmah (Aminoil)	1	4	2
	Republic	1	2	--
	Republic GI	2[a]	--	--
	Shell	1	3	2[b]
	Union	1	--	--
Nevada	Chevron	1	--	--
	Southern Union	1	--	--
Utah	Phillips	2	8	5[b]

[a]Not yet issued; awaiting Environmental Assessment (EA) and letter of approval.

[b]Plus one being drilled.

The paucity of geothermal well drilling cannot be attributed to government regulatory delay. Federal regulation is confined to federal lease tracts and would not sharply restrict the number of applications for drilling permits. State and local regulatory time spans for drilling are brief, especially outside of California. Power plant regulation problems occurred in the Geysers and in consequence of development practices.

The costs of regulation involved in obtaining federal permits appear to be greater than those for permits from some western states. While these permit costs may be burdensome to smaller firms, the larger entities with leases should be able to overcome this barrier.

Lessees seeking to drill on federal lands are required to post compliance bonds. The Bureau of Land Management (BLM) requires a $10,000 bond for each lease, to be filed prior to bidding. Alternatively, a lessee may post a $50,000 statewide bond, or a $150,000 nationwide bond. Explorationists seeking to proceed under a notice of intent to explore federal lands need not have a lease. However, they must file a protection bond.* Holders of national or statewide oil and gas bonds can meet this requirement by simply amending their bonds to include geothermal exploration. The obligation may otherwise be met by a $5,000 bond for each exploratory activity or by holding a pre-entry $25,000 statewide or a $50,000 national bond.[19] Small enterprises, and particularly those lacking prior gas or oil exploration experience, bear a heavy burden in meeting lease bonding requirements. They may seek bonds on security proffered by substantial participants in their own enterprises.

Between leasing and exploration, there is a planning lag of up to one year. The USGS requires federal lessees, prior to beginning work, to submit a plan of operations, which takes three to four months to be reviewed.[20] A Notice of Intent to explore on the public lands,[21] which takes up to 30 days to process, must be filed with the BLM. In addition, there may be state and local permits and authorizations to be obtained before drilling can begin.

A federal lessee is required to submit exploration plans to the USGS[22] and spends about a year collecting data for an environmental analysis which is then prepared. If and when a reservoir is defined, a permit is required for surface use by a power plant and transmission lines. Then a plan for production must be filed with USGS[23] where, reportedly, the time required for review of such plans has increased to several months.[24] Because of the dovetailing of filing lead-time and engineering-planning-time needs, regulatory delay does not appear to be severe.

The United States does not appear to employ lease rental provisions to require rapid prosecution of lease exploration by its lessees. In order to have acreage produced so that royalties will be paid, the United States imposes due diligence requirements on its lessees. A major incentive to diligent exploration of lease holdings is the structure of lease rental fees.

The General Accounting Office (GAO) states that leasing regulations can be strengthened to promote early exploration and development of leased lands:

> The [Interior] Department's leasing regulations do not require lessees to drill exploratory wells to evaluate an area's potential for heat, power, minerals, or fresh water. Under the geothermal leasing regulations, however, each geothermal lease is to provide for the diligent exploration

---

*As of November 1974, there were 51 exploration permits outstanding having terms ranging from six weeks to two years.

> of the leased resources until there is production in commercial quantities. Failure to perform such exploration may subject the lease to termination.[25]

The structure of lease rental fees affects the incentive of lessees to explore leases, rather than just hold them speculatively. Section 5 of the Geothermal Steam Act specifies minimum annual rental payments of one dollar per acre, and the BLM has set annual rentals at two dollars per acre for competitively let tracts. With production, a lessee may substitute a minimum royalty of two dollars per acre in lieu of rent.[26] Rental fees on the leased acreage can be increased after the fifth year if there is no production, or be eliminated once production begins. After the fifth year, certain expenditures for diligent exploration may be credited against rental fees.

A USGS official responsible for supervising geothermal leases told the GAO that USGS had not established a firm guideline on the required level of diligent exploration in the first five years of a lease and would probably not terminate leases if, during that period, no exploration activity were undertaken. GAO found no record of any lease being terminated for lack of due diligence. Subsequent inquiries at BLM and USGS reflect the same state of affairs.

For the sixth and succeeding years of a geothermal lease term, Interior Department regulations provide a formula for computing minimum expenditures necessary for diligent exploration. The minimum rents and expenditures necessary to maintain a lease for 2,560 acres if no commercial production takes place during the ten-year lease are as follows: total rents of $12,800 for the first five years and $64,000 over the first ten years; and no minimum expenditures in the first five years, and a total of $102,400 in ten years.*

The GAO recommended that due diligence requirements[27] now totaling but $166,400 in ten years be tightened to require expenditures during the primary term to approximate more closely the $800,000-plus costs of a deep well, and it urged that more specific requirements be imposed for minimum development during the initial five years of a lease.

As noted in a Federal Trade Commission (FTC) staff report, competition within the geothermal industry would help insure that the federal government receives fair value for its lands and, if geothermal energy is to reach its full potential, leases must be available to com-

---

*The law provides that a geothermal lease shall embrace a reasonably compact area of not more than 2,560 acres.

petitive enterprises that would foster development of alternate energy sources. "The fact that the leading geothermal developers are also leading West Coast petroleum suppliers suggests that interfuel competition may not be greatly enhanced."[28]

The paucity of geothermal drilling is particularly evident in the case of leased state lands. Wildcatting has been conducted by a number of enterprises, most notably such specialized firms as the Magma Companies, Republic Geothermal, and Geothermal Kinetics, Inc. On the other hand, several large firms that had acquired extensive acreage and were actively exploring are now, after some cold holes, apparently just holding on to their land. In this group are Sun Oil, Gulf Oil, and Getty Oil.

The development stage has been actually reached in only one area, the Geysers in California, where a great deal of drilling has been done by Burmah Oil* and Pacific Energy Corporation (PEC) to extend the defined resource area. Even there, uncertainty and what the industry perceives as long delay in getting generating units licensed and on line have greatly curtailed drilling. PEC and Burmah wells there have been shut in for several years.

Some industrial energy users, for example, Dow Chemical, Anaconda, Weyerhaeuser, and to some extent AMAX, have become involved in geothermal resource development activities. Dow, which now owns part of Magma, also purchases and transports natural gas for its own use in northern California. AMAX, 20 percent of whose equity ownership was recently acquired by Standard Oil of California, owns large reserves of coal, and uses a substantial amount of purchased electric power at its works near the Mt. Princeton geothermal area of Colorado. AMAX is exploring with Petro-Lewis for geothermal resources in this area. The industry participation is parallel with and in the alternative to other joint enterprises between utilities and small geothermal companies.

The limited amount of drilling in most western states is consistent with either a lack of financing, high risk perception, speculative leaseholding, a limited number of prospects, anticompetitive restraints on development, or some combination of the above. If good sites are limited in number, the possibilities for speculative withholding and anticompetitive practices are enhanced. Withholding

---

*Burmah, which acquired its Geysers interests in the 1974 acquisition of Signal Oil & Gas (the parent company), is a British petroleum firm which has held shares of BP and Shell Transport Co. and owned a number of tankers. It is currently being bailed out by the British government. Aminoil, owned by the Reynolds Company, recently acquired Burmah's domestic holdings.

in turn delays technological innovation, increasing risk and creating financing problems. The large number of leased tracts that have not been drilled clearly evidences some speculation, as parties know their financial situation when they take leases, and have developed their risk perceptions. Because development is largely financed through joint ventures with utility or petroleum firms, speculators may have assurance that development will not occur in ways that upset energy prices.

DEVELOPMENT ARRANGEMENTS

Smaller geothermal firms either associate with large enterprises that finance their activity, or they seek to finance exploration drilling by joint venturing with another energy company or energy consumer. The larger firms demonstrate a more go-it-alone attitude by joint venturing only where their acreage overlaps. They avoid action that would commit them to a subsequent sale of energy, such as joint venturing with consumers, asserting that they intend to prove up their acreage before offering its output. While delaying entering into sales may assure a vendor that he knows more about what he is selling, it also may be a means of reaping locational values or rents. The dependence of some geothermal enterprises on joint venturing to secure capital can result in a community of interest with energy users who wish to secure their future supplies at lower prices. These users obtain needed capital.

Joint ventures have previously occurred between Geothermal Resources International (GRI) and SCE; GRI and the Los Angeles Department of Water and Power (LADWP); Sierra Pacific Power Company and Magma; Petro-Lewis and Public Service Company of Colorado; and Dow Chemical with its affiliate Magma and with Chevron. More recently, there have been other joint ventures of consumers and energy firms.

In the area of financial contributions, securities holdings, and similar arrangements Union Oil owns 300,000 shares of common stock of Magma Power Co.; Hughes Aircraft provides capital for the use of Pacific Energy Corporation; and Dow owns a portion of Magma Power Company.

Separate business enterprises established by two or more entities for defined purposes include the following:

Pacific Energy Company and Burmah--well drilling operation.
Petro-Lewis and AMAX--exploration in Colorado.
Geothermal Power Corp. and Natomas--exploration, in Modoc County, California.

Republic Geothermal and city of Burbank--exploration, drilling in California.
Anschutz and Gulf Oil.
American Thermal Resources and Mapco, Dow, Gulf, and Standard Oil of California (Chevron)--separate joint enterprises.
Chevron and Geothermal Resources International, American Thermal Resources, Utah International, Mountain States Resources, and Phillips Petroleum--separate joint endeavors.
Earth Power Co. and AMAX.
Geothermal Resources and Occidental Petroleum and Natomas--separate ventures.
McCulloch and Utah Power & Light Company--negotiating joint venture.
Magma and Union Oil--well drilling, Nevada.

Land deals--farm-outs or pooled acreages--include the following:

Sunoco: Land farm-outs to Al-Acquitaine and Signal (now Aminoil).
Southern Pacific Land Company: Phillips evaluating 1.5 million acres.
Dowdle Oil: Farm-out to Natomas.
Mapco: Farm-out to Republic Geothermal.
Magma, Chevron, and the New Albion subsidiary of SDG&E: Pooled acreage at Heber, California.[29] (In its 1975 annual report, Standard Oil Company of California reports owning 75 percent of its Imperial Valley Project.)

Magma has received drilling contributions from Union Oil and from SCE. It has an agreement to earn a contribution from AMAX in Oregon. The Los Angeles Department of Water and Power has made a $25,000 bottom-hole contribution toward a well drilled by Republic Geothermal in a joint enterprise with the city of Burbank, California. On hydro project lands the Water & Electric Board of the Northwest Public Power Council also owns some geothermal prospects. In Idaho, the Raft River REC and the Northwest Public Power Council, with ERDA funding, have drilled two wells and are planning a 10mw binary-flashed steam pilot plant.

Utilities have participated directly and through fuel company subsidiaries in geothermal exploration in Utah, Idaho, Arizona, Oregon, and southern California. See Table 2.17.

There has been a small but significant number of joint ventures between industrial companies and geothermal firms. Table 2.18 sets out industrial company participation in geothermal exploration, identifying three such joint ventures.

TABLE 2.17

Utilities Recently Participating in Geothermal Endeavors

State	Utility	Associate
Arizona	Arizona Public Service	Geothermal Kinetics, Inc. (GKI)
	Tucson Gas & Electric	Geothermal Kinetics, Inc. (GKI)
	Salt River Project	Geothermal Kinetics, Inc. (GKI)
California	City of Burbank,	Republic Geothermal
	(Mono Power) SCE	Getty
	San Diego Gas & Electric (New Albion)	Magma
	Imperial Irrigation District[a]	Chevron
	Northern California Power Agency	RFL Ltd.
Oregon	Pacific Power & Light	Weyerhaeuser Lumber Co.
	Portland General Electric	Weyerhaeuser Lumber Co.
Idaho	Raft River Cooperative[b]	--
	Fall River Rural Elec. Coop.	--
Utah	Utah Power & Light	GKI McCulloch

[a]The Imperial Irrigation District is reported to have leased geothermal prospects, although District personnel did not confirm this report.

[b]In Idaho, the Raft River REC and the Northwest Public Power Council, with ERDA funding, have drilled two wells and are planning a 10mw binary-flashed steam pilot plant.

Source: Authors' survey of state geothermal leasing agencies, geological surveys, and utility commissions.

The tendency of utilities and industrial firms to join with specialized geothermal firms may reflect the matching up of capital and managerial talent with firms having specialized skill and knowledge. Geothermal joint ventures are reportedly not favored by larger firms which do not need financial assistance, and which believe that joint venturing merely adds to time needed for a project.

TABLE 2.18

Industrial Firms Recently Participating in Geothermal Endeavors

State	Company	Associate
Montana	Anaconda	GRI
Colorado	AMAX	Petro-Lewis
California	Dow	Magma
Oregon	Weyerhaeuser	(no partner)

Source: Compiled by the authors.

Because joint venturing is a significant means of financing geothermal development, it is important to understand the roles of carried or carrying partners in development, as well as the relationships among various firms.

Joint enterprises, either joint ventures or utilizations, are sought for exploration and development when acreage holdings of several companies intersect; when smaller firms want to market their geothermal prospects by associating themselves with enterprises seeking opportunities to drill and acquire a working interest in development programs; when the expertise of one firm is sought by others; to spread drilling risks; or to secure control of the output of a particular resource.

Joint venture agreements provide a means for, inter alia, financing the costly drilling of geothermal wells. In some cases an enterprise primarily engaged in obtaining geothermal leasing will enter into a joint venture with a large company which may hold adjoining acreage. In such a venture the oil company may earn an interest in the lessee's acreage and production, therefore, by completing a specified drilling obligation. The lack of a new-issues market, absence of the tax benefits available to oil and gas joint ventures, and the fact that investment needs are larger and time to payout longer than for comparable oil ventures, all appear to have impaired, but perhaps not destroyed, the ability of geothermal enterprises to raise venture capital by methods other than by association with another firm or by joint venturing.

The joint venture agreements studied involve smaller enterprises as one party, and frequently involve utilities. Joint venture agreements among major enterprises have not been reviewed; however, the impression is that such agreements are few in number and may generally involve sites where acreage holdings overlap.

The agreements reviewed are patterned after those in the oil and gas industry. They use similar terminology, and similar concepts for ownership of shares, areas of interest, and arrangements for cash participations and operating decisional choices.[30] The smaller enterprise is usually contributing know-how and/or land, and the larger enterprise is providing the financing and thereby earning an interest in the tract.

Geothermal joint venture agreements designate an operator and an area of interest in which the parties will operate as a joint venture. They frequently include provisions under which a party may elect whether it wishes to participate in a well, and provisions permitting the separate sale of product by each venturer. If an energy consumer is a party to the venture, there will be an option to purchase output according to a pricing formula.

Because the cost responsibilities of a joint venturer may otherwise exceed its cash flow capabilities, smaller lessees may seek to retain only a royalty interest, while farming out their acreage. Smaller entities may be required to assign their interest to others, when the point comes at which they can no longer carry their share of joint venture expenses or outlays.

When a joint venture is entered into, the smaller partner may be required to proffer a share of any interest it may have or acquire in any tract of land in the area of interest designated by the agreement. Such areas can involve wide geographical expanses. Joint venture drilling schedules are, as a practical matter, tied to decisions of the carrying party as to when and at what rate it will expend funds. As in oil and gas ventures, provision may be made for one party to drill if another "nonconsents" to participate in a well. The nonconsenting party may be allowed to "back in" to a share of any production after the drilling party has recouped its investment several times over.

Drilling is usually done by an independent contractor retained by the operator of the venture, although a few specialized geothermal firms have their own rigs.

TECHNOLOGICAL AND ENVIRONMENTAL PROBLEMS

Technological problems, delay, and uncertainty currently plague the geothermal industry.[31] The goals of the research program established "for the purpose of resolving all major technical problems inhibiting the fullest possible commercial utilization of geothermal resources in the United States" include:

> (9) the identification of social, legal, and economic problems associated with geothermal development (both locally and regionally) for the purpose of developing policy and providing a framework of policy alternatives for the commercial utilization of geothermal resources; . . .
> (11) the establishment of a program to encourage States to establish and maintain geothermal resources clearing houses, which shall serve to (A) provide geothermal resources developers with information with respect to applicable local, State, and Federal laws, rules and regulations, (B) coordinate the processing of permit applications, impact statements, and other information which geothermal resources developers are required to provide, (C) encourage uniformity with respect to geothermal resources development, and (D) encourage establishment of land use plans, which would include zoning for geothermal resources development and which would assure that geothermal resources developers will be able to carry out development programs to the production stage.

Locating the resource requires the use of geological, geophysical, and hydrological prediction methods yet to be tested by extensive drilling. Moreover, the proper mix of survey methods appropriate for different areas is still under development.

Industry personnel have resisted efforts to perform federal research on geophysical prospecting techniques on federal acreage. They have sought to have such work limited to further delineating areas already explored by others. One explanation for this position has been that federal exploratory work reveals the worth of the acreage and may drive up the lease costs on the federal land and on adjoining tracts. This point needs no response. Another explanation has been that experimental techniques are best evaluated where actual holes have already been drilled. This overlooks the fact that lessees have not allowed researchers to work on drilled tracts, as an alternative to testing unleased government acreage. Further, government testing to "prove up" private lands may lead to unfortunate political consequences.

Location and commercial use of hot water sources present problems of safety and environmental protection as well as of efficiently harnessing the energy. The hot water type of geothermal resource is most commonly found as a fluid of varying and sometimes high temperature and salinity. Technology must be developed that will make available such resources of varying temperatures and salinities in an environmentally acceptable manner.

Areas of technical uncertainty in geothermy are mirrored in ERDA's Definition Report for the Geothermal Energy Research Development and Demonstration Program (October 1975). This report states that the required hydrothermal technology development effort includes the following activities:

Reservoir engineering and field development to improve the economics and reduce the risk, for example, of premature reservoir failure. Tasks include improving technology for drilling,* well digging, reservoir modeling, well simulation, and downhole pumps.

Development of technology to utilize geothermal energy including heat exchangers, condensers, total flow systems, and absorption refrigerator systems.

Work in brine chemistry, scale control, and corrosion- and erosion-resistant materials.

Construction of field test facilities to test components.

Development of advanced systems and applications.

Study of environmental effects of field production and development of technology to control emissions, efficiency, noise, and land subsidence.

Demonstration of pilot plants and commercial-scale units, with two such commercial-scale facilities tentatively planned.

This program encompasses the area of technical uncertainty affecting the geothermal industry. To develop and disseminate better techniques for locating, evaluating, and producing geothermal energy, field data are essential; many of the problems require empirical solutions. In view of the expense and the limited number of ongoing geothermal exploration programs, federal acquisition (for example, purchase or direct exploration) and disclosure of field data might be very worthwhile.

Geothermal development can be expected to go forward only if environmental problems can be solved. Solution of these problems

*Apparently the programs of Southern California Edison Company and San Diego Gas & Electric Company.

would eliminate uncertainty and delay in the granting of permits. A partial inventory of environmental problems includes:

Topographical alterations due to the number and spacing of drilling pads. While closer spacing gives faster rates of withdrawal, impact on the land surface could be reduced by wider spacing and use of more expensive directional drilling for several wells from one pad.
Erosion and siltation, especially around drilling pads.
Water pollution, for example, from spills of drilling mud or steam condensate, or by impairment of aquifer water quality.
Subsidence, requiring reinjection of fluids.
Dust and particulate matter.
Emissions from cooling towers, which may include hydrogen sulfide, ammonia, boron, chlorine, and other chemicals.
Noise, such as occurs when wells are vented to the atmosphere.
Some seismic uncertainty.

Noise and hydrogen sulfide emissions are the most significant problems to local people in the Geysers. The effects of carelessness of prior operators are evident in current proceedings.[32]

In California, first-line environmental control decisions are divided among different jurisdictions and a variety of agencies. Well siting regulation is a county function, as well as a function of the state Division of Oil and Gas and, for state domain, the state Lands Commission. Power plant regulation is currently being transferred from one state agency to another; air pollution control is under another state agency;[33] water pollution control is under a district agency.

Pollution control agency personnel and environmentalists feel that cooperation between industry and environmental agencies has not been all that it might be.

Under the provisions of the California Environmental Quality Act, issuance of a geothermal prospecting permit or a lease constitutes a project having significant environmental effect. Accordingly a draft and a full Environmental Impact Report (EIR) are prepared either by the State Lands Division or a consulting firm, at the applicant's expense, and are circulated. After a minimum of 30 days the final EIR is presented to the State Lands Commission. At this time the commission in a public meeting will hear comments on the proposed project; determine if a final EIR has been prepared by the Lands Division following evaluation of comments and consultation with public agencies which issue approvals for the project; certify that a final EIR has been completed in compliance with the California Environmental Quality Act (CEQA) and that the commission has reviewed and considered the information contained therein; determine whether the project will or will not have a significant effect on the environment; and consider authorization of the proposed project.

Pollution control requirements and development delays increase when geothermal resource areas are rural and not industrialized. The local population may require that it be shown why proposed developments should be allowed. Leaving temporary pipe lines in place for protracted periods and failing to develop and adopt emission controls do not promote local approval for geothermal growth.

Public acceptance of geothermal development is also affected by the degree of public confidence in governmental regulatory bodies. While other areas may have fewer or different environmental problems, activity in the Geysers will affect public attitudes elsewhere. The Geysers Plan currently is being operated under a variance from air pollution control regulations. Installation of equipment for the control of hydrogen sulfide emissions from generating unit cooling towers is not scheduled to be completed until 1981.[34] At that time, emissions from venting wells will still be uncontrolled.

The California Public Utilities Commission,[35] granting certificates for PG&E for Geysers Units 14 and 15, states:

> PG&E claims that it is now capable of meeting the air quality standards prescribed by the Northern Sonoma County Air Pollution Control District as such standards apply to Units 14 and 15. Based on his evaluation of the record the examiner, as stated in the Final EIR, noted that the "Stretford process is an effective hydrogen sulfide abatement system now practicable for installation on Units 14 and 15. If installed, as proposed, on Units 14 and 15 emissions of H_2S from the units will meet the requirements set forth in the California ambient air-quality standard."

In its exceptions the state Air Resources Board stated that even with abatement Units 14 and 15 will ". . . emit substantial additional amounts of H_2S into the air in the region where concentrations of this pollutant already exceed the California ambient air quality standard." The ARB proposed the following substitute findings:

> The Stretford process, while of measurable benefit, is not a totally effective H_2S abatement system. When H_2S emissions from the other

> Geysers units are added to those of Units 14 and 15, even with controls operating, the California ambient air quality standard will not be met. Additional emissions from Units 14 and 15 will exacerbate this problem.

The decision reports that the air pollution control officer of the Sonoma County Air Pollution Control District testified that he would not grant authority to construct further units unless abatement action were undertaken on existing units.

While the PUC found that granting the applications will have no significant impact on the environment due to H_2S emissions, it went on to note that its approval could be negated by the Northern Sonoma County Air Pollution Control District.

Applications to construct Units 14 and 15 were filed with the Public Utilities Commission on July 24, 1973 and March 1, 1974, respectively. Certification required an environmental impact statement entailing a long procedure. Decisions were not made until the fall of 1976. Delays encountered in developing an environmental impact statement for Geysers Unit 12 were attributed by Tom Cordell, Sonoma County Environmental Officer, in part to a decision by Union Oil personnel to ask the limited county staff to devote most of their efforts to assessments of additional development wells and not to Unit 12.

Assertions of supervening development needs for reasons related to the energy crisis have not to date been persuasive. Courts have forbidden pollution control agencies to deviate from their statutory mandates, and legislative loosening of pollution control laws has, generally, been quite limited.[36]

In the future, geothermal development may encounter severe new environmental restraints. If California state proposals to limit emissions to 150 pounds of a compound per day are adopted, geothermal practices in regard to drilling, and field testing of wells, and field piping could be severely affected.

Utilities, petroleum companies, and geothermal enterprises must consider the effects of technological uncertainties in their investment planning. Holders of geothermal acreage must weigh the carrying charges on leases against the benefits of waiting for extraction technology to improve and fuel prices to rise. They must predict future lease acquisition charges to determine whether to lease more or to drill on currently owned leases. Smaller firms with pressing cash flow requirements must consider whether they should develop properties, assign acreage and retain only royalties, or seek joint venture development.

Technical uncertainty and delay in turn breeds further delay. Large petroleum companies with high rate-of-return criteria for investments do not wish to sink investment dollars into projects for which payout is five years away and will occur only if high prices are realized. Although such companies are in a position to warehouse projects by leasing acreage for future development or subleasing, they lack the incentive to supplant government research efforts or to try to solve geothermal utilization problems.

Innovation involves investment in risk-bearing activity. Utilities facing large investment requirements for power supply may be reluctant to embark upon unfamiliar, risk-bearing courses--they may seek to shift risks to others, and they will tend to be protective of their markets.

Firms which are experiencing good financial results are more likely to extend their holdings into new but related product lines, for example, from petroleum to geothermal. However, the service firms on which petroleum companies rely for much of the innovative work in petroleum field exploration are not heavily capitalized, and oscillate between very good times when they are fully occupied, and slack conditions. These service companies are not geared for long delays in realizing the value of work done.

Current practices for leasing and developing prospects reflect concerns for avoiding drainage by others and for reliably securing an adequate resource supply.

Discussions with geothermal industry participants indicate that it is considered necessary to obtain leases for an entire prospect before development either alone or in a joint enterprise. Their reasons include not providing free information to adjacent lessees, avoiding drainage of a property by rival wells, and relatedly, avoiding problems of coordination and allocation among lessees. Small developers are concerned that a field purchaser might ignore them in favor of a sole source vendor. Firms conducting exploration activity are concerned that their work might lead to adjacent areas being classified as KGRAs. These concerns have led to joint enterprises, and have created capital and time barriers to development, as firms feel it necessary to lease to effectively utilize a prospect before development.

Geothermal development is in a regulatory quagmire; multiplicity of problems has spawned a multiplicity of agencies and interagency groups. One way of lessening the problem might be to reduce the number of problems which necessitate extensive inquiry. If field developers and utilities on a project coordinate how questions are presented for regulatory action overtaxed agencies might make bottleneck decisions in a more timely manner, while decisions less critical to a project are put off.

Each stage of regulation reflects legitimate public concerns about protecting the environment, the public domain, and the property of others.* The number of agencies involved reflects to some extent the variations in land ownership and the different interests and apprehensions of persons at varying degrees of remoteness from proposed operations. The totality of regulatory procedures reveals multiple points at which environmental impacts are assessed. This indicates that there is redundant work in an area where the limited number of component personnel should concentrate on substantive issues. Consequently, developmental decisions are often delayed or made so expensive as to raise economic barriers to smaller enterprises.

Exclusion of enterprises which could develop geothermal resources while complying with pollution control requirements, could result in adverse consequences to the human environment, which statutes such as the National Environmental Policy Act are concerned to protect. This results not only from the direct nonproductive use of dollars, but also from the effect that increased concentration of industrial ownership may have on the governability of business enterprises.†

The investigations for and writing of environmental impact statements have required considerable time periods. At times this process has been lengthened by applicants' slow responses. In addition, waste and delay have resulted from a lack of focus on the material issues to be decided.[37]

A failure to focus on and resolve material issues leads to a failure to protect the environment. Such a failure makes the environmental impact statement an unguided missile--an end in itself, of little use in drawing reasoned conclusions from an adequate record. An example of this problem is found in the licensing of the Geysers units.

---

*The interests forwarded by environmental agencies can only be encouraged by regulatory or fiscal methods. Unless regulation or taxation imposes costs of pollution control on developers, these costs are shifted to others, and the stimulus to innovative pollution control is lost along with the market for pollution control equipment. Regulatory methods are needed in a new industry because the cost data needed for a fiscal control system are lacking.

†The adverse effects of this concentration on the human environment may be best understood by taking a deep breath in many "one-mill" towns.

While considerable time and effort were apparently expended on air pollution questions in recent Geysers licensing proceedings, the decision rendered fails to contain explanatory findings of fact, and basically concludes that resolution of air pollution problems is a matter for other agencies. The impact reports fill up hundreds of pages. But they do not lead to decisions that cope with major problems. In failing to make findings and resolve disputes, the decisions reflect a process of trial by attrition, not by reasoned judgment.

The purposes of the National Environmental Policy Act (NEPA)[38] and its state law progeny[39] might better be served by the presentation of probative evidence at a hearing where it would be subject to cross-examination and rebuttal. Most importantly, at a hearing such evidence would, of necessity, be focused on the major issues, without speculation on matters with little relationship to the physical processes at hand.[40]

Delay problems plague geothermal leasing at the federal as well as the state level. In particular, the Department of Agriculture's Forest Service has been a serious bottleneck to geothermal exploration in some areas. Applications for geothermal leases within lands administered by the Forest Service are not affirmatively acted upon by the Interior Department until the Forest Service concurs.[41] There appear to be instances where the Forest Service first determined that an environmental assessment is required prior to its concurrence, and then failed to proceed to develop such an assessment.[42]

As before noted, the detrimental effect of delay weighs especially hard on smaller enterprises which must plan their power supply in advance and which want to proceed with geothermal development, but are not likely to be able to bid against parties with speculative front-end money. Delay increases the likelihood of an overlapping filing and consequent designation of an area as a KGRA, and competitive bidding.

While the Forest Service is delaying issuance of an environmental assessment of proposed schedules under a lease of federal lands, prelease exploratory work by a prospective developer is apt to stop. If the developer generates information indicating the presence of a geothermal resource, it is obligated to supply this information to the Interior Department. If the Interior Department determines a geothermal resource exists, it is obligated to seek competitive bids. The party generating the information then may lose the competitive bid, and has no means of recovering the costs of work done from the value discovered as a result of the work.

As Mr. Barnett, consultant to the rural electric cooperatives, testified:

> If we were to proceed without those leases being issued to collect additional geological information, that very information could be used against our purposes and have the area described as a known geothermal area and open the area to competitive bidding to whoever might find an interest after our initial exploration. . . .[43]

Caution as regards exploration impacts on the environment is readily understandable. Difficulty in setting priorities for limited agency budgets is understandable. But a continuation of present budget priorities seems likely to stymie potential geothermal development work. Delay in evaluating proposals to develop a resource usable by small public utilities, while pressing ahead with coal or gas leasing used by major companies which often have dominant market positions, seems likely to add a bias toward greater economic concentration in the electric utility sector.

POTENTIALS FOR PRECLUSION OF GEOTHERMAL PROSPECTS AND SERVICES

Entry into geothermal development can be substantially restricted when services of firms most familiar with an area are exclusively contracted for by a limited set of potential buyers. Geothermal firms are in large part service enterprises, grouping prospect tracts and performing exploration services. Such firms are limited in number. Some are, of course, more familiar than others with prospects and problems of particular areas.

In at least one major instance, purchase and sale agreements and joint venture agreements contain, respectively, red-line or area-of-interest provisions giving the purchasing party, or the financing party other than a purchaser, the option to purchase output, or participate in all prospects of the second party located within the designated area. The areas covered by such agreements can be quite large, encompassing entire states in the case of joint ventures.* Purchase and sale agreements typically cover smaller areas.

It can be argued that designation of large areas is required to protect a prospect from drainage, or to prevent a service company from prematurely going off to other jobs, or to reduce the possibility that information obtained in one project will be improperly applied

---

*Areas of interest are often designated in petroleum joint venture agreements.

to benefit another. These provisions can, however, restrain trade, particularly if the number of prospects is limited, or a contracting party is a leader in an area, or the most likely acreage is preempted in an infant industry fraught with uncertainty. Such competitive issues have been raised in a formal court proceeding.[44] While the court found reason to be concerned with such issues, it left their determination to a public utility commission, which discovered no anticompetitive impact in the facts before it.

The Northern California Power Agency raised antitrust issues in a California Public Utility Commission proceeding on a PG&E application to build generation units in the Geysers area. * The commission refused to afford a forum for the contentions. On appeal, the California State Supreme Court remanded the order of the California Public Utilities Commission certificating the generating units, so that the commission might adequately consider and make findings on the antitrust issues raised regarding steam purchase contracts. The court noted, in holding that the issues required examination, that the red-lined area was of substantial size and economic consequence. Upon remand, the commission rejected the allegation of anticompetitive effects, as it has done on subsequent licensing proceedings. However, the commission has never made more than bare conclusory findings on this subject.

Although alternative prospects are available in the Geysers area, it is not clear what the effect of contractual practices has been upon other would-be entrants. In the context of other power coordination practices, it appears, as will be discussed later, that such practices, and practices such as Utah Power & Light Company's extensive filing of water appropriation claims, have a chilling effect on entry.

Joint venture agreements tying up a development company for all acreage in a state do not seem justified on the basis of a need to appropriate value of services in a particular field or pool. If service companies and sites are plentiful there is little need to be concerned about a particular service firm's severing relationships. If knowledgeable service companies or prospects are rarer, as appears to be the case with current technology, tying up service firms may create a real problem for later entrants. The services of service

---

*Units 7 and 8 which were the first served under the 1970 Contract between PG&E and Magma, Thermal Power, and Union Oil Companies. The first well was drilled by Magma in 1955 and in 1958. Magma and Thermal had entered into a contract with PG&E. In 1960, the first unit (12.4 mw) went into operation, and additional units went into operation in 1963 and 1968.

firms can be assured, perhaps more reliably, by other types of contract provisions.

Petroleum companies, and land companies hoping to sell leases to them, have acquired substantial acreage on geothermal prospects. Firms such as Gulf Oil, SOCAL, Union Oil, Getty, and Phillips Petroleum have amassed substantial holdings. As the information in the section on exploratory activity shows, these firms, and other firms leasing geothermal acreage, appear to have undertaken little drilling. For instance, AMAX (SOCAL) has not drilled on its extensive Geysers area holdings. A would-be entrant will find substantial acreage withdrawn from the market, and must compete for future acquisitions with firms having deep pockets and, apparently, warehousing practices.

STATE PRODUCTION CONTROL

Western states have legislative provisions regarding well spacing and encouragement of production up to maximum economic rates.[45] These provisions reflect concern about equitable treatment of owners of adjacent tracts. The statutory provisions related to conservation, creating authority to affect output levels, bring into being a device used by some state conservation commissions in the petroleum industry to restrict production so as to maintain prices.

An example of such provisions in geothermal legislation is found in New Mexico's Geothermal Resources Conservation Act which prohibits waste which is defined to include production in excess of the reasonable market therefor.[46]

The Oil Conservation Commission is authorized[47] to limit, allocate, and distribute the total amount of geothermal resources which may be provided from a reservoir. The commission is authorized, after notice and hearing, to allocate production among geothermal wells, including in the allocation any well it finds is being unreasonably discriminated against through denial of access to a geothermal resources transportation or utilization facility which is reasonably capable of handling the geothermal product of the well.[48] Similarly, the commission is authorized to allocate production well spacing among owners of a geothermal reservoir,[49] and can mandate unitization. Owners of land within a well spacing unit are to be compensated for production, either under a voluntary pooling agreement or in a like manner.[50] In the event that some owners of a working interest in a reservoir do not pay in advance a pro rata share of drilling costs, there is to be a deduction from revenues attributable to such part owner of drilling and completion costs, plus an allowance for risk set by the commission. The charge for risk may be set on an amount up to 200 percent of the nonconsenting owner's share of these costs.

As regards discrimination by a purchaser toward producers, the act provides[51] that one taking or purchasing geothermal resources in a geothermal reservoir is to be a common purchaser of all tenders at a reasonable point without discrimination among producers, even if a purchaser is also a producer, subject to a number of qualifications that leave substantial discretion to purchasers and the commission.

Any common purchaser taking geothermal resources produced from wells within a geothermal reservoir shall take ratably under such rules, regulations, and orders, concerning quantity, as may be promulgated. Such rules, regulations, and orders may consider the quality and the quantity of the geothermal resources available, the pressure and temperature of the point of delivery, acreage attributable to the well, market requirements, and other pertinent factors.

Nothing in the Geothermal Resources Conservation Act shall be construed or applied to require, directly or indirectly, any person to purchase geothermal resources of a quality or under a pressure or under any other condition by reason of which such geothermal resource cannot be economically and satisfactorily used by such purchaser by means of his geothermal utilization facilities then in service.

Geothermal developers appear to rely primarily upon contracts to allocate responsibilities, rather than the provisions of state regulatory law, or their enforcement. Fear of being left out, or of not achieving an all-embracing voluntary pool, may have led to such statutory provisions, and in particular situations may lead to hurried and inequitable arrangements, or delay exploration. These problems emphasize the need to carefully lay out government tracts, and to carefully watch drilling proposals, so as to treat all lessees equitably without unduly delaying exploration.

Long lead times, the close relationship between field and generating plant development, and inability to store produced geothermal energy, require advance joint planning between producers and generating entities. Parties not involved in that planning will likely be excluded from sales as a practical matter. As in electric power supply coordination, access to planning is vital.

The provisions of state law allowing state prorationing of production, noted at the outset of this section, could conceivably prove a problem to efficient willing producers. Prorationing has tended to favor inefficient producers.[52] It offers each state an opportunity to attempt to function as a little OPEC,[53] maximizing its economic rent. Oil prorationing has never been fully effective as a price controller, but it has significantly affected oil structure and performance.

As noted by the attorney general,

> In sum, the conservation regulatory system, accomplishes only an indirect and partial balancing of overall crude supply, which permits higher prices to be established than would normally prevail if no controls existed. At its utmost, the effectiveness of control is limited to insuring only that no supplies enter crude oil markets outside the regular market channels maintained by the industry. . . .[54]

There is no federal enactment regarding geothermal energy equivalent to that which sanctions state prorationing in the oil industry. Under federal law,[55] it is illegal for a private party to ship in interstate commerce petroleum or petroleum products produced, transported, or withdrawn from storage in excess of amounts permitted under state law.

This restriction on shipment can be suspended by the President if he finds that the amount of petroleum or products moving in interstate commerce is so limited as to be a cause of lack of parity between supply and consumptive demand resulting in an undue burden or a restriction of interstate commerce. State regulation of economic as opposed to physical waste may be constitutionally suspect as well as unworkable if there are a number of prospects in various states.[56] Smaller enterprises principally located in only one state may be disadvantaged by the state's production controls, compared to other larger enterprises that can trade in several states, and in the case of generating firms, that have access to interstate power lines.

MARKETING GEOTHERMAL POWER

There are two major marketing patterns for geothermal energy: sales from field developers, as in the Geysers, and developer-user joint ventures. The field developer (including a joint venture) is ordinarily responsible for delivering the working fluid to the generating plant intake. The price received will probably be related either to generating unit output or the quantity of energy made available to the generating unit. Prices may be set either on the basis of costs (including return on investment) or by the commodity value of energy. Some utility systems have sought to tie price to the costs of development, while developers seek price provisions that are tied to alternate fuel costs and contain escalation clauses. Developers sometimes seek take-or-pay or other minimum charges.

Geothermal prices for resources developed without utility participation are likely to reflect costs of alternative fuels, while prices of energy produced with utility participation in development will more often be based on costs incurred. A firm generating electric power for sale to others usually must be a utility that files tariffs with either a state utility commission or the FPC. Thus, except for industrial entities engaged in generation of their own power needs, generation will be by utilities, not by oil/geothermal entities which seek higher rates of return on investment.

A combination of factors may incline utilities to prefer to own sources of geothermal energy-supply generation rather than to purchase energy produced by others. Rate base advantages may accrue from ownership. Bulk power market positions are not threatened by ownership. Risks to operating security may be minimized by ownership. As against this, utility financing and risk avoidance problems might be alleviated if geothermal units were to be installed by others* and their output purchased.

An arrangement intermediate between utility construction and separate ownership of a geothermally powered generating station is to have a separate entity construct such a station and the utility purchase it after the unit proves itself. Institutional problems may interfere with such postoperative sales. The return earned on the unit would be subject to regulatory review under the Federal Power Act, as the seller would be a public utility having to file tariffs for its sales in interstate commerce. If the selling firm is owned by another enterprise or person, an exemption will have to be sought under Section 3 of the Public Utility Holding Company Act.[57]

PG&E presently generates more electricity from geothermal energy than any other entity in the world. However, as was shown above, large utilities cannot be expected significantly to increase their involvement in geothermal projects. A primary reason is that the scale of a project is too small to warrant the managerial resources required. Other reasons may include public relations concern over boosting what might be construed by the voting public to be a viable alternative to nuclear power, and the reduction in concern over escalation in fuel prices with the advent of fuel adjustment procedures.

---

*Costs to a utility purchasing unit output would be an operating expense not providing a rate of return. However, purchasing power under a long-term contract, in lieu of building units, provides off-the-balance-sheet financing needed by utilities having poor capitalization and interest coverage ratios.

Also, utilities having large alternative sources of power may have less need to expose themselves to the risks involved in a new technology.

It will be shown in Chapter 3 that a hostile environment for small utilities exists in the western United States and that this environment has the double effect of creating an urgent desire to develop an independent generating capability and frustrating present attempts to fulfill that desire. Many smaller public power systems in the West are seeking to enter into self-generation. Heretofore many of these systems have relied on federal hydro generation or purchases from large private systems. Federal hydro power can no longer meet their expanding needs and reliance on wholesale purchases can be very detrimental to system growth and survival because of anticompetitive problems, described hereinafter.

The small scale of geothermal generation is well within the reach of small utilities. (Larger fossil or nuclear units entail joint ownership arrangements.) The load growth possibilities of geothermal capacity loom far larger for a small than for a large utility, and make better use of managerial time for a smaller utility. The smaller system often does not have simultaneous competing opportunities for large nuclear or fossil units under its management control.

To utilize geothermal units successfully, however, small utilities need access to reserve capacity, transmission, and bulk power supply coordination services. These can be sought from large investor-owned utilities and the federal power agencies. Transmission is needed to connect geothermal fields with load centers,* and uncertainty as to coordination and wheeling or exchange services could rob small systems of their already constricted ability to plan and finance geothermal development within the time span that a field developer is willing and able to wait for a market. It is noteworthy that the larger public power systems, having generation and a capability to build geothermal or other units, are the same systems that can best anticipate obtaining participation in alternate joint coal or nuclear units, with related coordination and transportation.

Small systems are generally not interested in geothermal energy if its price is tied to that of oil. They want a cheaper energy source, and their development plans will likely reflect this desire. There is a belief in some public power quarters that pricing problems in the geothermal area are created by petroleum company control of geothermal resources.

---

*For instance, development of 500 mw, which would be the goal of a SMUD geothermal project, would entail new transmission lines.

It has been shown in a previous section that petroleum companies have obtained most of the leases to federal geothermal lands and that there is little evidence of development. The large petroleum companies' failure to engage in rapid development efforts apparently is due to the existence of more advantageous investment opportunities elsewhere. Geothermal activities take seven to ten years to result in production and revenue. Investments in gas and oil operations generate revenue earlier, involve more reliable technology, receive more favorable tax treatment, and have higher anticipated rates of return than investments in geothermal generation. In addition, petroleum companies have much more flexibility in dealing with oil and gas than in dealing with a geothermal field. A geothermal field can only yield heat that would be valuable to facilities installed in close proximity to the field. The market most likely would be confined to one of the few electric utilities in the area.

Moreover, geothermal energy so developed would compete with other energy products of the same petroleum companies such as residual oil, natural gas, coal, and nuclear fuel. Sullivan, McDougal, and Van Huntley noted that an inverse relationship between petroleum company activity in offshore drilling and geothermal development on the west coast could be observed before and after the Santa Barbara oil spill incident.[58]

Warehousing of geothermal prospects by major and large independent petroleum companies is often accomplished by being the high bidder at sales in which there is competition to secure leases and by being the only bidders for those tracts whose commercial worth is so speculative and remote that bidding is not practical for smaller enterprises bent on development. The low prices and few bids received indicate that leasing of poorer prospects is going on at a rate in excess of markets for development, yet ERDA and petroleum companies continue to lobby for leasing by BLM and the Forest Service. As of April 30, 1977, the federal government had 1,612,463 acres under geothermal lease to others, mostly undrilled. *

Independent geothermal developers have been active. However, they are limited in the degree to which their activity could be expanded

*At the 19 federal geothermal lease sales held between mid-1975 and mid-1977, the ratio of bids received to units bid on has exceeded two at only two sales, and has been one at nine sales, and zero at three. Prices per acre have been under ten dollars per acre at all but five sales, and have been under three dollars per acre in several recent sales. The Bureau of Land Management continues to offer more than double the number of leases that even speculators will bid on.

by financing problems and speculation--withholding development in anticipation of price increases.

Unlike major oil companies, the independent developer has to obtain external financing. Given the technological uncertainties, the absence of tax advantages, and the limited market for geothermal heat, such financing has been forthcoming primarily from the electric utilities that would be the purchasers of the geothermal heat, and from joint enterprises headed by large oil companies. Large investor-owned utilities do not have the incentive to significantly expand their present contractual commitments to finance independent geothermal developers. Small utilities such as municipals and cooperatives, have the loads* and an intense motive to participate with independent geothermal developers in rapid development of geothermal energy resources, but are held back by the hostile environment in which they exist.

Joint enterprise between independent geothermal developers and small utilities for the development of geothermal generation might be a promising strategy for small utilities. To such joint enterprises--such as joint venture arrangements for finding and developing geothermal resources--a small utility would bring much of the financing and a capability to convert geothermal energy into marketable electricity. The independent geothermal developer brings to the partnership some financing, technical expertise and manpower for development and operation of geothermal wells, and the ability to use investment tax credits, and other available write-offs. Again, before this strategy can become widely used, certain changes will have to occur in the environment for small utilities, and the small utility and the developer will have to see eye-to-eye on energy pricing.

Industrial firms are constrained as developers and users of geothermal energy because they are unaccustomed to, and unenthusiastic about, being subject to utility regulation. As geothermal developers, they might be forced into unfavorable sale and repurchase agreements by uncooperative utilities, and bulk power coordination might be available, if at all, only on harsh terms.†

Industrial energy consumption is heavily concentrated in a relatively small number of industries--primary metals, chemicals,

---

*NCPA cities alone have loads of 580 mw and receive 230 mw from the federal Central Valley Project (CFP), and 350 mw from PG&E. They have 52 percent load factor and will require 500 mw over their federal allotment by 1980.

†Most industrial processes require firm power, which cannot be obtained from an isolated generating unit; hence the need for reserve arrangements.

petroleum refining, food processing, paper, and ceramics. These industries are most likely to be attracted by the economies of a system of self-generation that also affords process heat or cooling water. Geothermal process heating requires low salinity fluids occurring near plant sites.* Joint production of process heat and power requires coordination of generation and process heat use. Self-generation can be operated in parallel with a utility system with which sales and exchanges are effected if regulatory and competitive problems† are overcome.

Self-generation requires capital outlays while purchase, on the other hand, involves current expenses that reduce taxable income. The return on plant is unlikely to equal that earned in production. Furthermore, with self-generation, if intracorporate transactions are made on a cost basis, there are lower gross revenues for depletion.

Industrial self-generation could, like municipal self-generation, attract enterprises for which a 100mw resource is significant. This is particularly so if the industrial load is interruptible, so that standby arrangements are not critical.

Rapidly rising industrial power rates may make the use of waste heat at industrial sites more attractive. As energy costs become a larger proportion of total costs, more management time and other corporate resources may be used to limit increases in energy costs. Also, the point has been made that the possibility of reducing power costs while obtaining a secure power supply has been a factor inducing industrial investment in geothermal power.

In regard to self-generation, utility personnel have argued that because they have a general charter to supply the public, utilities should have a monopoly. However, failure to develop cheaper energy sources, and requiring industries to pay more than needed, have harmful effects on the economy. A monopoly would eliminate diversity of approaches and rates of development, and would preclude coordination by utilities with generators paid for by a new source of financing. It would also eliminate a source of pressure on utility rates and cost consciousness. In California, air pollution control regulation may prevent the use of coal-fired boilers, oil is very expensive, and the natural gas supply is being curtailed. These conditions make geothermal self-generation more interesting.

---

*Sites for the listed energy-intensive industries are limited by transportation and, relatedly, raw materials requirements.

†Problems encountered in recent efforts to obtain total energy systems will be described later.

Industrial public power generation will not be feasible without receipt of credit for geothermally driven generation as firm capacity not requiring unusually large reserves. Before utilities will treat the capacity of a generating unit as firm, they will insist on its assured availability, and may insist on controlling the operation of the unit. In the absence of a capacity credit, an industry or utility may not save demand charges on its bill, and the price for its power will be at a dump-energy rate, only a little above incremental operating costs. If capacity sales are made only after a substantial discount for reserve requirements, the kilowatts credited as sold are reduced, and recovery of capital may be jeopardized.

To the extent that geothermal capacity is not marketed as firm capacity, energy sales may occur only at economy or dump rates. Prices for economy transactions are usually midway between the seller's incremental energy costs and the buyer's incremental energy costs.* (For example, if a seller has energy costs of 7 mills [not including recovery of capital charges] and a buyer has energy charges of 23 mills, the sale price is 15 mills.) Dump sales are for only a few mills. If capacity is not treated as firmed up, dump sales seem much more likely than economy sales. Sales to other power systems--made at split savings or on a cost plus basis--would be profitable. Also, if use of geothermal energy lowered average costs, a utility would profit as a result of regulatory lag. The profit potential of regulatory lag in bulk power sales is dependent upon the price and terms at which geothermal energy and capacity is proffered to utilities.

PURCHASE AND SALE AGREEMENTS

The use of geothermal energy to produce electric power necessitates a close, long-term relationship between steam producers and utilities; and because a number of wells will be required to provide energy for a generating unit, coordination of producer field activity will also be required.

Terms of agreement for the purchase and sale of geothermal steam determine the distribution of risks, the ability to sell or purchase from alternative parties, and the cash flows a developer will experience. These factors both reflect and determine the market power of parties, and they define capital requirements of field development.

---

*At times such sales are made at incremental energy costs plus 15 percent.

If the market for geothermal energy were such that sellers competed against each other to make sales, then prices would be determined by the cost of producing the increment of energy being sold. However, for a market perceived as encompassing all types of energy for base-loaded units, vendors will seek a price reflecting the buyers' other choices and will compete only if one vendor believes another's transaction would cost him a sale. Vendors would of course compete for control of resources, for example, drilling prospects. Because of the novelty and risk of geothermal energy, vendors perceive that the value of the geothermal commodity is somewhat less than that for alternative, more conventional energy types. Consequently, their offering prices will start at a discount from the prices of energy alternatives.*

Geothermal developers of all types are seeking such a discounted price. Interviews conducted have indicated that the cash flow requirements and specialized nature of some smaller firms make them more vulnerable, and hence more willing to give larger discounts than some of the larger enterprises. Similarly, some smaller firms are particularly receptive to the idea of joint venturing with customers that would own part of the working interest.

As noted, geothermal energy vendors want prices based on the amount of heat delivered by them, rather than on kwh generated. They believe a cents-per-million-Btu price would encourage efficient use of the steam and enhance its value. Their concern about the consequences of being tied to a particular generator has led at least some producers to seek take-or-pay provisions that require the purchaser to pay for a minimum amount of energy, if tendered, even if the generating unit is not operating. Take-or-pay provisions limit the buyer's flexibility in choosing between energy sources.

Buyers who are seeking to reduce prices to costs (with a return of less than the 20-25 percent sought by a number of developers) must, first, assess their ability and willingness to enter into a geothermal venture on the scale necessary to ensure evaluation and testing of a sufficient number of geothermal prospects.[59] Second, they must assess the impact of a price on future geothermal investment both by

---

*The value of energy never exceeds that of alternative energy sources and prices are governed accordingly. When development began at the Geysers, PG&E insisted on paying no more than its alternative average costs of generation. It should be noted that sales agreements at the Geysers were not set at incremental costs of energy but at average rates, and that the prices based on alternative average busbar costs have no take-or-pay provisions, and are not set at a price for Btus delivered.

their suppliers and by their shareholders or governing boards. Finally, they must try to find a way out of the present seller's market for energy. *

Some preliminary appreciation of the problem may be obtained by comparing company heat costs (inclusive of return on investment) with prices. If a drilling and development program costs $10 to $16 million, as estimated by an FTC staff report, yielding a capital cost per kilowatt of $262,[60] the amortized cost per mill rate at an 80 percent capacity factor would be:

$$\frac{262}{7000} \text{ X (Fixed Cost Rate)}$$

Fixed costs (or an amortization rate) of 20 percent yield a cost of 7.5 mills; and fixed costs of 25 percent would yield 9.35 mills. (Well operation and maintenance costs, which are expected not to exceed a mill per kwh are ignored.)

Using field development costs of $31.5 million as presented in a recent paper by Greider of SOCAL,[61] it is found that for an 80 percent capacity factor operation, and a rate of return on investment of 20-25 percent, costs not including operation and maintenance range from 4.5 to 5.6 mills per kwh. By comparison, discounts of 20 percent from the price of $12 a barrel for oil yield fuel costs of 16 mills per kwh. This kind of cost calculation gives a utility considerable incentive to push for cost-based pricing, or to consider developing its own geothermal resource.

Agreements such as those entered into by NCPA, Burbank, and UP&L are likely harbingers of the future pattern of development for smaller geothermal enterprises. Such enterprises will probably look to utilities for financing, while providing prospect tract aggregation services and/or expert geothermal exploration services.

The small firm in a utility joint venture can seek to make sizable profits on its share of production and still leave a large share of production for the utility; also, the small firm could make a large portion of these profits from project tax deductions: for example, abandonments, accelerated depreciation, investment tax credits, and perhaps depletion and intangible drilling costs, which utilities are unlikely to be able to use because they are tax exempt or have very low effective tax rates.†

---

*Receptiveness to cost sharing (and product sharing) arrangements and promotional zeal vary not only between but within enterprises.

†Utilities use their low effective rax rates to assist their financing. In lease financing, the utility pays for financing, in part by allowing the lessor to have title to the leased property and to deduct investment tax credits and accelerated depreciation.

Non-joint-venture marketing will be attempted by larger petroleum companies. These firms seek prices based on the value of the energy as determined by costs of alternative utility fuels. They have the wherewithal to submit higher bonus bids in anticipation of later selling geothermal energy at such prices. These firms also do not make money solely by providing services and have the cash flow to permit warehousing of geothermal prospects.

In offering sales of steam without requiring prior buyer participation in field development and production, larger firms may be offering utilities a less risky avenue into geothermal development.

On the utility side of the market, utilities looking for a low cost alternative fuel seem unlikely to be able to participate in highly competitive bidding for lease tracts. High lease bonus bids would tend to drive up project costs, and raise the prices of energy produced. Also, smaller firms are not likely to have the resources to pay large bonuses. Thus sales from competitively let properties may be expected generally to be from producers to utilities; on other tracts, producer-utility joint ventures are to be expected. While utilities are not high risk-taking entities, and prefer to participate in projects on prime sites, possibilities for leasing private land, and low bid prices may permit them to find some suitable prospects.

Smaller power systems are accustomed to lower federally generated power costs and are very cost conscious. They cannot simultaneously participate in many projects to avoid risks in delay of one or two. Accordingly, they seek participation in development programs in such a way as to give them an active role in decisions governing timing of development and providing lower cost energy.

Larger utilities having a number of generating resources may feel less restrained in dealing with large petroleum companies in a mutually dependent arrangement such as geothermal development than they would with smaller firms. In purchasing steam, utilities may prefer to deal with substantial firms that have the assured financial capability to develop a field over a period of years. Their participation in fuel resource projects will necessarily be directed first toward securing coal and uranium resources for their future large base-load units. Combination utilities, such as PG&E also are engaged in gas supply ventures.

Overall, the current situation, in which large utilities deal primarily with large petroleum firms (or large firms grouped in joint ventures with small ones) and smaller utilities deal with smaller firms, is likely to persist. Vendors of geothermal energy thus may generally tend to be either large companies and independently engaged primarily in sales of petroleum, or vertically integrating utilities.

Frederick Scherer has stated that "the more prone input markets are to a breakdown of price competition, the stronger is a firm's incentive to integrate upstream."[62] Vertical integration by utilities will only result in cost-competitive entry if the utility has reason to be cost-conscious. Otherwise, it may merely provide an opportunity to pass along to consumers the price charged by a utility's unregulated fuel affiliate, and preclude markets. If pressure from regulators, or better yet, competition in bulk power markets induces a utility to be cost-conscious, utility fuel affiliates could provide competition to oil companies.

Vertical integration, by both large and small utilities, is unlikely to be on such a scale as to present severe diminutions of competitive potentials. Smaller systems which vertically integrate are unlikely to present market preclusion problems, by virtue of their limited size. Because of other cost disadvantages--in part a result of anticompetitive coordination and rate setting by larger utility systems--the smaller systems are more likely to pass fuel savings along in order to keep their rates competitive.

At the same time, these small systems do provide an alternative mode of geothermal development seeking cost-based pricing--not prices based on alternative fuels or what the traffic will bear. Were geothermal developers to come forward with cost-based prices, there is no reason to believe that they would find the market closed to them.

Geothermal firms are concerned about the range of markets available to their production. A variety of customers could provide joint venture opportunities, while avoiding the influence that a single customer may bring to bear. Furthermore, buyer competition may bid up field prices.

Geothermal developers contacted were sanguine about having at least one market for their energy from proven fields, but not about having multiple outlets. They were aware of possible wheeling problems, generally expecting to offer the product only to the local utility. Such limited marketing expectations may be attributed in part to the reluctance of one private utility to erect plants in the service area of another.* This reluctance has been reported to us by utility personnel, energy companies, and by a California Public Utility Commission staffer who has been "informed" it is commission policy.†

---

*An employee of Southern California Edison told us that SCE considers the Geysers to be "PG&E's resource."

†Publicly owned utilities involved in the Raft River Project are confident about their being able to obtain wheeling over Bonneville Power Administration power lines.

Geothermal energy sales are, to date, all made to PG&E in the Geysers. Vendors there are (1) a joint enterprise made up of Union Oil Company of California plus a joint enterprise of Magma Power Company and Thermal Power Company (now a subsidiary of the Natomas Company); (2) Aminoil (successor to Burmah Oil and Gas Company); and (3) Pacific Energy Corporation. Also, the NCPA has entered into a joint enterprise to explore for and develop geothermal resources on acreage in the Geysers. In this agreement, NCPA agrees to carry its leaseholder partner.

The contracts under which steam is purchased and sold in the Geysers are noteworthy in several respects. In each case, the seller developed steam to support a unit, prior to the contract, and is obligated to develop steam well in advance of receipt of any revenue. In each case, all present and future reserves of the seller in a larger red-lined area are dedicated to PG&E, which has access to producer field data.

Producers have a duty to continue to explore and develop their acreage. The obligation of PG&E to install units to employ developed steam is loose. PG&E has a substantial amount of time within which it may elect to construct a generating unit. In the Union-Magma-Thermal agreement, PG&E may defer a unit for up to 11 years.*

Producers can only sell steam to third parties if such steam is free of PG&E's purchase rights, and only if it can first be established in an arbitration proceeding that such sales will not detract from the steam supply of PG&E units within their expected remaining life. In such arbitration current lack of a generally accepted reservoir theory might prove an insuperable barrier to outside sales.

In each case, the price of steam is tied either to average costs for fossil and nuclear generated kwh, or (for PEC) the average annual cost for gas and oil. For all these contracts average fuel costs are weighed by the lowest PG&E unit heat rates.

For a geothermal unit to be economical it must be able to employ the high voltage transmission lines of others, to sell its output as a part of the bundle of power services required by most users, and to find alternative markets in which to sell output in excess of demands of regular customers.

Thus a certain level of cooperation among power systems is requisite. How power systems compete with one another and how larger systems decline to cooperate with smaller ones seeking self-generation will be described later. This refusal to coordinate constitutes a major barrier to development of geothermal energy.

---

*The contracts provide that 20 pounds of steam is to be delivered per kilowatt hour in contemplation of, and perhaps freezing, the use of only condensing turbine generating units.

NOTES

1. See, for example, Wall Street Journal, June 7, 1976; and Edison Electric Institute, Nuclear Fuel Supply (1976).

2. FPC Staff Report, Annual Summary of Cost and Quality of Steam Electric Plant Fuels, 1973 and 1974.

3. The distribution of costs among various elements of a 55mw isobutane binary-cycle plant have been estimated by C. H. Bloomster of Batelle Pacific Northwest Laboratories in Economic Analysis of Geothermal Energy Costs, Report to ERDA (November 1975).

4. Conversation with Mel Schrecongost, California Division of Oil and Gas, Sacramento, California (Summer, 1976).

5. The risks of drilling a "cold" hole are primarily exploratory, and not development risks. Burmah Oil's Annual Report for 1974, page 20, indicates that four of the five steam wells drilled by Burmah Oil and Gas in California in 1974 were successful and brought the number of completed wells to 13. ERDA research plans regarding drilling of "slim" holes could greatly assist in reducing the costs of exploration and resource delineation.

6. In the Geysers area, taxpayers prevailed in United States v. Reich, 454 F.2d 1157 (CA 9, 1970); and see Geo. D. Rowan, 28 ICM 797 (1969), which held that intangible drilling expenses and depletion deductions are available. However, the Internal Revenue Service has not acquiesced in this holding.

The Senate Finance Committee recently (May 1975) voted to proffer intangible drilling deductions and a 22 percent depletion allowance to geothermal development. These provisions passed the Senate but were dropped in conference.

7. P.L. 84-455, 26 USC §465 (1976). Knutson employs these six hypothetical reference cases:

REFERENCE CASE DATA

Category Represented	Wellhead Temperature (degrees C.)	Flow Rate (thousands of lbs. per hour per well)	Well Cost (thousands of dollars)	Cost of Generation (mills per kwh)
Dry steam	182	100	300	14.5
HTLS* 1	250	500	300	15.6
HTLS 2	225	500	300	18.9
HTLS 3	175	700	300	24.0
HTLS 4	175	600	300	25.9
HTLS 5	175	500	300	28.9

*High Temperature Low Salinity.

Source: Battelle Pacific Northwest Laboratory, Richland, Washington, 1976.

Treating tax reductions as net reductions in costs, he presents the following table, showing "cost reductions due to lower taxes":

THE GENERATION COSTS FOR DIFFERENT TAX SUBSIDIES

	Cost of Generation (mills per kwh)			
Resource Category	No Subsidy	15 Percent Depletion Allowance	Intangible Write-off	Both Allowed
Dry steam	14.5	13.2	13.3	12.2
HTLS 1	15.7	14.5	14.8	13.7
HTLS 2	18.9	17.2	17.6	16.1
HTLS 3	24.0	21.7	22.4	20.3
HTLS 4	25.9	23.3	24.0	21.7
HTLS 5	28.9	25.9	26.6	23.9

Source: Battelle Pacific Northwest Laboratory, Richland, Washington, 1976.

8. April 1975 report of West Systems Coordinating Council to FPC.

9. ERDA, Definition Report: Geothermal Energy Research, Development & Demonstration Program (ERDA-86) (October 1975) p. I-12.

10. Ibid., p. I-13.

11. See Dow Chemical Company, Draft Report "Energy Industrial Center Study," NSF Grant OEP 74-20242.

12. 30 USC §1001, et seq. If either the federal or the California state government prevails in pending proceedings to settle the question of title to geothermal resources when mineral rights are reserved, the significance of government leasing will be increased even more.

13. Telephone conversation with Jack Kennedy, Minerals Division, New Mexico Land Office, April 1976.

14. Rules and Regulations Relating to Geothermal Resources Leases, New Mexico State Land Office, April 1976.

15. Telephone conversation with John Mitchell, Idaho Department of Water Resources, April 1976.

16. Under the Idaho Geothermal Resources Act, Idaho Code Sec. 42-4001.

17. Telephone conversation with Ed Schlender, Manager of Raft River Electric Cooperative, May 1976. Raft River's lands are checkerboarded by federal lands so that the Cooperative may be confronted with the choice of either facing offset wells or trying to bid against oil companies. Schlender favored preference leasing for public bodies, and noted that in leasing state lands, an adjacent developer has priority. He praised the USGS and their open-file reports.

18. Telephone conversation with Vern Newton, State of Oregon, May 1976.

19. Conversation with Bob Pablovich of Uplands Mineral Division, BLM; and see 42 CFR, part 3206.

20. 30 CFR §270.34; a second exploratory hold would take a much shorter time to process.

21. 43 CFR §3209.

22. CFR §240.34 (k).

23. 30 CFR §240.35.

24. Telephone conversation with T. Reid Stone, USGS, and conversation with Barry A. Boudreau, USGS, Menlo Park, California, May 1976.

25. GAO, Problems in Identifying, Developing, and Using Geothermal Resources, March 6, 1975.

26. 30 USC 1004 (d).

27. 43 CFR 3202. 5.

28. FPC Staff Report, supra.

29. Bureaus of Competition and Economics, Federal Energy Land Policy: Efficiency, Revenue and Competition, report to the Federal Trade Commission (October 1975), reprinted by Senate Interior Committee, 94th Cong., 2 Sess. as Serial No. 94-28 (92-118), 1976.

30. See Rocky Mountain Mineral Law Institute, Proceedings of Offshore Exploration, Drilling and Development Institute (1975) for a good discussion of oil joint ventures.

31. The Geothermal Energy Research, Development, and Demonstration Act of 1974, §104, 30 USC §1124.

32. Telephone conversation with Tom Cordell, County Environmental Officer, Sonoma County, California, June, 1976.

33. The California Code provisions regarding air pollution control were overhauled in 1975 (Stats. 1975 ch. 957, §12).

34. PG&E, Amended Environmental Data Statement, Geysers Units 14 and 15 (1974).

35. California PUC, Decision 85720, April 20, 1976.

36. In Clean Air Constituency v. California Air Resources Board, 423, P.2d (Calif., 1974) the State Supreme Court held that the Air Resources Board lacked authority to delay a vehicular emission control program "for reasons related to the energy crisis," as such reasons did not relate to the purposes of the Board's enabling legislation. 523 P.2d at 624-26. Also see Natural Resources Defense Council v. EPA, 489 F.2d 390 (CA 5, 1974). The court in rejecting a state implementation plan, held that considerations of economic costs were to be subordinate to public health considerations.

37. Irrelevance of portions of federal environmental impact statements has also been criticized. Henry R. Myers, "Federal

Decisionmaking and the Trans-Alaska Pipeline," vol. 4, Ecology Law Quarterly 915 (1974-75).

38. 42 USC §4321, et al.

39. For example, the California Environmental Quality Act, California Public Resources Code §21000, et seq.

40. NEPA was intended to be not only an environmental full-disclosure law but also was intended to effect substantive changes in decision making. See Environmental Defense Fund, Inc. v. Corps of Engineers, 470 F.2d 289 (CA 5, 1972); cf. Scenic Hudson Preservation Conference v. Federal Power Commission, 453 F.2d 463 (CA 2); cert. denied, 407 US 926 (1971). The obligation NEPA imposes on agencies is to consider environmental consequences while requiring (1) rigorous inquiry, Calvert Cliffs Coordinating Committee, Inc. v. Atomic Energy Commission, 146 US App. D.C. 32, 449 F.2d 109 (1971); Greene County Planning Bd. v. Federal Power Commission, 455 F.2d 412 (CA 2), cert. denied 409 US (1972); (2) a reasoned decision, Monroe County Conservation Council, Inc. v. Volpe, 472 F.2d 693 (Cal., 1972); NRDC v. Morton, 148 US App. DC 5, 548 F.2d 827 (1972); and cf., Hanley v. Kleindeinst, 460 F.2d 640 (CA 2) cert. denied 407 US 990; and (3) a statement which is not a brochure (cf., Brooks v. Volpe, 350 F. Supp. 269 [D.C. Wash., 1972] supplemental 350 F. Supp. 287); see, EDF v. Armstrong, 382 F. Supp. 50 (D.C. Ca., 1972); EDF v. Froehlke, 473 F.2d 346 (CA 5, 1972); Sierra Club v. Froehlke, 345 F. Supp. 440 (D.C. Wash., 1972), Greene County, supra. (This case and Morton supra called for a single coherent statement and rejected substitution of written testimony, but this testimony had not been circulated to afford other agencies a real opportunity to come forward.)

While questions of broad significance may be required to be addressed, as in Swain v. Brinegar, 517 F.2d 766 (CA 7, 1975) and Sci. Inst. for Public Info. v. AEC, 581 F.2d 1078 (CADC, 1973), procedures should assist judgments: Sierra Club v. Morton, 510 F.2d 813 (CA 5, 1975); Lathan v. Brinegar, 506 F.2d 677 (CA 9, 1975); Sierra Club v. Morton, 395 F. Supp. 1187 (DCDC, 1975) and Union of Concerned Scientists v. AEC, 499 F.2d 1069 (CADC, 1974); Sci. Inst. for Public Info., supra; cf. NRDC v. TVA, 347 F. Supp. 128 (DC Tenn), 502 F.2d 852.

41. This situation is reflected in testimony presented in the fall of 1975 to the Subcommittee on Energy Research and Water Resources of the Senate Interior Committee. U.S., Congress, Interior Committee, Geothermal Development: Hearing, 94th Cong., 1 Sess., 1975.

An administrative tribunal in the Interior Department, the Interior Board of Land Appeals (IBLA), has determined that the department has the authority and duty to develop environmental impact statements on forest lands for proposed oil leasing, in a case where the Forest Service has not developed such a statement: Chevron Oil Company, IBLA 76-424, decided March 15, 1976. This decision may provide, by parity of reasoning, authority for the BLM to prepare environmental impact statements so that it might make determinations upon applications for geothermal leases, and not be bound to await Forest Service action.

42. U.S. Congress, Senate, Interior Committee, *Hearing on Geothermal Energy Development*, 94th Cong., 1 Sess., 1975, pp. 75-76. Testimony of J. Barnett.

43. Ibid.

44. Note *California Power Agency v. PUC*, 5 C.3d 370, 96 Cal. Rep. 18, 486 P.2d 218 (1971), and cf. allegations in Letter of Advice of Attorney General in regard to PG&E's application for the Mendocino Plant, 37 F. R. 1642 (1972).

45. See, e.g., Utah Code Anno. §73-1-20, 14 Colo. Rev's Stat., 1973, §34-70-101, 35. et seq.

46. New Mexico Stats. 9, Part 2, Sections 65-11-4 and 65-11-5 (C).

47. Ibid., Section 65-11-9.

48. Ibid., Sections 65-11-10 and 65-11-11.

49. Ibid., Section 65-11-11.

50. Ibid., Section 65-11-13.

51. Ibid., Section 65-11-14.

52. Stephen L. McDonald, *Petroleum Conservation in the United States: An Economic Analysis* (Resources for the Future Ser.) (Baltimore: Johns Hopkins Press, 1971); John Vafai, "Market Demand Prorationing and Waste--A Statutory Confusion," *Ecology Law Quarterly* 2 (1972): 118.

53. New Mexico has created a body under its Energy Resources Act, New Mexico Stats., 9, Part 2, §65-13-1, to consider how the state might capture economic rents from resources located in the state; Montana has sought to capture such rents by imposing a large severance tax on coal, rising to 30 percent of gross sales for strip mined coal over 900 Btu per pound heating value. Revised Codes of Montana, Part 3, 1975 Interim Supplement for Title 84 to 95, Sections 84-1312-1818.

54. Attorney General's Report, 1967.

55. Connally Hot Oil Act, 15 USC 715.

56. As regards oil prorationing, "the key to the effectiveness of State regulation, then, was a federal consent to these market controls, and federal enforcement of prohibitions on interstate shipment of oil produced in violation of them," according to the memorandum for the Attorney General, accompanying *Report of the Attorney General* pursuant to Section 2 of the Joint Resolution of September 6, 1963,

consenting to an Interstate Compact to Conserve Oil and Gas (July 1967). This report was one of an annual series of Justice Department Reports regarding the Interstate Oil Compact Commission. The June 1964 report describes state market demand control, while the 1969 report contains an extensive history as to the Connally Hot Oil Act.

57. 15 USC 79(c).

58. Michael Sullivan, Steven McDougal, and F. Van Huntley, Patterns of Geothermal Lease Acquisition in the Imperial Valley: 1958-1974, Report to the State government, University of California, Riverside, August 1974, pp. 102-05.

59. Because of the risks of exploration, planning a drilling program should encompass risk analysis. By improving analyses and analyzing more structures, the probability of locating a commercial resource improves. See Megill, An Introduction to Exploration Economics (1971).

60. Table 11.6 of the FTC Staff Report on Federal Energy Land Policy (October 1975).

61. Robert Greider, "Geothermal Statement: Based on Testimony Presented to Lieutenant Governor Bynally's Committee," Geothermal Energy Magazine 4 (April 1976): 27.

62. Testimony of F. M. Scherer, U.S. Congress, Senate, Judiciary Committee, Subcommittee on Antitrust and Monopoly, Hearings on the Petroleum Industry: Vertical Integration, Part III, 94th Cong., 2 Sess. (1976), p. 1834.

CHAPTER

3

ELECTRIC UTILITIES AND THE CLIMATE FOR GEOTHERMAL DEVELOPMENT

The electric power industry's basic functions are the generation, transmission, and distribution of energy to ultimate customers. These functions are coupled through interconnection and coordination of power systems. Some utilities engage in all of the industry functions while others only perform one or a few functions, for example, distribution. Figure 3.1 provides a description of the industry.

Technical advances in generation and transmission have made possible enormous economies of scale in the electric power industry. Few, if any, existing investor-owned utilities are alone large enough to exhaust these economies. Private utilities have often sought to expand through acquisitions and mergers. They have frequently used the holding company device. To a significant degree, consolidations have been aimed at achieving scale economies. Particular goals have included the use of a united organization to coordinate planning, to share reserves, and to achieve operating economies. Utilities have also sought to cooperate in varying degrees with one another to achieve these objectives.

The working of power systems is referred to as integration where it proceeds by means of uniting ownership, and coordination where the expansion involves cooperation between firms.

THE ROLE OF POWER POOLING

Interconnection and joint operation and planning of electric power systems (pooling) are necessary to find outlets for the output of generating units. Reaching these outlets is necessary to realize the economies of scale that large, high capital cost equipment can provide if fully utilized. It also permits systems to employ cheaper

FIGURE 3.1

The Electric Power Industry, 1970

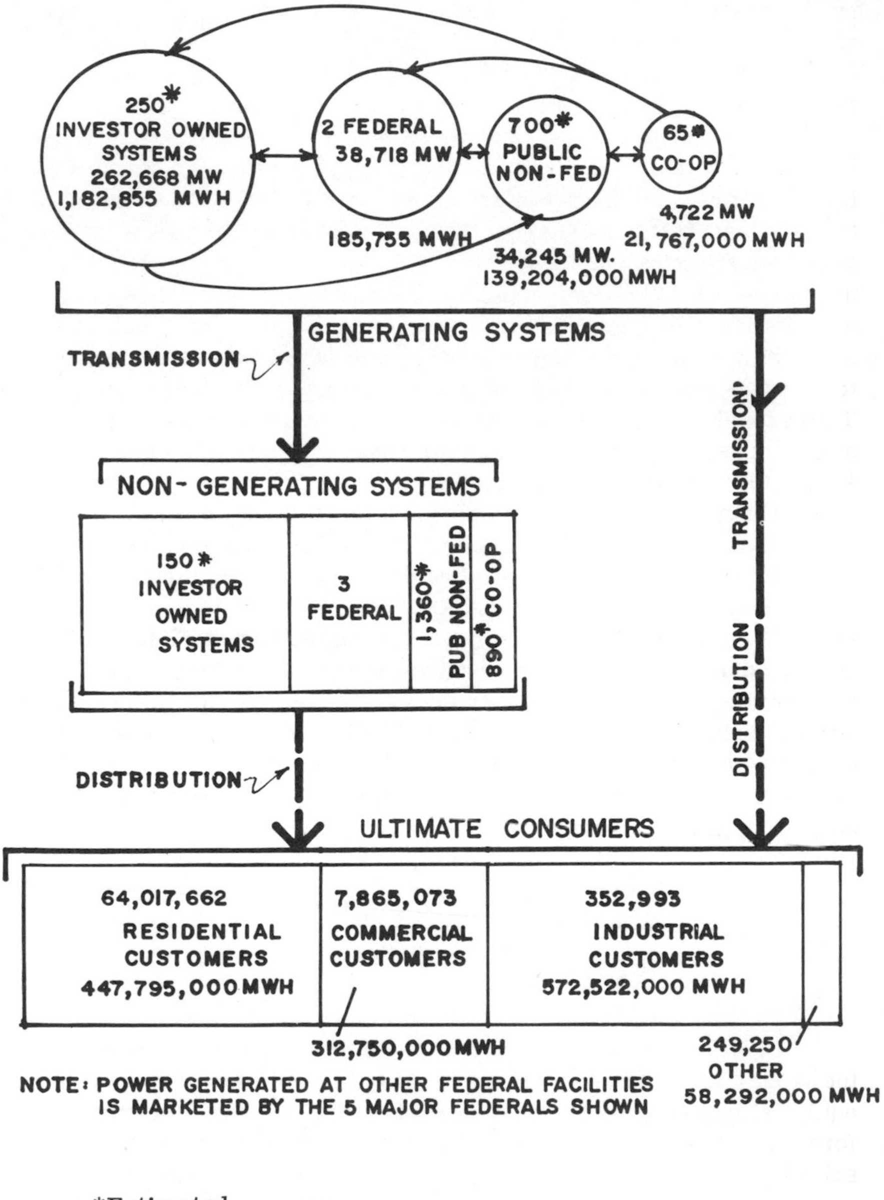

*Estimated.

Source: Federal Power Commission.

source units before more expensive ones, thus minimizing operating costs, and to share reserves.

Isolated systems must maintain generating capability sufficient to serve forecast loads and to provide adequate reserves in the event of forced or scheduled outages of generating units. They also cannot share reserves or economies of scale obtainable from units too large for one system to install alone. Similarly, an isolated system cannot engage in trading with other systems to take advantage of intersystem variations in operating costs.

The structure of the industry today is the product of an evolutionary process that occurred over the past century.[1] Small isolated utilities combined into groups of interdependent systems through transmission tie lines that permitted the flow of energy between the systems during periods of equipment outages. Utility managers found that they could reduce operating costs through capacity sharing transactions and economy energy exchanges. As the size of the most economical units increased, utilities were prompted to move to more extensive and longer-term coordinated planning. Table 3.1 describes the functional stages of this evolution. Each stage represents an increase in the degree of sophistication, but the stages are not mutually exclusive.

The actual degree of planning and operating coordination among utilities varies. Some pools have a far higher degree of coordination than others; for example, under some coordination arrangements economic dispatch is limited to periodic transactions, and not done on a moment-to-moment basis. Similarly, planning coordination may be limited to sales of capacity excess to needs of a unit owner's system, and may not encompass coordinated construction of generating units of sizes designed to serve needs of several coordinating utilities.

An explanation of some of the more important features of coordination in pools follows.

KEY FEATURES OF ELECTRIC POWER POOLS

Reserve Sharing

Utilities plan to have sufficient generating capacity to meet their peak loads when provision is made for reserve capacity to cover equipment outages, frequency regulation, load swings, errors in forecasting loads, and slippage in planning and construction schedules.[2]

The sharing of reserves is the cornerstone of power pooling and accordingly will be discussed in some detail. In the absence of

TABLE 3.1

Function Stages of Coordination within the Electric Utility Industry

Stages	Purpose of Interconnection	General Description	Types of Transactions/Service
I	Reliability	Decrease probability of loss of load due to forced outage or other equipment failure (for a given mix of generation and transmission).	Reserve sharing and/or emergency support.
II	Reliability and operating efficiency	Minimize cost of serving a given load at a specified level of reliability with existing mix of generation and transmission.	Economy exchange; daily diversity exchange; maintenance scheduling and energy exchange; short-term power exchange.
III	Reliability, operating efficiency, and joint planning	Determine the mix of generation and transmission voltages and configurations that minimizes the cost of serving electric loads over time at a given level of reliability.	Seasonal capacity exchange; staggered construction, including unit power and long-term power exchange.

Source: Based on David W. Penn, et al., Coordination, Competition, and Regulation in the Electric Power Industry, Nuclear Regulatory Commission, Economic Analyses Section, Office of Antitrust and Indemnity, June 1975 (NUREG-75/061), p. 30.

reserve sharing, an entrant generating firm using geothermal power may be unable to sell firm power.[3] Reserves may be classified in terms of the immediacy of their availability (for example, spinning reserves, ready and standby units*) and in terms of their roles (for example, reserves for scheduled maintenance as opposed to reserves to be used in the event of an unscheduled operating event).

Spinning reserve is that portion of operating reserve capacity which can pick up in five to ten minutes. Ready reserve is that portion of reserve capacity which can take load within 10 to 20 minutes. Generally, hydraulic turbine generators, quick-starting diesel engine generators, and gas turbine generators can be used for ready reserve. Standby reserve is the remainder of a utility's reserve capacity. It generally consists of older units with relatively high operating costs.

Reserve sharing arrangements may also vary according to the duration of the service to be provided. Separate provisions will be made for short-term operating contingencies, and for reserves to be provided for month-long periods to meet problems of planning or long-term unit outages. The amount and type of reserves a utility must maintain depend on the number, size, and type of its generating units and the configuration of its transmission lines.

To understand the purposes of reserve sharing, consider first the situation of a simple system operating in isolation. If the system is to have 10 mw of firm load-carrying capability, Figure 3.2 shows three ways in which firm power might be provided. Spinning reserves must always equal the capacity employed on the largest unit generating. Accordingly a choice must be made between use of larger units having lower costs per unit of capacity and the rising reserves required by such units use. This trade-off is illustrated in the lower portion of Figure 3.2.

Interconnected systems can share reserve capacity. Reserve sharing requires both adequate interconnected power transmission capacity and contractual arrangements to provide emergency power and to maintain sufficient reserve capacity so that agreed upon reliability criteria are maintained. Under such contracts, reserves are to be sufficient to cover all combinations of unit outages except those whose likelihood of occurrence is remoter than an agreed upon level--usually one service interruption in ten years. The system reserve is the difference between the system capability inclusive of

---

*Geothermal wells cannot be quickly stopped and started, nor can they be kept in operation when not being harnessed for generation. Accordingly, a geothermally powered system must go to others for its ready reserves.

receipts under power purchase contracts and the supply requirement of a system. Installed reserve requirements of electric systems generally range from 15 to 20 percent.[4]

FIGURE 3.2

Three Ways to Provide Firm Power

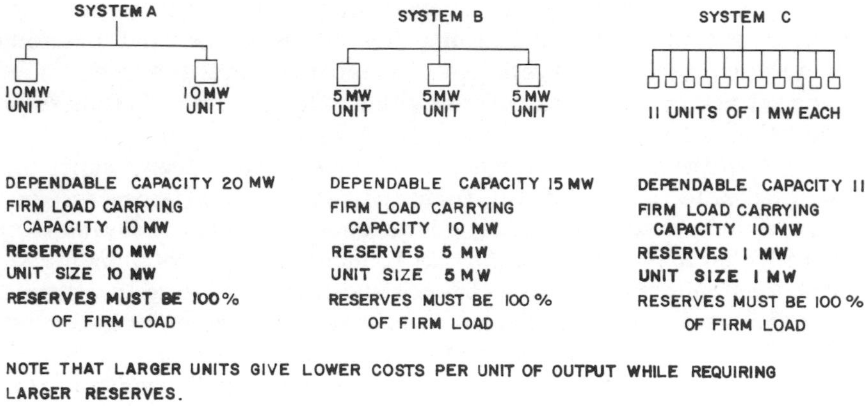

There is active trading of operating reserves among interconnected systems where the cost of capacity is customarily quoted in dollars per megawatt-day and the price varies with sources. Interconnection agreements frequently provide for a fixed service charge to be paid irrespective of any actual deliveries of power; emergency power and energy may also be provided on the basis of subsequent return in kind.

As a system grows larger, the benefits conferred on the system by entering into such coordination with smaller systems declines progressively. Smaller systems, however, benefit greatly by this coordination. A smaller system contributes progressively less reserves as a percentage of load to a larger system than it receives, while the smaller system's load-carrying ability (with existing plant) rises.*

---

*If a system A having two 10mw units coordinates with a system B having four 10mw units, system A's reserves decline from 10 mw to 3 1/3 mw and its load-carrying ability rises from

Coordinated Development

Utilities also coordinate the installation of capacity. To understand the purpose of this type of coordination, consider again the case of an isolated system. While system loads will rise fairly smoothly over time, new units must be installed in discrete lumps. This results in excess reserves until load again catches up with capacity. See Figure 3.3.

When several systems coordinate load growth, the number of units installed in the coordinated network increases. Each unit is a smaller proportion of the total network load. The step-like installed capacity curve comes closer to the line that corresponds to the curve for required capacity for the network, so that excess capacity is reduced.*

Coordination of development can be achieved by joint unit ownership, by each system building units in turn and making short-term or long-term sales of power, or by a jointly owned separate generating company. With such coordination, the size of planned units can be based on the aggregate annual load growth of the coordinating systems. Achievement of scale economies and more intensive use of plant sites can result. Again the larger of unequal-sized systems generally gains less from load growth coordination than the smaller.†

---

10 mw to 16 2/3 mw, while system B's reserve needs decline from 10 to 6 2/3 mw and its load-carrying ability rises from 30 mw to 33 1/3 mw.

If system B has 6 units of 10 mw each, its reserve need in coordination with system A declines from 10 mw to 7.5 mw and its load-carrying ability rises from 50 to 52.5 or 5 percent. System A's reserve needs decline from 10 mw to 2.5 mw and its load-carrying ability rises 75 percent from 10 mw to 17.5 mw.

If a 100mw system shares reserves with a 1500mw system carrying 15 percent reserves, the small system increases its load-carrying ability 30.4 percent while the larger system may add nothing to its load-carrying ability.

*Moreover, with coordination, parties having excess capacity are more likely to be able to arrange to sell and transfer output to third parties.

†As another illustration, consider the situation of load growth coordination between two systems, A and B. Each has a 10 percent per year load growth rate: A, a 700 mw system, has a 70 mw annual load growth increment; and B, a 1500mw system, has a 150 mw annual load growth increment. In the absence of load growth coordination, Systems A and B would separately install units whose size

FIGURE 3.3

Capacity and Loads

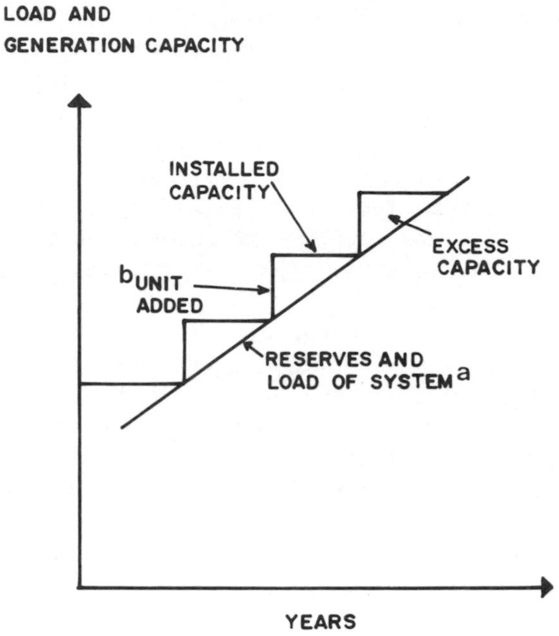

[a]The slope of this line is set by annual load growth.

[b]Planning coordination can be employed to increase the size of units added, as well as to reduce excess energy.

Coordination to Match Resources and Loads

Different types of capacity have different load-carrying characteristics. The FPC defines dependable capacity as "the load-carrying ability of a station or system under adverse conditions for the time interval and period specified when related to the characteristics of the load to be supplied. Dependable capacity of a system includes net firm power purchases."[5] Systems attempt to

---

would be based on three years' load growth, that is, 210 mw for A and 450 mw for B. With coordination, larger, more economical units are possible.

match capacity load-carrying characteristics with load durations. Linking hydro and thermal units permits some economic possibilities not otherwise available. Coordination between electric utility systems allows additional scope for such link-ups to occur.

For example, the dependable capacity of a system using hydro units can be maximized by allocating the hydro use to expectable peak demand periods; and by providing thermal unit capacity back-up for dry years. The situation is illustrated by Figure 3.4, representing a system load duration curve. A unit capable of generating 50 mw at any given time, but with a limited source of energy (for example, water supply), could be assigned only 30 mw of dependable capacity, were it required to serve at Area B on the load duration curve of Figure 3.4. It could be credited with 50 mw of dependable capacity if it were planned to serve at Area A, representing a shorter time period.

FIGURE 3.4

Load Duration Curve

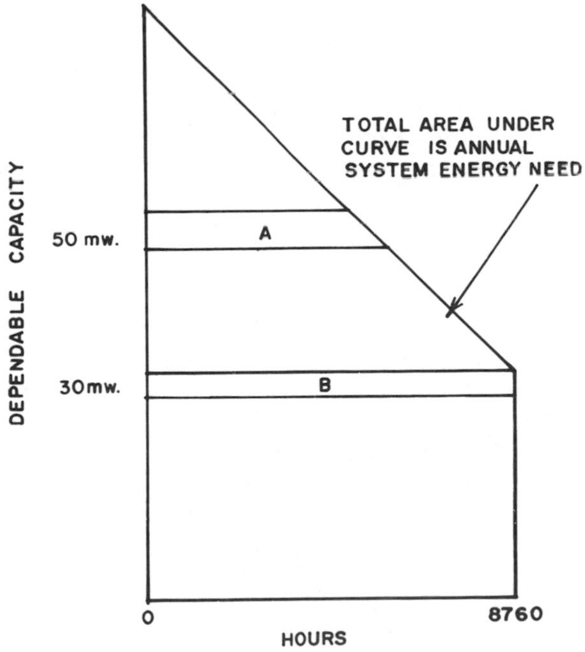

Note: With energy in Area A = that of Area B, unit loaded at A will have 50mw dependable capacity, while at B only 30mw.

Transmission Coordination

Coordination of power systems requires interconnected transmission. The feasibility of interconnecting transmission facilities depends upon distance and load, as illustrated in the set of examples set out in Figure 3.5.

Installation of larger transmission lines can result in lower transmission losses and reduced carrying costs per kw. Transmission coordination reduces the effect of loss of a line or transformer. Space can be conserved by coordinated planning to permit more complete use of the limited number of transmission corridors. This may be essential for construction in an environmentally concerned era. A system that dominates transmission may be able to choose whether smaller systems in its region have opportunities to use the various forms of coordination, or are prevented from realizing such possibilities.

Economic Dispatch

A utility generally attempts to bring existing generation units into operation as load increases in the order of the units' marginal delivered costs--from least cost toward higher cost units. System operators often seek to adjust output from each generation source so that the marginal cost of output from each source is equal. By interconnecting, it is possible for two or more contiguous utilities jointly to practice this economic dispatch. As the combined load of the several utilities increases, that unit with the lowest marginal cost is brought into service first, regardless of which of the utilities owns it. In this manner, production costs of the several utilities are jointly minimized.

Diversity in incremental costs for different generating sources may arise because of differences in the loads of various units at any given time, different effects of climate, other seasonal differences, differences in fuel supply, and costs or benefits associated with alternative disposition of the units.

Potential benefits of joint dispatch increase in proportion to the diversity of the load patterns of the utilities. Without transactions of this sort, a utility would not make available a large unit with low marginal costs to another utility approaching its peak load and having only its least efficient units to put into service. With economic dispatch, the utility with cheaper idle generating resources can sell energy to the other utility at a small fraction of the other utility's internal generation cost.

FIGURE 3.5

Economics of Transportation Coordination

Case 1

CASE I ECONOMICS FAVOR CONSTRUCTION OF TRANSMISSION LINE WHERE COST OF DEVELOPING ISOLATED GENERATION MORE THAN OFFSETS THE COST OF TRANSMISSION INTERCONNECTION.

Case 2

IN CASE 2 IT MAY NOT BE FEASIBLE TO BUILD TRANSMISSION SYSTEM WHERE CONTINUED OPERATION OF ISOLATED GENERATION IS LESS THAN COST OF OPERATING LARGE SCALE GENERATION AND LENGTHY TRANSMISSION SYSTEM.

Case 3

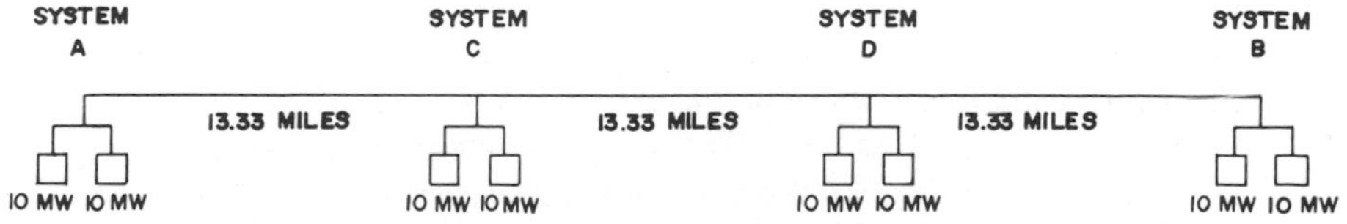

IN CASE 3 IT IS FEASIBLE TO CONNECT ALL FOUR BECAUSE OF SHARING TRANSMISSION COSTS AND COORDINATED DEVELOPMENT OF GENERATION. YOU CAN INTEREST C AND D BECAUSE THEY HAVE MUCH TO GAIN.

PRINCIPLE NO. 5
THE ECONOMIC FEASIBILITY OF COORDINATED BULK POWER SUPPLY DEPENDS ON THE SIZE OF THE SYSTEMS AND THE DISTANCE BETWEEN THE SYSTEMS.

Although less effective, periodic longer term transactions for exchanges of energy can capture some of the economies of joint dispatch.

POTENTIAL COORDINATION OBSTACLES IN POOLS: THE PROBLEM OF TRUST

The benefits of power pooling are obtainable only to the extent that a variety of problems are overcome by participating utilities. Each member of a pool is interested, understandably, in getting the most benefits it can. A member may be inclined to take advantage of other members, which might be termed opportunistic behavior. The problem of trust in pools is that a wide variety of situations that frequently occur in pools could lead to opportunistic behavior on the part of one or more members. Consequently pool members may be inclined to limit the degree of their coordination. Instead of attempting to achieve the full economies of coordination, a pool member might enter into only piecemeal agreements on the sharing of reserves, economic dispatch, and capacity expansion in order to minimize the potential for opportunistic behavior by others at its expense.

Achieving economies of reserve sharing and reliability in a pool is not a sure thing. Broad agreements and long-term contracts can be made, but there is no assurance that they will be kept. Trusting a neighboring utility to keep an agreement to provide reserve at a particular time is not as safe as having one's own.

Imperfect information permits opportunistic behavior as to reserve sharing and deficiency purchasing obligations. Such obligations are based on the capacity of a member. The unit capacity available cannot be measured without an expensive performance test. Consequently, each member must rely on the integrity of the others in reporting capacity. It is not uncommon for members to report capacity at 5 or 10 percent below nameplate rating. This problem is magnified when utilities in certain regions experience shortages of condenser water. If a member with 2,000 mw of capacity is obligated to provide the pool with reserves equal to 15 percent of its capacity, and if the member understates its true capacity by 10 percent, the member could, in the course of a year, avoid paying the other members over $1 million. If a pool member overestimates his available capacity, he can avoid paying deficiency charges. If deficiency charges are less than new unit costs, a member may seek to rely on capacity of more conservative members--until the contract is refused. The potential for this sort of activity could motivate pools not to exploit entirely the economies of reserve sharing.

In the area of economic dispatch, problems of trust are also present. Probably the greatest deterrent to formal pool operation is skepticism regarding the fairness of determining and allocating the savings resulting from economic dispatch. In a pool, the true saving from economic dispatch is the difference between the sum of the costs that would have been incurred by the individual members without interchange and the sum of the costs incurred in actual operation. Three general methods are used to allocate the savings to members: equal distribution of all pool savings to all participants, distribution on a split savings basis as a series of two-party transactions, and distribution of savings in proportion to member net generation. Measurement problems plague these allocations.

Suppose Utility A experiences an increase in load for one hour. The pool central dispatcher will satisfy this extra load from the lowest operating cost unit not in use. Suppose this unit, B-3, is owned by Utility B. The true savings which result to the pool would be the difference in operating costs for one hour between the lowest cost unit not in use in Utility A, say, A-6, and unit B-3. Suppose unit B-3 were nuclear. There is no known method for accurately measuring nuclear fuel costs for a period as short as one hour. Suppose unit B-3 were hydraulic. An argument could be made that the cost of one hour's generation was negligible. An argument could also be made that the cost was higher than the corresponding cost of one hour's generation from unit A-6. The true answer is that the cost was the opportunity cost of the potential energy lost by operating unit B-3. Since water can be stored for prolonged periods, the opportunity cost depends on the magnitude of the pool peak, equipment availability, and natural resupply of water over a period of several weeks in the future. By using the water now, the pool might have to bring its most expensive peaking unit into operation to meet peak demand in two weeks. Standard methods can be stipulated in the pool contract, but there is much opportunity for disputes, contract renegotiation, and exercise of escape clause privileges.

Other potential problems include accounting for semi-variable costs such as start-up costs, station lighting, maintenance costs, and labor. Since union contracts vary from member to member, Utility B might have to pay the station crew overtime or even a full day's pay for, say, the last 15 minutes of operation because the business day for the full crew might end during the hour of generation.

All of the above possibilities could conceivably be stipulated in a pool contract, but there are always new possibilities arising that are not currently covered. Moreover, the basic problem of trust is involved, a great deal of detailed information is required to compute savings and allocations, and much of one member's reported information cannot easily be verified by the other members. Note that these same costing problems must be solved, also, in an integrated system.

Trust is challenged in the area of long-term capacity expansion programs. If all economies of staggered expansion are exploited, the member who delays building its own new capacity runs a risk that the pool might dissolve before its turn arrives to build an optimal scale plant. In that event, it will have to supplement its own generation with much more expensive firm power from other utilities for a period of several years. Contract stipulations could cover this, but not completely. The very procedure of trying to cover every possibility of fraud or default breeds suspicion in the pool. A tendency might be present for suboptimal expansion programs.

The Problem of Size Disparities with Concomitant Benefits and Bargaining Power

As heretofore noted, small utilities can provide only limited benefits to larger utilities.[6] The larger utility might prefer that the small utility abandon its capacity and buy all its requirements as firm power.* If the larger utility did pool with the small utility, the lowered production costs of the small utility would permit it to compete with the larger utility for industrial customers.

Another problem is that small systems often cannot provide other small systems with substantial benefits. Many are in small enclaves in the enormous service areas of large utilities far from the service areas of other small utilities. The cost of buying rights-of-way and building transmission facilities to the closest small utility would be prohibitive. Adjacent large utilities are reluctant to coordinate with, or wheel power for, small utilities; it may be in their own interest to retain small systems as captive wholesale customers or to acquire them.

Small utilities find themselves frequently unwelcome in pools, unable to form pools on their own, and operating in an environment in which larger utilities may impose price squeezes so as to cut off the small chances of attracting and retaining industrial customers.

THE ROLE OF COMPETITION IN THE ELECTRIC POWER INDUSTRY

In providing electric utility services, there are a number of opportunities for competition, some latent and some realized. Potential and actual competition occurs at the retail level, in wholesale

---

*Current financing problems have apparently prompted some utilities to drop their resistance to jointly constructing projects with municipal or cooperative systems.

bulk power sales, and in trading among bulk power supplies. As one would expect, this potential and actual competition affects relationships among electric utilities.

The nature of the rivalry among systems can determine their operating and capital costs, and consequently their rates. Sometimes it affects their very survival. Rivalry also provides comparative standards for judging the performance of power systems and may lead to more economic performance.

Retail Load Competition

Opportunities for retail competition are limited in the electric power industry. Local franchise requirements, service area certification by state regulatory commissions, and state antipirating laws frequently result in exclusive territories for retail distribution.

Retail load competition can occur for new customers that have not yet chosen a location, or on the borderline between adjacent utilities when residential, commercial, or industrial developments are built in previously undeveloped areas or when a municipality with its own distribution system annexes surrounding areas already served by another utility. Although such occurrences appear to be infrequent in any locality, they are important when summed over a region or nation, and over 5- to 10-year time spans.[7]

Retail load competition to attract industrial customers can be vigorous:

> Although electric power is a minor expense for the majority of manufacturers, the industries where it is important (more than 3 percent of sales) account for about half of total industrial load. Industrial electric rates are generally low because industrial demand is very elastic. That, in turn, reflects the ability of industrial customers to locate in any of many service areas or to generate their own power. The rates charged very large industrial customers are commonly negotiated between the company and the utility and then filed with the public utility commission. The competitive efforts of utilities to get new industrial load are often very vigorous.[8]

While power costs are a small part of the total costs of most firms, this is not the case for certain heavy industries: chemicals, metals, paper, petroleum, and stone, clay, and glass.[9]

A recent examination of 49 cities served by two competing utilities, in most of which cities a consumer had a choice of being served by one firm or the other, found that up to the point where annual sales reached 222 million kwh, average costs were lower for municipal systems that faced competition than for those that did not.[10] There was no net effect of competition beyond that level. It was also found that innovation (installing underground cable) by the municipal was quickly imitated by the competing investor-owned firm.

Utilities can limit and stifle retail rivalry in a variety of ways. Often, competition for industrial load is between the small utility and the large utility that supplies the small utility's energy. Power system managers, whether of large or small systems, actively seek industrial customers because of their beneficial effects on system load factor. Indeed, it is possible for a large utility, which has succeeded in capturing the retail industrial market to the exclusion of small systems that are also its wholesale customers, to argue that its wholesale rates should be raised because of the poorer load factor of the smaller systems. This argument can be used to defend a price squeeze. The large utility can raise the wholesale rate for the small utility to a level above the large utility retail rate that the small utility must offer to land a prospective industrial customer. Such a price squeeze effectively prevents the small utility from gaining the customer and generally results in the prospective industrial firm becoming a customer of the large utility, possibly at a retail rate higher than the small utility would otherwise have charged.[11]

Large utilities can reduce opportunities for retail competition by refusing to wheel power from others to small utilities. If a large utility refuses to wheel power to a small utility in its service area, the small utility will be deprived of a cheaper source of power and is then unable (without resorting to illegal rate discrimination) to offer potential industrial customers retail rates lower than those possible under its own wholesale rates from the large utility. (This form of conduct also reduces wholesale competition.)

Sometimes the larger system may refuse to interconnect with a smaller system except upon onerous terms. A more subtle practice is the discouragement of even minor ventures by wholesale customers into the construction of transmission facilities. Some utilities, for example, refuse to consider instituting a discount for delivery to the wholesale customer at a higher (transmission) voltage, even though under such arrangements the customer would assume the cost burden of building and owning the necessary stub lines and transformer stations. Without the economic incentive to make a start in transmission, the wholesale customer is in a poorer position to seek to connect to a different bulk supplier or to join in a bulk power project with other small systems.

Large utilities sometimes reduce opportunities for retail competition by refusing to permit small utilities to seek to lower their bulk power costs by joining coordinating groups or by participating in the construction of new generating units. These forms of conduct will be discussed in more detail hereafter.

Finally large utilities can reduce opportunities for retail competition by inserting restrictive provisions in wholesale or wheeling contracts with small utilities. Such provisions may limit resales of power to existing customers, or they may prohibit the resale of the power to wholesale customers. Also, de facto market-sharing provisions in coordinating arrangements--for example, provisions for sharing new facilities or for allocating ownership of transmission--affect the potential for retail competition.

On-Site Industrial Power Installations and Utility Opposition Thereto

Industrial, and occasionally commercial, power consumers sometimes individually generate their own electric power. This is a form of retail competition.

About 4 percent of the installed capacity of electricity generation in the United States is at nonutility complexes--mines, mills, electric railways, office buildings, and so on. Utilities have historically resisted such projects. For example, parties attempting to install on-site units providing heating, cooling, and electric generation (total energy installation) have found it necessary to install their own standby equipment.[12] Similarly, obtaining supplementary power has been a problem for total energy users.[13] Consequently, most such units provide space heating and cooling, but not electricity.

According to Mr. Edward A. Myers, Jr., vice-president of SCE:[14]

> . . . utilities have generally been concerned about the potential loss of base load, about the use of utility transmission facilities for wheeling, and about any idea that the utility system might be expected to provide backup to on-site generation without adequate compensation.
>
> . . . if a project is both technically and economically feasible, then the utility will itself give serious consideration to installing the generation.

> . . . Concern was expressed about frequency control, the danger that proliferating small units might pose to system dispatching problems, and the need to be "fully compensated" in standby service.
>
> . . . "very large" customers with multiple plants in a utility's service territory may want to have wheeling and combined metering. This concept is not cost-effective or practical from an electrical stability viewpoint. Additionally, the concept poses numerous liability problems in the event of transmission problems. The customer also may not be able to afford his fair share of the costs of wheeling as do the various interconnected utilities.

Franchise requirements can thwart self-generation moves. For example, an effort to obtain coordinating sources for a total energy system in Utah was thwarted because service outside of a single building was required, and this was deemed to require a franchise. No franchise was available because none would be issued in the service territory of an existing utility.[15]

Competition in Bulk Power Supply Markets

Electric power supply competition generally occurs in what will be referred to as bulk power markets. Bulk power transactions are made among suppliers to retail markets. Competition in bulk power supply markets requires that a utility system have a choice among sources of power and energy and of coordinating services. With such choices, it can enter into sales, purchases, or exchanges with the systems offering the lowest costs, and it in turn can sell or exchange power, energy, and service when its costs are lower than others. Likewise, with competition a utility can acquire access to and develop a variety of new power supply resources.

Utility system planners continually evaluate available power supply options in order to assure themselves of dependable, efficient, long-term sources of power and energy. These options include wholesale purchases for resale, self-generation, purchases of reserves and other back-up support, a share in a jointly owned unit, an entitlement to output from another system's generating unit, power on an interruptible basis, power on a firm basis for a defined period, and various other types of arrangements that can be lumped under the "coordination" heading. The ability to pick and choose among these various options, by type and source, is a competitive process. This process can stimulate efficiency and progressiveness.[16]

Integrated utilities engage in active trading of capacity and energy, in order to minimize the costs of bulk power supply by preferential use of lowest cost generation resources regardless of ownership. Similarly, integrated utilities coordinate the planning of high voltage transmission and new generation jointly to reduce their costs. Transactions are made in this market for emergency or stand-by power, for ready and spinning reserves, for economy and dump energy and for economy power, for transmission services, for unit power or deficiency power, for joint enterprises to install large base-load generating units. Opportunities for wholesale and coordinating services competition have improved with advances in transmission technology and with the growth of interconnected transmission grids.* Coordinating services may be separately traded, or, as is usually the case, bundled in a coordination agreement.

Because of system stability, transmission cost, and administrative burden problems, interconnection and transmission service arrangements are generally coupled with provisions regarding reserves and emergency services. Agreements for more extensive coordination, including those for poolwide economic dispatching, short-term or unit power sales, and for planning almost invariably provide for the reserve and emergency coordination needed to keep the trading systems interconnected.

Grouping of various coordination services into a market, while recognizing that some of the discrete services so bundled can constitute distinct submarkets when separately traded (for example, short-term power and energy) conforms with business practice. Such grouping of services is recognized in antitrust law as having a distinct value and constituting a market.[17] A market is characterized by the substitutability of the items traded therein.[18] Coordinating services as a group are obtained in one agreement along with other economy transactions, short-term purchases and sales, and unit or seasonal power arrangements.

---

*Wholesale loads are loads of utilities served by exchanges or purchase of power and energy from another utility. Transmission costs are directly proportional to distance and inversely proportional to the square of transmission voltage. Use of extra high voltage transmission lines has made it economical to transmit power to Los Angeles from the Four Corners plant in New Mexico 600 miles away and from the Columbia River 850 miles away.

The pool created by coordinating agreement constitutes the "area of effective competition" for many transactions particularly where a pool encompasses a wide geographic area.[19] Treating coordinating services as one market, rather than as a series of markets, fulfills such practical indexes of market as industry recognition, the grouped products' peculiar characteristics and uses, the assemblage of production facilities required, customers distinct from those buying wholesale, firm, or retail power and energy, the relationship of terms of trade to the existence of a package, and the lack of alternate vendors.[20]

It should also be noted that by controlling access to bulk power markets through control of transmission or by direct and contractual control of sources of back-up power, a firm or group of firms controls coordination. The submarket for high voltage transmission and back-up is quite local, being limited by the costs and environmentally related inability to implicate high voltage grids, and the necessity to obtain back-up from nearby sources not subject to large transmission losses and to outages. Transmission facilities and back-up facilities are both bottlenecks through which the commerce of bulk power must flow in any geographic area.

A firm with monopoly power in one market may not lawfully employ selective refusals to deal so as to expand that monoply into another market.[21] Similarly, a firm or group controlling a bottleneck resource that must be employed to reach a market cannot refuse to grant access to this resource to competitors.[22] Exclusion from a bottleneck need only confer a "competitive disadvantage" to be actionable.[23]

Refusals to deal may also constitute unfair trade practices. Such refusals may occur both by outright means and by imposition of unfair or unworkable terms in contracts. The case law referred to earlier teaches that refusals on the part of large utilities and power pools to afford coordination and transmission--essential for access to bulk power and retail markets--are not in keeping with the policies of the antitrust laws.

Without the development of smaller-scale generating resources such as geothermal units, small systems must obtain their new base-load generation from large coal or nuclear-fueled units. Small system participation in the ownership of these large units would generally entail cooperation between large and small systems. In the light of the history of anticompetitive conduct in bulk power markets, such cooperation cannot be relied upon to permit small systems to continue to exist as a competitive force. Dependence of small systems upon large ones would be fostered by reliance on large units. Especially with other forms of coordination being denied them, small systems are generally unable, alone or in association, to build large units.

Coordination need not preclude competition. With coordination, power systems are able to and do make choices among resource possibilities when they assemble their overall power supply package.

> For a large area, there are often many ways of developing an efficient overall bulk power supply plan or pattern of development. The existence of a diversity of approaches and the freedom to shop for options provide a degree of competitive stimulus to search for new and better power supply alternatives.[24]

While some components of coordinating services are interchangeable and some are complementary, the different combinations of these components are substitutable, one for another.[25]

The geographic scope of markets for wholesale bulk power and for coordinating services is defined both by the cost and availability of transmission and by the boundaries of interconnection and coordination agreements. Just as product markets, geographic markets reflect the limits of power pools within which trading occurs. The pools may contain the nearby load centers within economical transmission distance; pools may also be the areas within which facilities for the trading of bulk power and coordinating services are most developed.

The geographic area within which a small system can shop for coordination and bulk power is not as large as that for a larger power system. The transmission cost limits on geographic markets are more constraining for these systems, with their limited ability to install their own transmission and their difficulties in obtaining transmission services from other systems. These limits make the geographic areas in which small-system marketing occurs those of their surrounding large utility and the compact pool, if any.

However the bulk power/coordination market is defined, access to it may generally be blocked by exploitation of bottleneck facilities (that is, transmission) in a small geographic area surrounding the system that seeks access both to power supply and to use of such bottleneck facilities for coordination purposes.

The first--and in many cases the only--market definition or market share investigation needed is to establish the (narrow) boundaries of such local transmission and subtransmission market, and to show whether the large utility controls it. Relatedly, to show the large system's arrangement with neighboring utilities to prevent them from furnishing transmission which they might otherwise offer is to show crucial anticompetitive conduct.

Wheeling--use of one utility's transmission facilities by one or more other entities for energy transmission--is the key in bulk power competition. Leonard W. Weiss states, regarding wheeling arrangements that,

> The purchase of energy by the owner of the transmission line and its resale to a bulk power customer is not the same thing. If the energy is wheeled, any bulk power supplier in the region can compete for the business of the customer, but where it is purchased and resold, the owner of the transmission line continues to have a monopoly. At present wheeling agreements are voluntary and wheeling charges are not really regulated. About 3 percent of the energy generated by privately owned utilities is wheeled.[26]

This percentage has been expanding in recent years. Wheeling is of particular convenience in the distribution of federally generated power from hydro projects to utilities having marketing responsibility. Such an operation is exemplified by Bonneville Power Authority which distributes power to a number of private municipal and public utilities. Similar arrangements exist in the distribution of power into a number of nongenerating cooperatives. The wheeling of power by privately owned utilities is also widely practiced.

The current structure of the industry, however, limits the development of wholesale competition along the lines set out here. Large utilities generate, transmit, and distribute power. Their vertical integration precludes wholesale competition. Approximately two-thirds of the sales to final customers involve electricity distributed by the utility that generates and transmits it, or a like amount of power and energy as an interconnected grid. The distribution network of a large investor-owned utility could be viewed as a set of smaller distribution systems interconnected by transmission lines, each system possibly serving a distinct populated area. Each of these smaller distribution systems could, perhaps, operate independently, obtaining its power requirements as a wholesale customer from any of a variety of bulk power suppliers. The fact that it is not independent of the large investor-owned utility eliminates it as a potential wholesale customer.

The conduct of the industry also limits bulk power supply competition. The customary distinction between structure and conduct can be somewhat misleading if applied rigidly to the electric power industry. Structure, in the electric industry, is largely affected by the size, ownership, and interconnection of bulk power facilities. The economic use of these facilities requires a high degree of joint action and agreement among firms in the industry, and the firms themselves are interdependent to a degree not common in other industries. The relationships among utilities are governed by contracts, as in many instances are the growth patterns of, and investments made by, individual firms (for example, when a coordination agreement provides for staggered construction of facilities to match capacity growth to the load growth of the entire pool). Under these circumstances, the conduct of industry members in including or excluding entities from participation has a clearly determinative effect on structure. Looked at from the perspective of the excluded segment of the industry (that is, generally, the small systems), their structure as essentially small scale distribution enterprises dependent on larger integrated firms for bulk supply is largely determined by the contractual and other conduct of the integrated enterprises.

As noted in one discussion of retail competition, arrangements among pooling utilities or among large utilities and their wholesale customers sometimes contain provisions explicitly or implicitly allocating customers. Wholesale transactions occur between supplying system sellers and buyers who distribute at retail and either own no generation or whose generation can only partially carry their loads. Wholesale contracts for all of a buyer's requirements typically contain separate rates for the purchase of energy and for the service of instantaneous load. Contracts for partial or supplemental service involve purchases of a block of firm power, or the purchase of transmission and back-up services so as to permit purchasers to use entitlements in third-party power sources. Additionally firm power may be purchased.[27]

Some wholesale tariffs restrict the buyer's ability to acquire power from others or to install its own generation. Frequently such tariffs contain all-requirements provisions explicitly preventing use of alternative generation, and long notice periods are provided for cancellation of agreements. Agreements to provide partial requirements service may not be offered. For instance, if interconnection is afforded, wheeling may not be offered and reserve sharing may only be offered on a basis that is very onerous to small systems. Power pool agreements may forbid or sharply restrict participation by new members. New power systems have been refused interconnections and coordinating services in efforts to prevent their operation.

Anticompetitive obstructive tactics are not unknown in the electric utility industry. These tactics have involved political interference in competitive systems, filing of baseless legal proceedings, preemptive acquisition of power sites, and political resistance to federal construction.

Predatory features have been employed in the design of rates for services to systems competing for other trade with the supplying utility. To understand these, it is first necessary to note that wholesale tariffs differ from coordination agreements in that wholesale customers usually pay rates based on average system costs while coordination agreements provide for trading on usually lower incremental cost bases. Coordinating utilities may purchase discrete blocks of firm power and other services to shape their supply to their loads. Wholesale tariff demand charges are based upon customer responsibility for systemwide peak loads; customers pay peak charges for power and energy which, in unbundled transactions, might otherwise be separately acquired, off-peak supplies. Under "ratcheting" provisions in wholesale tariffs, peak demand charges are levied for each of a series of succeeding monthly billing periods based upon the customer's peak load incurred during its, but not necessarily its supplier's, peak. The tariffs for wholesale and coordinating services are filed with the FPC, while rates for retail trade are filed with state or local bodies. Wholesale rates have been set, and rate increases have been so timed as to effect price squeezes on wholesale customers.[28]

Yardstick Competition

The concept of yardstick competition in the electric power industry has its roots in the public power movement of the early 1930s which culminated in the creation of the Tennessee Valley Authority in 1933. Supporters of public power placed great weight on its potential role as a yardstick for private power rates. Today, yardstick competition involves performance comparisons between and among various utilities; it occurs at all levels of the electric utility industry. The intense rivalry between public and private systems is manifest at the political level. Kahn notes that "competition-by-example or by threat of displacement by public enterprise has greatly improved the performance of the industry."[29]

Use of such comparisons is essential in regulatory proceedings where, usually, absolute criteria for the prudence of costs incurred or practices followed are not available.

Penn, Delaney, and Honeycutt summarize the role of yardstick competition as follows:

> The process of yardstick competition can provide regulators and the public with information about the range of feasible utility performance. In addition, it offers managers a means of evaluating their own market performance. How and to what extent regulators and managers use yardstick

comparisons varies, but they are a potential tool for improving performance in the electric utility industry.[30]

Comparative Procurement Practice, Important Aspect of Yardstick Competition

The electric utility industry uses tremendous amounts of capital, procures major quantities of fuel and heavy equipment, and must attract management capable of dealing with its manifold problems.

Rapidly escalating equipment and fuel costs, and high capital costs have caused rates to rise substantially in recent years, while securities ratings of utility systems have in many instances declined, and returns on investment have been less than those allowed by utility commissions. Both these facts result in upward pressure on rates, and both offer opportunities for management to gain a competitive advantage. By reducing as much as possible their outlays for fuel and other purchases, managements can maintain or increase return on investment. Lowered operating costs will increase sales of economy power and energy, thus improving the system load factor. The most effective utility in the competition to reduce these costs, and to acquire low-cost supply factors, will furnish a yardstick for regulators (and investors) to judge the performance of other firms.

Short-term financing has traditionally been obtained through sales of commercial paper, bank lines of credit, and revenue anticipation notes. Long-term external financing has traditionally been obtained through sales of debt and equity, leasing, and long-term contracts. With rising capital needs and high rates for hiring capital, utilities have embarked upon more novel forms of financing, such as nuclear fuel leases, creation of new special-purpose subsidiaries which issue and sell debt, joint enterprises with bodies issuing tax exempt securities,* and various forms of joinder in the creation of large-scale generation facilities.

Utilities compete in attracting financing in their traditional capital markets, and in developing new forms of financing and access to other capital markets. If a utility competing to sell commercial paper or debt performs poorly, the ratings of its issuance may be lowered. Lowered ratings lead to higher costs of capital, can preclude purchases by certain classes of investors, and can adversely affect the value of common stock. Lack of success in capital markets

---

*Currently applicable tax legislation places limits on the use of this device.

appears to be a major factor in determining the size of some utility construction programs and willingness to enter into joint enterprises for plant ownership.

In recent years, the potential for comparing utilities has increased because of more frequent requests for rate increases by utilities and more frequent financings. The potential investor interest in comparative studies has also increased due to erosion of earnings, escalating capital costs, and rising fuel prices.[31]

Competition in Fuel, Equipment, and Management Personnel Markets

The costs of fuel have risen substantially as a share of overall costs of energy. Utilities can compete to control fuel costs. They can seek better contract terms for purchases, a better mix of long- and short-term purchases, and vertical arrangements to locate and produce energy.

Fuel contract and fuel procurement practice terms vary; some contracts have terms far more favorable than others. In some parts of the country, utility enforcement of coal supply contracts has varied, with enforcement deficiencies being alleged. Procurement practices differ in regard to the size of fuel contracts sought, and the extent of reliance on competitive purchasing methods.

A number of utilities have entered into the production of raw fuel. Entry into raw energy provides opportunities to compete for resources. Lower fuel costs also increase the number of bulk power sales. The incentive to employ fuel holdings to compete on a cost basis may be subverted by greater opportunities available in passing high costs from captive mines on to consumers through fuel adjustment clauses. Fuel acquisition practices can be compared by utility commissions seeking to hold costs down.

There are only a small number of firms manufacturing the very large capital equipment used by utilities. Equipment procurement and quality maintenance are tremendous cost items for utilities. In the past, the heavy power equipment industry has been subjected to price fixing. Several large utility systems have made material efforts to encourage new entry by European firms. By flexible design practice, some competitive bidding can be arranged.

There are wide variations in the costs (per kw) for new equipment among utilities, indicating opportunities for competition at the architect-engineer procurement level, in quality control over deliveries, and somewhat less in purchasing. Utility systems employ a mix of in-house and contracted-for architect-engineering services. Some systems contract out most of their work to one firm year after year, while others have developed a substantial in-house capability. Often these latter firms realize far lower costs for new capacity. With rising scheduling problems and tight capital budgets, utilities will need to seek and train management capable of holding down costs. Sleepy procurement practices[32] should become a thing of the past if inflation is to be controlled.

Interfuel Competition

The foregoing has dealt with competition within the general framework of electricity services. Such competition proceeds within a larger framework of competition among forms of energy. At the consumer level, fuel competition occurs for new installations of space heating, air conditioning, and appliances. In times past, competition has been quite vigorous between oil, gas, and electricity for space-heating load.[33]

Electric utilities with summer peak loads have sought to promote use of electric heating as a way of improving their annual load factors. The heating load does not contribute to annual peak loads and helps keep expensive base-load generation more fully employed.

Several factors affect the shape of interfuel competition in the future. The rising cost and declining availability of natural gas limit what was heretofore a major factor in appliance (including air conditioning), and space-heating markets. Pollution control costs and declines in the rate of scale economy realization tend to limit electric utility competition capacities but more directly limit use of coal for space heating.

In one area, the potential for interfuel competition has been, many suggest, severely limited--that is, the combination of gas and electric utilities. Where a utility owns both gas and electric distribution systems, it has the opportunity to discriminate in favor of one or the other service. Such combination utilities are considered to be generally anticompetitive under the Public Utility Holding Company Act of 1935, Section 11(b), and the Securities and Exchange Commission (SEC) has ordered a number of such systems to divest themselves of their gas properties.[34] While divestiture may be deferred in times of gas shortage, the basic policy of Section 11 remains in force.[35]

A number of differing positions have been put forward on the comparison of performance of combination and straight utility systems. With the advent of Alaskan production, gas systems may again seek new loads. Even when curtailing, gas systems may seek to shift from industrial gas to higher revenue customers. There are circumstances in which a combination electric and gas utility would have the economic incentive to promote the use of one form of energy at the expense of another. As long as the relationship between costs and benefits is different between the two markets, the utility may very well wish to promote one at the expense of the other.

> If, for example, the costs for providing additional electricity were falling more rapidly (or rising less rapidly) than those associated with providing additional energy in the form of gas, the utility would, in the presence of regulatory lag, maximize profits by promoting electricity and retarding the use of additional gas. Likewise, if the utility can earn a high rate of return on additions to capital, it may promote that service which results in the largest accumulation of capital. In fact, this temptation to distortion will be present as long as allowed rates of return exceed the costs of capital. Where service improvements in one service might lead to a loss in sales to the other, the combination company has less incentive to undertake the service improvement than do single service utilities. Extensions of competing service may likewise be retarded to maximize profits. Certainly, one would expect a combination company to promote its products less vigorously than would a utility facing a competition in a wide range of its sales. In fact, one can argue that a combination company may very well lead the "quiet life" of a monopolist and relax its concern with costs and service.[36]

Geothermal energy might in some circumstances provide interfuel competition. If an independent developer of a geothermal resource were to market it only to direct heating and similar uses, he would provide a local competitive alternative to oil, gas, coal, and electricity in a limited market. If the resource were controlled by a firm selling other forms of energy for the same end uses, such competition might be unwelcome, and the geothermal resource might not be developed. A geothermal resource exploited for power generation might provide a salable amount of waste heat for local use. This product might not be perceived as a threat to a utility's power sales. If the cost of a by-product is very low, so that it might be sold in markets that electric energy cannot penetrate, it would open a new market for the utility rather than cutting into an existing one.

Utility systems will be primarily interested in encouraging uses of geothermal energy that complement existing power systems (for example, in generation applications). They can be expected to be unenthusiastic about uses that would displace their services and may seek to design tariffs to influence the path of development.

Some have pointed to the future possibility of an electric economy. Given the very large market niche occupied by electric energy, and the limits on competing energy sources, interfuel competition at retail will not compensate for lack of competition in the bulk power sector.

ACCESS TO MARKETS FOR BULK ELECTRIC POWER AND COORDINATING SERVICES

Geothermal energy will be harnessed for electric power generation only if the resulting kw of power and kwh of energy can be produced at a price less than that of alternatives. For the production, or busbar, price to be competitive, a developer utility must be able to receive power and energy on a continuous uninterrupted basis, that is, as firm power and energy.

Entry into the business of generating electricity for firm power and energy sales is possible only if an entrant is able to enter into trading relationships with other power systems. Intersystem trading is necessary to obtain coordinating services to permit firm service without constructing smaller units and redundant capacity. This trading is also necessary for access via high voltage transmission to other buyers and sellers of bulk power and coordinating services.

Coordination arrangements depend upon contracts among utilities. These arrangements, pooling resources, vary as to their comprehensiveness and mode of execution. Systems excluded from them are at a competitive disadvantage and can be effectively precluded from installing their own generation. If a system is purchasing part of its requirements from others, addition of a generating source will require at least limited coordination of planning with its supplier. Access to coordination is thus essential to systems seeking to install new generating units. They need reserves to back up their units, supplemental power, outlets for excess capacity, and transmission services. To develop a geothermal field with several units, they require planning coordination with other pooling systems over a period of time.

Reserves and transmission services cannot be obtained without either trading among systems or duplication of facilities.

Duplication is nearly always undesirable, and may be impossible, from an economic or environmental standpoint. The need for trading of services among utilities and for wheeling places large utilities in a very advantageous position vis-a-vis their smaller rivals. Larger systems control transmission and power pooling and have often acted to exclude smaller systems. Larger systems have not coordinated planning and operations with smaller ones, but have used the economic position obtained by their control of high voltage lines in a noncompetitive manner.

Competition in bulk power supply has long been recognized as an important public goal. Unfortunately, federal power operations of the U.S. Bureau of Reclamation, intended by statute to help smaller public power systems, have become subservient to a few large power systems. The reluctance or refusal to trade on the part of larger utilities, coupled with the subservient power operations of the Bureau of Reclamation, create very difficult barriers to would-be public-power entrants.

The problems of these entities are increased by government policy and practice regarding geothermal leasing, utility regulation, and loan guarantees--all of which favor large private entities.

An attitude favoring central planning by big government and big industry, rather than competition in markets, permeates federal utility relationships. It should be replaced by one favoring competitive entry if innovation and "hustle" are to characterize efforts to meet bulk power needs.

THE POWER INDUSTRY ON THE WEST COAST

On the West Coast are located the major load centers in which geothermal energy would be marketed as electricity. These load centers are coupled with generation through a high voltage grid (see Figure 3.6). The high voltage grid in the western states is shaped like a giant doughnut centered around Nevada. The lines of this grid are heaviest in capacity in the populated areas of central and southern California, and in the Pacific Northwest, while lines running north and south through the mountain states are lighter. The heavy lines connecting California and the Pacific Northwest include a unique direct current (d.c.) line.

Nuclear and hydro generating stations are found along the Pacific coast and in the great river valleys of the Pacific Northwest; coal-fired stations are found in the mountain states extending from Montana to New Mexico.

An extensive amount of oil-and gas-fired generation is found in the dense load areas of southern and central California. Oil- and gas-fired units and future nuclear units are located near the growing load centers of southern Arizona, and along the Front Range in Colorado. There are a series of large hydro developments on the lower Colorado River.

FIGURE 3.6

Western High Voltage Lines (A.C.)

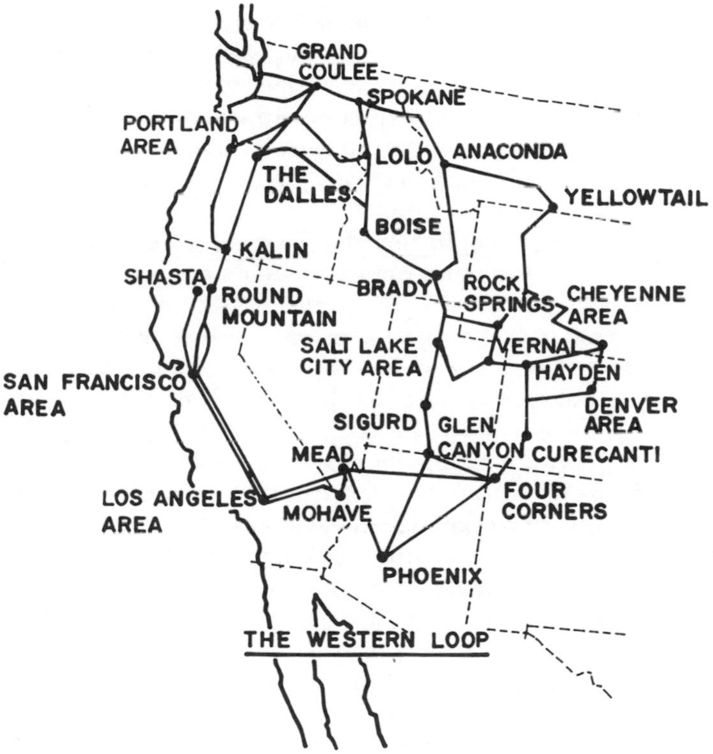

Source: Federal Power Commission, National Power Survey, p. 111-3-179.

The western leg of the doughnut (consisting of two 500kv alternating current [a.c.] lines and one 750kv direct current [d.c.] line) is known as the Pacific Northwest-Southwest Intertie. The Pacific Intertie was a significant step in transmission development of the region. Its two 500kv lines supplement the backbone transmission grid in California. Together with a 750kv d.c. line, they provide interconnections between the hydro generation of the Northwest, and the steam-electric units in California. Because these lines are larger than parallel circuits, there are problems, not yet fully resolved, in realizing full use of their capacity.

The intertie's high voltage lines offer lower impedance to the flow of energy than parallel, connecting low voltage lines in the area. This difference in impedance between these larger and smaller parallel lines, combined with an inability fully to schedule energy flows, results in unscheduled spillovers of flow over the larger lines. Thus, not all their capacity is available for scheduled flows. These frequent unscheduled flows generate, over time, a counterclockwise flow of electric power. The net effect is to reduce the power transferable from the Northwest to California. The future high voltage grid in the West will serve to transfer power from generating sites to load centers, to back up large nuclear plants built near load centers, and to permit exchanges of economy power and energy and peaking service.

Proposals are afoot to improve transmission links across southern Nevada, connecting Utah coal fields to southern California, and to construct a second d.c. line from the Pacific Northwest to either southern California or to Arizona. The overall developmental pattern reflects a tendency to develop transmission linking hydro and thermal resources, and to develop coordination among pools and firms.

The North-South alignment of the grid will be affected by construction of coal-fired capacity in the Mountain States to serve West Coast loads. Stronger lines will connect eastern Montana with the BPA. The large power units and long transmission distances involved in western power system planning will encourage transactions to acquire reserve generation and alternative transmission routes for back-up purposes,* and will encourage interruptible or short-term sales from systems having excess output capability with relatively low opportunity costs.

The evolution and increased importance of very high voltage transmission and large nuclear and coal-fired plants has increased interdependence and trading among power systems. This makes access to such trade necessary if smaller systems are to retain their identity and competitive viability. With the increasing need

---

*In this regard the recent multiparty transaction in which municipal power systems acquire power from the Pacific Northwest, over the Pacific d.c. Intertie, backed up by reserves acquired from the Nevada Power Company, shows what can be done.

to conserve transmission rights-of-way, coordinated expansion and higher line voltages will become increasingly important.

The doughnut-shaped grid traverses many of the sites where geothermal resources are believed to exist, while other areas such as in northern Nevada are not near to these lines.

The Western Systems Coordinating Council (WSCC) has forecasted that 33,623 mw of new generating capacity would be installed during the five-year period 1975-79. This will be approximately 10.6 percent above the 93,544 mw of installed utility capacity at the end of 1976.[37] Thirty-seven percent of the total scheduled additions are hydro, 34 percent coal-fueled, 15 percent nuclear, 11 percent various forms of gas- and oil-fueled units, 2 percent geothermal, and 1 percent undefined. Schedules have subsequently slipped for many units.

For the longer term, the WSCC prospectus was:

Year		Peak Load*	Generating Capability†	Reserve Capacity before Maintenance	
Order	Calendar	mw	mw	mw	Percent of Peak Load
10th	1983	124,000	162,000	38,000	30.6
15th	1988	168,000	214,000	46,000	27.4
20th	1993	223,000	279,000	56,000	25.1
Ten-year increase		99,000	117,000	18,000	

*Net firm December peak load less any firm transfers into WSCC from areas outside of WSCC.

†December generating capability (not reduced for scheduled maintenance).

Conceptual plans for the 11th through 20th years show resource additions by type as follows:

Type of Resource	Percent of Additions 1983-92
Hydro and hydro-pumped storage	20.3
Nuclear	57.4
Fossil-fueled (including combustion turbine, diesel, and combined cycle)	19.8
Other (including geothermal)	2.5

As can be seen, under present conditions in the electric power industry planned additions of geothermal generation are quite small.

WESTERN ELECTRIC SYSTEMS

The configuration and arrangements among western power systems reflect their varying efforts to achieve the benefits of coordinated operations and planning. Table 3.2 lists the western power pool members. The power pooling arrangements of the mountain and Pacific states, and among California utilities, will first be reviewed. Each of the pools discussed is linked to the others and transactions occur between pool groups. The pools discussed are those whose geographic area includes likely sites for geothermal development.*

It should be noted that power pooling is more comprehensive and more accessible to public power systems in the Pacific Northwest than it is in the mountain states or in the Southwest. In the Pacific Northwest, the high voltage grid of the BPA ties the area together, while traditionally light load density has adversely affected power pooling prospects in the mountain and southwestern states. In the mountain states, large private systems have resisted cooperation with publicly owned systems.[38]

ELECTRIC UTILITIES IN CALIFORNIA

The sheer magnitude of electric consumption in California and the location of major geothermal resources within its borders make the rate at which development of this resource progresses in that area critical to its overall employment. The electric power industry of California consists of six investor-owned utility companies, 26 municipal systems, five irrigation districts, the Central Valley Project, and the State Water Project. They perform all or some of the usual public utility functions of providing generation, transmission, distribution, and power exchange services. California utilities also obtain substantial amounts of energy and dependable capacity from outside the state, particularly from the Bonneville Power Administration, other northwestern sources, and the Hoover Dam in Nevada. Table 3.3 shows the dependable capacity of electric systems in California.

---

*Development of geopressured resources might also take place in Texas. Power pooling in Texas is in a state of flux. Heretofore, the power pool in much of Texas was isolated from other systems. However, the Central and Southwest utility holding company system must either interconnect its system in the Texas power pool with its interstate system or face divestiture proceedings.

TABLE 3.2

Organization of the Utility Industry: Western United States Power Pools and Their Members

Geographic Area	Power Pool	Members
California	California Power Pool (CALPP)	Pacific Gas and Electric Co.
		San Diego Gas and Electric Co.
		Southern California Edison Co.
	Southern California Municipal Group (SCMG)	City of Los Angeles
		City of Burbank
		City of Glendale
		City of Pasadena
Pacific Northwest	Pacific Northwest Coordinating Agreement (PNCA)	Bonneville Power Administrator
		Division Engineer, North Pacific Division
		Corps of Engineers, Department of the Army
		The United States Entity, designated pursuant to Article XIV of the Treaty
		City of Eugene, Oregon
		City of Seattle, Washington
		City of Tacoma, Washington
		Public Utility District No. 2 of Grant County, Washington
		Public Utility District No. 1 of Chelan County, Washington
		Public Utility District No. 1 of Pend Oreille County, Washington
		Public Utility District No. 1 of Douglas County, Washington
		Public Utility District No. 1 of Cowlitz County, Washington
		Puget Sound Power and Light Co.
		Portland General Electric Co.
		Pacific Power and Light Co.
		The Washington Water Power Co.
		The Montana Power Co.
		Colockum Transmission Co., Inc., a subsidiary of Aluminum Company of America
	Northwest Power Pool (NWPP) (Operating Committee, Voluntary Participating Systems)	Bonneville Power Administration
		British Columbia Hydro & Power Authority
		Eugene Water & Electric Board

(continued)

TABLE 3.2 (continued)

Geographic Area	Power Pool	Members
		Idaho Power Co.
		Montana Power Co.
		Pacific Power & Light Co.
		PUD No. 1 of Chelan County
		PUD No. 1 of Douglas County
		PUD No. 2 of Grant County
		Seattle Department of Lighting
		Tacoma Public Utilities, Light Division
		U.S. Army Corps of Engineers, North Pacific Division
		USBR--BPA (Southern Idaho)
		Utah Power & Light Co.
		Washington Water Power Co.
		West Kootenay Power & Light Co.
	The Intercompany Pool (Interpool)	Pacific Power and Light Co.
		Portland General Electric Co.
		Puget Sound Power and Light Co.
		The Washington Water Power Co.
		The Montana Power Co.
		Idaho Power Co.
		Utah Power & Light Co.
Mountain States	Associated Mountain Power Systems (AMPS)	Idaho Power Co.
		The Montana Power Co.
		Pacific Power and Light Co.
		Utah Power & Light Co.
		The Washington Water Power Co.
	Rocky Mountain Power Pool (RMPP)	Public Service Company of Colorado
		Pacific Power & Light Co. (Wyoming Division)
		USBR - Region 4
		USBR - Region 7
		Montana Power Co.
		Consumers Public Power District
		Southern Colorado Power Division of Central Telephone and Utilities Corp.
		Colorado Springs Department of Public Utilities

		Utah Power & Light Co.
		Black Hills Power & Light Co.
		Tri-State G & T Assn., Inc.
		Colorado-Ute Electric Assn., Inc.
		Cheyenne Light, Fuel & Power Co.
New Mexico	New Mexico Power Pool (NMPP)	Community Public Service Co.
		El Paso Electric Co.
		Plains Electric G & T Cooperative, Inc.
		Public Service Company of New Mexico
		USBR-Rio Grande Project
West	Western Energy Supply and Transmission Associates (West)*	Arizona Public Service Co.
		Department of Water & Power, city of Los Angeles
		El Paso Electric Co.
		Nevada Power Co.
		Public Service Company of Colorado
		Public Service Company of New Mexico
		San Diego Electric & Gas Co.
		Sierra Pacific Power Co.
		Southern California Edison Co.
		Tucson Gas & Electric Co.
		Utah Power & Light Co.
		Arizona Electric Power Cooperative
		Arizona Power Authority
		Burbank Public Service Department
		City of Colorado Springs
		Colorado-Ute Electric Association, Inc.
		Glendale Public Service Department
		Imperial Irrigation District
		Pacific Power & Light Co., Wyoming Division
		Pasadena Municipal Power & Light Department
		Plains Electric Generation & Transmission Cooperative, Inc.
		Salt River Project
		Southern Colorado Power Division of Central Telephone & Utilities Corp.

*Planning only.

Source: Based on Federal Power Commission, The 1970 National Power Survey (Washington, D.C.: U.S. Government Printing Office, 1971).

TABLE 3.3

Electric Power Industry of California, Winter 1975-76 Capacity

System	Dependable Capacity before Sales and Purchases		Net Dependable Capacity	
	Number of mw	Percent of Total	Number of mw	Percent of Total
Private				
SCE	12,276	37.2	13,349	35.7
PG&E[a]	10,143	30.8	13,537	36.2
SDG&E	2,106	6.4	2,279	6.1
Sierra Pacific Power[b]	550	1.7	708	1.9
Public				
CVP	1,200	3.6	320	0.9
LADWP	3,960	12.0	4,698	12.6
SMUD	1,295	3.9	724	1.9
San Francisco[c]	315	1.0	10	--
Imperial Irrigation District	270	0.8	318	0.9
Burbank	252	0.8	257	0.7
Pasadena	220	0.7	261	0.7
Glendale	207	0.6	225	0.6
Turlock-Modesto Irrigation Districts	165	0.5	275	0.7

[a]The dependable capacity stated for PG&E includes that derived from the Oroville Wyandotte Irrigation District (67.5 mw installed), the Merced Irrigation District (89 mw installed), and the Yuba County Water Agency (370 mw installed). The entire output of these entities is dispatched by and sold to PG&E.

[b]Primarily in Nevada.

[c]Hetch Hetchy project dispatched by PG&E and obtaining power via exchanges with it.

Source: Compiled by authors from material filed with the Federal Power Commission.

PG&E, SCE and SDG&E have collectively 78 percent of net dependable capacity, and 74.4 percent of dependable capacity in California. Their shares in the northern and southern sections of the state where they are respectively located are greater. Municipal systems have 16.5 percent of net dependable capacity (dependable capacity net of firm sales) and 19 percent of dependable capacity in California. The Central Valley Project of USBR has 0.9 percent of net dependable capacity and 3.6 percent of dependable capacity. Irrigation districts, not captive to a utility, have 1.3 percent of capacity and 1.6 percent of net dependable capacity. Total electric energy use was 12,708 million kwh in December 1974: 39 percent was supplied through PG&E (including from PG&E, to San Francisco, the California State Department of Water Resources [DWR] water project and CVP loads); SCE supplied 35 percent; LADWP 11 percent, and SDG&E 6 percent.

Installed capacity, excluding a portion of the Sierra Pacific System (system totals 550 mw), in California totals 33,071 mw. Of this PG&E has 31 percent and contracts for output of an additional 2 percent (excluding capacity from SMUD and the California State Department of Water Resources); Southern California Edison has 35 percent, LADWP 15 percent, and SDG&E 6.4 percent. 1974 generation totaled 156,578 million kwh, of which PG&E produced 36 percent, SCE 35 percent, SDG&E 5.4 percent, and LADWP 11 percent. 1974 peak loads in California totaled approximately 30,100 kw. Of this, PG&E had 30 percent, SCE 31 percent, LADWP 11 percent, and SDG&E 5 percent.

All the municipal generating systems engage in transactions for power with the private utilities as well as the public agencies. In addition to the generating municipal systems, there are 15 municipal systems, three rural electric cooperatives and one utility district that engage only in distribution, purchasing all of their power and energy. Table 3.4 lists these California systems that obtain all of their power from others.

Pacific Gas and Electric Company (PG&E)

PG&E is sole supplier to five systems that purchase all of their requirements. These systems have an aggregate peak of 149 mw and used 695 million kwh in 1974. PG&E serves 47 counties in northern and central California; it operates a 17,000-mile transmission grid and is directly interconnected with PP&L, CVP, the Sierra Pacific Power Company, SMUD, and SCE. Additionally, via the Pacific Intertie it is interconnected with the Pacific Northwest. Wholesale sales are made by PG&E to 11 municipals and one cooperative-owned distribution system, while wheeling services are provided for the CVP and several other public agencies including the California State Department of Water Resources' water project. PG&E controls most of the high voltage transmission (230 kv and over) in northern and central California.

TABLE 3.4

California Distribution Systems, 1974

System	Peak (mw)	Energy Use (millions of kwh)	Sole Supplier
Alameda	63	333	PG&E
Anaheim	305	1,431	SCE
Azuza	30	128	SCE
Banning	12	50	SCE
Biggs	2	6	USBR
Colton	22	87	SCE
Gridley	5	19	USBR
Healdsburg	7	35	PG&E
Lodi	57	192	PG&E
Lompoc	10	64	PG&E
Palo Alto	133	737	USBR
Redding	48	231	USBR
Riverside	233	912	SCE
Roseville	33	105	USBR
Santa Clara	162	1,037	USBR-PG&E
Shasta Dam PUD	4	18	USBR
Ukiah	12	71	PG&E
Anza Coop	2	10	SCE
Plumas Sierra	9	44	USBR
Surprise Valley	13	54	BPA

Source: Federal Power Commission.

The Central Valley Project of the Bureau of Reclamation (CVP)

The Bureau of Reclamation's CVP consists of a series of federal hydro projects and some transmission facilities. This project has a total machine capability of 1,471.2 mw.* In 1974, the total aggregate peak demand of loads served by CVP amounted to 1,114 mw (energy load of 5.3 billion kwh).

Since 1951, CVP has sold electric power not needed for project irrigation pumping to its statutory preference customers which include 23 federal facilities, five state facilities, several municipalities, an irrigation utility district and one cooperative. The remaining power goes to PG&E. The CVP facilities are integrated operationally with the PG&E system, and PG&E transmission is used for all project sales with the exception of sales to small federal agency load and to its two utility district customers, SMUD and Shasta Dam Area PUD (SDAPUD).

The Sacramento Municipal Utility District (SMUD)

SMUD is a quasi-municipal corporation supplying electricity in most of Sacramento County and a small portion of Placer County, California, an area surrounded by PG&E's service area. SMUD's generation resources consist of six hydro plants with a total of 649 mw of machine capability and 913 mw of nuclear capacity; SMUD purchases 360 mw from CVP, and also purchases 165 mw of Canadian Entitlement Power (CEP). It is interconnected and coordinates operations with PG&E.

Southern California Edison (SCE)

SCE serves a 50,000 square mile area in southern California. It operates a transmission network of 10,000 miles. SCE is interconnected with PG&E, Arizona Public Service Company (APS), Nevada Power Company (NPC), SDG&E, and LADWP. SCE serves at wholesale two cooperatives, six municipals having aggregate peak demand of 604 mw and energy loads of 2,614 million kwh. SCE supplied over 60 percent of the requirement of municipal wholesale power in California.

Los Angeles Department of Water and Power (LADWP)

LADWP is the largest municipally owned electric system in the nation, with an extensive transmission network. LADWP is

---

*CVP has purchased 400 mw from the Centralia, Washington, coal-fuel plant. This purchase expires in 1982.

interconnected with the DWR, NPC, SRP, SCE, and the cities of Burbank, Glendale, and Pasadena. It makes no sales at wholesale.

San Diego Gas and Electric Co. (SDG&E)

SDG&E serves southwestern California, is interconnected with SCE, and sells power at wholesale to the Escondido Mutual Water Company.

Other Electric Utilities in California

The cities of Burbank, Pasadena, and Glendale in southern California are interconnected with LADWP. They receive power from federal hydro projects and from the Northwest.

The cities of Anaheim, Banning, and Riverside are large wholesale customers of SCE. Some or all of them may enter into self-generation by participating in large coal-fuel or nuclear projects.

NCPA is a joint power agency established in 1968 that has 12 members: the cities of Alameda, Biggs, Gridley, Healdsburg, Lodi, Lompoc, Palo Alto, Redding, Roseville Santa Clara, and Ukiah, and the Plumas-Sierra REC. While none of these systems currently generates power, they are, as a group, considering construction of or participation in a variety of other generation facilities. Certain NCPA members receive power supply from CVP,* and others are wholesale customers of PG&E.

The State of California Department of Water Resources (DWR) operates the state water project which provides reservoir storage and water transportation, pumping and power facilities by which water from northern California is delivered to southern California. The project requires electric energy for pumping in excess of what its hydroelectric resources (1,041 mw) will produce. The department accordingly has arranged for power supply and has proposed to participate in the ownership of the proposed San Joaquin nuclear project.

Three irrigation districts, Imperial, Modesto, and Turlock, produce, transmit and distribute power. The Imperial Irrigation District, the largest, in 1974 had a peak load of 300 mw and a net system input of over 1 billion kwh, of which 340 million kwh were

---

*CPV customers are Redding, Roseville, Biggs, Gridley, Palo Alto, and the Plumas-Sierra REC. Additionally, Santa Clara receives a portion of its requirements from CVP.

purchased from several sources--the USBR, SCE, and SDG&E. Modesto had a net system input of 933 million kwh, almost 90 percent of which was purchased from public agencies; the Turlock Irrigation District had a net system input of 653 million kwh, of which approximately 60 percent was purchased (primarily from public agencies).

POWER POOLING IN CALIFORNIA

The three dominant concerns in California--PG&E, SCE, and SDG&E--have established a relatively loose pooling arrangement covering most of the state's area and population. These utilities collectively govern exchanges with the major external source of energy to date, the Pacific Northwest. Within the widespread network of the major utilities are a few public bodies disposing of significant energy supplies--Sacramento, CVP, the California DWR, and a group of municipalities around Los Angeles. There are also a number of irrigation districts (total municipal governments) that generate some electricity. The Los Angeles group has formed its own local pool.

The California Power Pool

Power pooling arrangements in California are somewhat complex. The three large power companies--PG&E, SCE, and SDG&E coordinate certain phases of planning and operation through the California Power Pool (CALPP). These three firms have approximately three-fourths of California dependable capacity and 59 percent of California high voltage transmission.[39]

Coordination

The pool provides for both operating coordination and coordinated planning. It provides for no new members and coordinates operation with the four private utilities in the Pacific Northwest, so regulating use of the Pacific Intertie.

Reserve Sharing.

Under the pool arrangements each member is initially responsible for serving its loads and providing adequate resources for its area system.* Required resources including energy, capacity, and spinning reserves† not so provided are acquired from the pool with a penalty on the deficient party.

Economy and Short-Term Transactions

The agreement provides for exchanges of emergency services for up to two hours, which are to be returned in kind, economy exchanges of capacity and energy, and transactions to purchase and sell short-term firm and standby power. Capacity charges for short-term firm power are based on the higher of 80 percent of power commitments or the maximum actual demand prorated for the portion of the month over which it was incurred. Attendant energy is priced at 115 percent of supplier's incremental costs.

"In order to protect the Parties from unknown and unreasonable risks, service to third parties outside its area system involving provision by a party of standby or protection of a supply of power from its Area System is prohibited, unless all Pool members agree."[40]

Pacific Intertie Agreement

California Companies Pacific Intertie Agreement, dated August 25, 1966 and filed with the FPC on January 18, 1968, is an agreement among PG&E, SCE, and SDG&E. Its purpose is indicated to be to facilitate building and operating the intertie, to transfer power and energy to and from the Northwest, to facilitate transactions

---

*The area system of a party consists of its system plus (1) each third-party system normally operated in parallel and which is by facilities and agreement, effectively integrated as to loads and resources, and (2) generation plants whose output is substantially all delivered to the party.

†Spinning reserves are, according to the pool agreement, 5 percent of the sum of daily peak loads and interruptible receipts. Energy resources are to equal a party's monthly needs less energy resources unavailable due to maintenance, plus 50 percent of the energy output of the largest system unit (if it is on line). Capacity resources must be satisfied or damages are assessed for the deficient month and for each of 12 months thereafter. Reactive power is the responsibility of each party.

among the companies, and to reinforce their own backbone grids.* Necessary coordination is sought to be accomplished through CALPP, through intercompany assignments of power and transmission, and by economy transactions. Benefits are to be shared among the company-parties.

Joint Obligations

CALPP members assumed certain obligations in the course of their effort to show that a privately owned set of linkages with the Pacific Northwest would provide as many benefits as would facilities owned by the federal government. These obligations include, besides coordination with BPA and other Pacific Northwest entities, the exchange of company energy for USBR hydro capacity, and transmission of Canadian Entitlement Power (CEP).† The intertie capacity is apportioned among the companies by prorating the intertie rated capacity (less assingments made under the agreement to other companies), in proportion to their relative size. This proportionality is specified in the contract to be 50 percent for PG&E, 7 percent for SDG&E, and 43 percent for SCE.

In the event of any modification by a party, primarily directed at increasing intertie capacity, other parties shall have the right to participate in the costs thereof, and to share in the increased capacity. Sharing would be in relation to the contract shares of each party. Each company may use vacant intertie capacity as it sees fit so long as rights and performance of obligations of others are not impaired.

Joint Purchasing and Scheduling

The companies agree to jointly arrange contracts for transmission capacity with PP&L and BPA. Each party has the right to

---

*The Pacific Intertie is defined to consist essentially of the 500kv a.c. circuit coming down from WSBR lines to the Oregon-California border; PP&L's 500 kv line linking this point to lines built by PG&E; a 750kv d.c. line from Oregon via Nevada to Southern California; and a 230kv a.c. SCE link in Southern California between the end of the 750kv line and the 500kv lines. The 500 kv lines were designed with a rated capacity of 2,000 mw, and the d.c. line at 1,350 mw (50 percent owned by SCE and 50 percent owned by LADWP).

†San Diego, being at the southern end of the intertie, would receive all intertie shipments through SCE power lines. SCE is obligated to wheel for SDG&E only such capacity as is within SDG&E's share of intertie capacity.

a share in sales of line capacity to the state when such capacity is sold to the state to permit it to receive from the Northwest. The contract is carried out through the Coordination and Joint Accounting Committees.

The Coordination Committee will schedule intertie operations. Transmission is to be scheduled for power to or from north of California, energy in exchange for USBR peaking capacity, CEP power resold to the California DWR, and power delivered to SDG&E by SCE as part of an exchange for power that SDG&E arranges to have delivered into northern California.

Allocated Marketing

The parties to provide interconnections to the intertie to serve nonparty entities are specified as PG&E to SMUD, USBR, and California DWR; and SCE to Metropolitan Water District and LADWP.

Limited Access

A party may dispose of its share of intertie facilities unless such disposition would reduce any other party's intertie capacity; materially reduce reliability of transmission service over the intertie; or impair any party's ability to perform its obligations to provide reserves to other parties, coordinate its use of power from north of California (and resales of such power to others) only in proportions that will provide maximum benefits to all of the companies, or to meet certain joint obligations as defined below.

Other than for transactions involving specified transmission of dump energy for the state, SMUD exchanges, and the metropolitan water district, transfer of intertie capacity to others by a company is subject to approval by the company owning the facility to be used, and is subject to a right of first refusal by other parties.

Similarly, capacity contracted to the USBR, the state, and SMUD, if unused, is to be offered to other parties which are given a further right of first refusal if one party does not wish to acquire its share of such capacity. Rejected shares are to be offered to other parties, subject to recapture by the rejecting party on five years' notice.

Seven-Party Agreement

Related to CALPP is the Seven-Party Agreement establishing coordination for interconnection by the Pacific Intertie among the pool members and four Pacific Northwest utilities (the members of the 1957 Intercompany Contract): Washington Water Power, Puget

Sound Power and Light, Portland General Electric, and PP&L.[41] Under this agreement, the Northwest utilities, trading as a group, sell the California utilities, also trading as a group, hydro energy at BPA rates and purchase excess steam-generated energy at seller's incremental cost plus 15 percent. Provision is made for new members.

If SDG&E is unable to use its full share of intertie capacity for Canadian or firm power, it may reduce its share of intertie participation or increase its exchanges of power delivered for it in northern California for power in southern California. This reduction is subject to arbitration at five-year intervals unless SDG&E's unused share exceeds 20 mw. Also, if SDG&E can generate for less than the cost of northwest power (other than Canadian, or Bureau of Reclamation capacity exchange for energy) it may reduce its share or participation in the intertie.

A schedule of curtailments for use in emergencies is provided.

The California State Department of Water Resources (DWR) Agreement

The DWR is constructing and operating an extensive project to bring water to southern California. As a project, DWR generates a significant amount of electricity. However, it has no transmission to integrate its facilities.

Transmission

PG&E, SCE, SDG&E, and LADWP have entered into a 1966 contract to provide electric energy and transmission services to DWR. Additionally, under the agreement, DWR obtains transmission services over the Pacific Intertie, at times when there is unused capacity in the 750kv d.c. line. This complementary service enables it to obtain hydro power from the Northwest. The California suppliers also sell DWR off-peak energy, firm capacity, and on-peak energy to supplement the energy available from DWR's generation, USBR, and Pacific Northwest facilities. Energy use must be scheduled five years in advance, except for changes of no more than 10 mw, or 10 percent (up to 40 mw) of capacity which may be made on two years' notice.*

---

*Rates are at fixed amounts, as follows: on-peak and off-peak energy, firm capacity--three mills per kwh, $1.477 per kw per month with a 12-month ratchet provision for the 12 prior or coming

Distribution of DWR Energy

The contract provides for the exchange of energy from DWR plants into suppliers' systems in return for delivery of like quantities of energy to DWR elsewhere.

Limitation on DWR Access to Other Energy Sources

The sources of power for DWR are specified in the agreement.* Before the DWR may obtain generation from additional facilities of its own or from the Oroville-Thermalito hydro power plants, it must give suppliers five years' notice and negotiate terms and conditions for such use.

Use of other sources requires six years' notice and negotiation as to terms. If, after one year, subsequent to notice, negotiations are unsuccessful and DWR still wishes to use another source, suppliers may terminate the contract on five years' notice. Any new DWR plant or other source "may be connected to suppliers' interconnected electric systems only if the parties agree that such power sources meet electric industry standards for reliability and agree upon suitable standby arrangements for such power sources."[42]

Specific provision is made for the possibility of a state thermal generating plant. In the event such a plant is proposed, suppliers are to be given six years' notice, and negotiations are to ensue as to transmission and standby services, plant operation and maintenance, sales of power produced in excess of project needs, and

---

months; extra capacity provided without notice--two dollars per kw per month. In the event extra capacity is provided for over one-half hour for reasons on suppliers' systems, state water project emergencies or curtailments of states' Northwest power, a 12-month ratchet is imposed. This two-dollar sum is payable only if the capacity is deliverable. Under this ratchet, DWR is to pay two dollars per month per kw of extra capacity in the month it is taken and for each of the succeeding 12 months.

PG&E is billing agent under the contract. The contract rates are to be effective until March 31, 1983, subject to review on or before March 31, 1978. Rate revisions are to be effective five years later and are to be based upon anticipated fuel costs, and suppliers' generating efficiency and economies experience.

*Plants are listed as denominated state hydro plants, Bureau of Reclamation power, Northwest power via intertie lines, Northwest dump power, purchases from suppliers per this contract, and generation from Oroville-Thermalito or a state thermal plant.

integration of the plant with suppliers' interconnected electric systems. If negotiations fail to produce agreement within a year and DWR intends to proceed with the plant, suppliers may elect to terminate the contract on five years' notice.

Each party is allowed to install and set as it sees fit protective relays for automatic separation of the parties' systems.

LADWP is obligated to provide only surplus, not firm, electric energy.

The arrangements between DWR and LADWP and the members of CALPP indicate how an entrant system is dependent upon the arrangements it must make with firms providing transmission and operating coordination. In this setting, PG&E, SCE, SDG&E, and LADWP, in acting together, have at their disposal options that could make it quite difficult for DWR to substitute a new source of energy in competition with them. This is the case whether the source is developed and operated by DWR, or by others. In either event, permission from its existing suppliers is required.*

Sacramento Municipal Utility District (SMUD)

SMUD is surrounded by PG&E's service area, and is dependent on PG&E for access to external sources of power, and markets for power. The Justice Department has alleged that a 1955 contract between SMUD and PG&E contained a clear-cut anticompetitive restriction, in Article 13(b), which prohibited SMUD from selling or wheeling power outside a geographic area delimited in the contract. Justice alleged that SMUD refused to wheel power in 1965 solely because of this restrictive contract.[43]

On August 1, 1967, SMUD executed a 38-year contract with pool members for access to the intertie. The pool will provide up to 400 mw of intertie capacity between April 1, 1971 and March 31, 1976, and up to 200 mw in other periods to wheel capacity and energy for SMUD to or from the Northwest. Imported Northwest power cannot be sold by SMUD outside its boundaries unless pool members agree. SMUD must provide the pool five years' advance notification of intertie shipments and must provide immediate notification of any contract with a Northwest entity. The contract also allows SMUD to resell excess Northwest firm power to pool members provided that

---

*Representatives of NCPA have alleged that the unusual long-term fixed rate provisions of DWR's power supply agreement may have been proffered as an inducement for DWR to forego construction of its own transmission grid or steam-electric facilities.

five years' advance notice is given before initial delivery. Once SMUD has notified pool members that it is invoking the resale provisions, SMUD may not amend its purchasing contract with Northwest entities without consent of pool members during a period less than 61 months in advance.

On June 4, 1970 SMUD executed an integration contract with PG&E, which provides for coordination of bulk power supply planning, and for provision to SMUD of the coordinating services required to support its first nuclear unit. The agreement arose as a result of SMUD's plans to construct nuclear units. The 1970 agreement requires SMUD to sell to PG&E all capacity and energy generated by the resources of SMUD's system, including power purchased from CVP, in excess of SMUD's load. In addition, the agreement appeared to place a 830mw limit on the second SMUD nuclear unit and to include all surplus energy from that unit in the power that must be sold to PG&E.* Justice has alleged that this 1970 agreement prevents small utilities from participation in the construction of the second unit or purchasing power from it.[44]

In November of 1975, PG&E filed an amendment to the 1970 contract which, among other things, provides that capacity and energy surplus to SMUD's load from up to 400 mw of planned SMUD geothermal capacity be sold to PG&E. The price is to be midway between SMUD's cost and the cost PG&E would have incurred in generating an equal amount of capacity and energy from its own geothermal units, but in no event can the cost exceed PG&E's cost (even though PG&E's geothermal capacity consists of dry steam units, whereas SMUD may have to use more expensive hot water units). This contract is discussed hereinafter. Similarly, PG&E will acquire peaking energy from 150 mw of proposed SMUD oil-fired combustion turbines at rates based on nuclear energy costs far below those for oil, paying full costs only for exceptionally long-duration use of such turbines.[45]

Southern California Municipal Group

The Southern California Municipal Group engages in operating coordination, including economy transactions, emergency services,

---

*In a proceeding before the FPC (FPC Docket No. E-7597) on this contract, NCPA objected to the agreement as being an anticompetitive preemption of bulk supply sources by PG&E. In its response to this allegation, PG&E contended, inter alia, that the contract's reference to unit sizes of 830 mw was not intended to limit the size of SMUD units so as to prevent others from obtaining unit power entitlements or ownership in those units.

and spinning reserves, and coordinates acquisitions of power from Hoover Dam and from Canada and the Northwest. Smaller systems in this group can obtain wheeling over lines of LADWP enabling them to acquire power and energy from the Northwest and elsewhere outside of southern California.

Central Valley Project (CVP)

CVP has a major set of coordinating arrangements with PG&E. The agreements provide for PG&E firming up CVP's hydroelectric capacity, bringing Pacific Northwest capacity and energy to CVP, and transmitting CVP power to publicly owned distribution systems. While these services are of substantial value to CVP, the terms of the agreements as such limit CVP access to new power sources, and its ability to increase service to existing and new customers and to trade with other systems. These matters are discussed further in the section on conduct that follows.

CONDUCT IN THE POWER INDUSTRY IN CALIFORNIA

Small utilities in California operate in a hostile economic environment. Large investor-owned utilities in the area use their substantial market power in a pattern consistent with maintaining their dominant positions to the disadvantage of smaller systems. The Department of the Interior operates several major power projects in the western states in a manner that has had the effect of reinforcing and extending the market power of the large investor-owned utilities. Significant proportions of project power and energy intended for preference (publicly owned) customers have been routed to large investor-owned utilities. These utilities, in turn, sell this power and energy to preference customers at higher prices and under restrictive contractual provisions. Given this situation, the small utilities have an incentive to develop their own generating capacity, to provide themselves with a secure source of power and energy at reasonable costs in order to insure their survival and growth.

Central Valley Project (CVP)

WSBR's CVP provides power to many small utilities. The evolution of a hostile environment for small utilities during the past 25 years has involved CVP and its relationships with CALPP members.

CVP is a major Interior Department (Bureau of Reclamation) water conservation development in California which commenced generation in 1944. At that time, it lacked transmission and coordination needed successfully to market its surplus power to preference customers.* No preference customer had a system capable of providing the transmission and coordination needed. Efforts for federal construction of transmission facilities were successfully resisted by investor-owned utilities.

The amount of firm power CVP could provide if operated in isolation is quite limited in relation to its total installed capacity. This is because many CVP hydro developments have sufficient storage to be run only as peaking or intermediate power sources. CVP must coordinate its operations with base-load resources if it is to obtain dependable capacity commensurate with the amount of its installed capacity. When the CVP first sought coordination, it found that it must either erect a federal steam-electric system or employ the transmission grid and generation of PG&E.

No federal steam-electric system has ever been authorized except for the Tennessee Valley Authority (TVA). While base-load generation might have been obtainable from the Pacific Northwest, that area's hydro resources were not as yet so developed as to permit export, and the transmission lines to BPA were resisted by PG&E before Congress.

Initial Contracts

After several years of negotiation, two contracts were executed in 1951 between USBR and PG&E, one for transmission and exchanges,[46] and one for sites and interchanges.[47] While providing CVP with needed coordination, the two contracts were so phrased as to create a situation in which PG&E would be able to preserve and extend its great market power, while the other small utilities in northern California would remain dependent upon it.

The transmission and exchange contract pertained mainly to the terms under which PG&E would transmit CVP power to preference customers. Under the contract a uniform rate was established for transmission anywhere within a limited service area.[48] The service area to which this contract effectively restricted CVP encompassed the Central Valley and several counties in the Bay area.[49] By so confining CVP sales to preference customers to a

---

*Publicly owned electric systems, and REA cooperatives, have a preferential right to purchase output from federal power projects.

specified geographic area, it became (and predictably so) difficult for CVP to market power in a broader area in future years when its capacity was enlarged. Bureau representatives were aware of the fact that the wheeling provisions would provide PG&E with a strategic advantage in future negotiations concerning any change in the sale and interchange contract.[50]

The contract was to run for a term of ten years. It provided for wheeling service only to those CVP preference customers served by PG&E on the effective date of the contract, which were located and used the delivered power and energy beyond the corporate boundaries of municipalities served at retail by PG&E, and which had a monthly maximum demand of at least 500 kw for three consecutive months in the 12 months preceding the date on which the requested service would commence.[51] PG&E was to decide within a 90-day period whether or not service would be available to new preference customers.[52] This provision provided PG&E with the ability to eliminate CVP as a potential wholesale competitor or as a vendor to municipalities not covered by CVP in 1951.

The economic leverage possessed by PG&E in consequence of its control of transmission is seen in a report that PG&E took the position that it would not enter into a transmission service agreement unless there was "every possibility that a purchase and sale agreement were to be entered into."[53] One PG&E draft contract called for the transmission contract to be ineffective if an agreement for the purchase and sale of CVP power was not reached by the bureau and the company.[54]

Sale and Interchange Agreement

The 1951 agreement, after a great deal of negotiation, provided essentially for the operation of the CVP as an element of the PG&E-controlled power system. With its system so operated, the CVP would receive certain credit for power and energy, which it might sell over PG&E lines to preferential customers. Any remaining output from CVP generation or any remaining power received by CVP from the Northwest would be acquired by PG&E in a rather peculiar manner hereinafter described.

Because preferential customers required firm power, the project's dependable capacity rating (PDC) defined its marketing ability. The PDC arrived at reflected a negotiating process, not an engineering calculation. The PDC that USBR was able to negotiate was only 300 mw. This figure is substantially less than the amount that would have been credited to the CVP using either the total nameplate ratings of the CVP generators (450 mw), as had once been suggested by PG&E, or the total CVP system capability, originally estimated to be 500 mw by the bureau and 400 mw by PG&E.[55]

As noted by GAO, PG&E reports to the FPC indicate that integration of CVP facilities into the PG&E system provided 518 mw of capacity.[56] There is no indication that the bureau was compensated for benefits conferred on PG&E's system by CVP. The bureau was not able to acquire the firming capacity it sought to obtain, either from PG&E or otherwise.[57] Rather, the company's position that PDC was to be limited to CVP and preferential customer needs was imposed.

Understating CVP project dependable capacity reduces CVP firm sales and thereby increases the purchases made by preferential customers from PG&E. Moreover, understatement of PDC results in PG&E's receiving free capacity.

The CVP hydro system provides nondependable capacity in excess of its dependable ratings. This capacity is taken by PG&E but paid for only in limited circumstances. The final contract calls for PG&E to pay for nondependable capacity made available to it only if PG&E receives at least 60 days' advance notice, and if the capacity is available for at least five consecutive calendar months.[58]

The GAO later charged that from July 1952 through December 1956 this payment arrangement enabled PG&E to take $963,261 worth of nondependable capacity for no payment since the capacity had been available for less than a five-month period.[59] To this charge USBR responded with a specious argument that the five-month period was necessary to allow PG&E to place plants on cold standby or to dismantle equipment for extended maintenance, and that nondependable capacity had value only as dump energy.[60] This argument neglects both hydrological forecasting and the practice of short-term capacity sales elsewhere.

It will be seen that preference customers would view this matter as an attempt by the bureau to subsidize PG&E at their expense by forcing them to buy federal energy and power at higher rates from PG&E.*

Sale and Interchange Contract

In the negotiations for this interconnection agreement, much consideration was given to specification of the capacity and energy that would be credited to the USBR and how these would be disposed of. The bureau did not receive credit for the amount of dependable capacity and associated energy derivable from CVP projects, which resulted from the operation of CVP as part of a unified PG&E-CVP

*Sales of short-term and economy capacity are well known among utility systems.

system in which CVP capacity could be used for peak periods. Instead, as noted above, CVP project dependable capacity was agreed to be 300 mw (as against the 518 mw PG&E reported to the FPC).

Implementation of the Initial Contracts

The implementation of the new contracts went forward after 1952 in a new administration, and with a new attitude on the part of the bureau differing from earlier efforts to build an expanded federal system in contention with PG&E. During the 1950s, CVP power was disposed of in a manner that violated the federal preference law. The House Committee on Government Operations found in 1960 that the bureau had misled preference customers by issuing repeated statements to the effect that all CVP dependable capacity had been allocated to preference customers and that, consequently, none was available for sale. The committee found that the bureau knew that between 40 mw and 98 mw of dependable capacity would be available between September, 1960 and August, 1964 because prior allocations exceeded preference customers' maximum simultaneous demand. At the time, the bureau was selling this power to PG&E, which is not a preference agency.[61]

The bureau denied requests from the California DWR for interim or nondependable power. In so doing, the bureau asserted that provisions of its contract with PG&E required preference customers to have generating capacity to replace the nondependable power when it became unavailable, a requirement the DWR and most other preference customers would not meet.[62] Ignoring the peculiar interruptible nature of the DWR water pumping load, the bureau asserted that supplying nondependable power to preference customers could burden PG&E, because PG&E might be called upon to supply capacity when the nondependable capacity was not available. The substantial interest in and need for dependable capacity on the part of public entities is reflected in testimony by cities, cooperatives, utility districts, and the state of California.[63]

The refusal to sell to the DWR is significant because of the potentially large size of the DWR power system and demand. The prohibition of sales of nondependable power was incorporated as a contract amendment only in 1959 and it was then made applicable only to new preference customers for such power.[64] This amendment curiously would not bar CVP sales of nondependable power to either PG&E or perhaps SMUD, located in the district represented by CVP's principal critic, Representative John Moss. The House Committee on Government Operations, in analyzing the requirement that preference customers provide assurance of a supplemental power source as a condition for purchasing CVP power, noted that this requirement was a contributing factor to a delay of several years in execution of contracts with preference customers for power they had been allocated; that the requirement forced preference customers to contract with PG&E for supplemental power; and that meanwhile PG&E purchased the subject CVP capacity at a price below that which the company would then sell power to preference customers contracting with it for a supplemental power source.[65] This requirement of supplemental power was dropped in 1959 after USBR Commissioner Dominy conceded that it was not the bureau's responsibility to compel customers to obtain supplemental power.[66] The committee also indicated its disagreement with requirements purporting to limit PG&E's responsibility for providing supplemental service only to those preference customers that would purchase all such power from that company.[67]

After the congressional inquiry,[68] USBR partially corrected its misallocations of capacity. The bureau also denied requests by certain municipals for CVP power on the grounds that they were outside PG&E's service area, even though other systems would be willing to take delivery from PG&E and, in turn, wheel the power to the municipals.[69]

Another instance of the bureau's unresponsiveness to preference customers involved SDAPUD. In the late 1950s SDAPUD had to curtail service to new customers and limit increased service to old customers. It was unable to obtain additional CVP power over existing USBR transmission facilities. An appropriation was requested by the bureau for FY1960 for construction of a line to remedy the situation. However, the request was withdrawn after PG&E offered to wheel bureau power to SDAPUD. SDAPUD protested that PG&E's offer was unacceptable because the wheeling rates would force the district to raise its own rates, and because the arrangement would force the district to become dependent on the discretion of its direct competitor, PG&E, for future expansion of its load.[70] These protests had no effect on the bureau's decision to withdraw the request for a transmission line.

Implementation of the 1951 contracts was also accompanied by pressure from PG&E to prevent construction of federal transmission lines. Wheeling of CVP power to customers by PG&E in lieu of construction of federal lines had been assertedly justified on the grounds that it was cheaper. However, the House Committee on Government Operations, in a study of transmission of CVP power during the 1950s, concluded that in many instances wheeling service cost the United States and its preference customers significantly more than direct delivery over government facilities would have.[71] USBR data showed that the United States realized savings of $2,947,750.81 from July 1, 1957 through November 30, 1959 following a changeover from wheeling by PG&E to direct delivery over government lines to SMUD Cumulative savings projected to 1994 were estimated by the bureau to exceed $95 million.

Northwest Intertie

Power from the Pacific Northwest might provide CVP with a source of firming power alternative to PG&E. Construction of Pacific Northwest-Pacific Southwest intertie transmission lines permits the exchange of water-generated electricity from the Northwest for steam-generated electricity from the Southwest. The intertie could also provide preference customers with an avenue to new sources of supply. Unfortunately, the arrangements entered into for the intertie prevent the choice of alternative power sources by CVP or its preference customers.

Authorization for federal construction of the intertie was conditioned upon the secretary of the interior's finding that such transmission could not be provided equally well over hired private lines. To compare federal versus private lines, the secretary sent a federal yardstick plan and a related set of criteria for evaluating nonfederal proposals to all utilities and other entities that had indicated an interest in building a part of the intertie.[72] The yardstick plan was to serve as a model of a transmission system that would carry out the intent of Congress with respect to an intertie system. Nonfederal proposals were to be compared with it.

Ten utilities and other entities responded with proposals which were evaluated by Interior. The secretary concluded in a report to the House and Senate Appropriations Committees that the proposal submitted by the California Power Pool would result in the most beneficial service to the Central Valley region of California.[73] In making its choice, the department was on notice that the proposal it was accepting limited CVP's ability to serve new preference customers,* and placed an effective ceiling of 1,050 mw on the amount of service CVP might provide existing preference customers. The agreement accepted by Interior did resolve the PG&E-public power struggle for wholesale markets, and appeared to resolve the power supply needs of preference customers then served by CVP until 1980.

The acceptance of the proposal was asserted by the Interior Department to cost less for CVP customers than would alternative approaches. However, these alleged cost savings are disputable,

---

*This results from PG&E's being obligated to wheel only to existing CVP customers in a limited area (delineated in the 1951 CVP-PG&E agreement), and load limits on CVP.

while the market of CVP was clearly limited.[74] The record makes it hardly disputable that PG&E sought and gained, through the agreement, the upper hand in determining system planning in northern California. PG&E is in a position to determine future access to high voltage transmission--the control link in all power system combinations--both within northern California and from external sources such as the Pacific Northwest.

Intertie construction started in 1964, involving 500kv a.c. lines with a rating of 2,000 mw, and a 750kv d.c. line, owned half by SCE and half by LADWP, with a rated capacity of 1,350 mw. The d.c. line runs through Nevada to southern California and cannot be economically tapped into en route. The price of the intertie was asserted by the National Rural Electric Cooperatives Association to be the preclusion of federal lines in California for 20 years.[75]

The Interior Department's acceptance of a privately controlled intertie did not end PG&E resistance to federal transmission lines. In 1966, PG&E lobbied for public works legislation that would prohibit construction of a federal line between CVP and its Marysville dam if a local public or private agency could offer distribution and transmission service at a low cost. In approving the intertie arrangements, the secretary of the interior waived standard provisions in federal transmission-line rights-of-way permits allowing the United States to transmit power over unused line capacity and to require expansion, at the government's expense, of line capacity.[76]

After approval of the intertie proposal, the USBR entered into negotiations with CALPP for a transmission services contract, and a CVP-PG&E contract was negotiated. The members of CALPP allocated shares of intertie capacity among themselves under an August 1966 agreement.[77]

Intertie negotiations found USBR having to argue for omission of a clause in the transmission agreement which would have allowed the pool companies to cancel their intertie agreement in the event that the bureau built or acquired an interest in any other high voltage lines in California.[78] After several years of negotiation, a contract was arrived at between PG&E and the bureau for the sale, interchange, and transmission of capacity and energy.[79] This contract will be referred to as Contract 2948A or the 1967 Contract.

The Justice Department has asserted that the 1967 contract is designed to restrict electric power sources alternative to PG&E's.[80] Justice alleged that this contract gave PG&E a high degree of control over all marketable electric energy generated in California or imported into California by CVP for a 40-year period; that it effectively limited the geographic area in which the USBR can market electricity; that it effectively limited the sources of capacity and energy that may be included in or wheeled in by the CVP system;

and that it effectively limited the maximum amount of capacity that CVP may maintain in its system.

While the contract does contain provisions that make it possible for some bureau power to be wheeled by PG&E, Justice inferred that PG&E made these concessions as the only alternative to eventual construction of a federal generation and transmission network in the heart of its service area, an action that PG&E had strongly resisted. Although the contract had the effect of temporarily relinquishing certain resale customers to the bureau, Justice noted that it secured PG&E's control over generation and transmission in northern and central California for many years.

Article 24(a) of the 1967 contract defines a geographic area in which PG&E has agreed to wheel CVP power to preference customers. This wheeling area excludes various preference entities that were and still are all-requirements wholesale customers of PG&E. Article 24(a)(1) provides that PG&E will wheel CVP power only to preference customers which were customers on April 2, 1951 and which, at the time of applications for wheeling, have no customers within their municipal boundaries which are served at retail by PG&E. Justice alleged that Article 24(a) also effectively precluded access to CVP power by municipalities whose service areas lie outside the wheeling area.[81] Bureau requests to PG&E to expand this area in 1967 were uniformly denied.

Article 12(a)(7) permits CVP, in lieu of making energy available from its own hydro plants, to substitute another source of power for delivery by PG&E to CVP customers. This source of power must be transmitted on the intertie from the Northwest, see Articles 12(a), 19(d), and 19(e). Article 19(g) provides that capacity and energy from a new source to be sold or used in PG&E's service area may not be delivered over CVP's system without PG&E's consent. PG&E has used Article 19(g) to prevent CVP from agreeing to wheel NCPA geothermal power.[82]

The contract limits CVP's marketable system capacity to 1,050 mw. In 1971, CVP withdrew part of its allocation of power from the city of Santa Clara in order to meet other demands upon its system. The cities of Biggs, Gridley, Palo Alto, Redding, and Roseville have commitments from CVP for supply of their entire load growth requirements until 1980, and to maintain the 1980 level of supply until 2004. However, if CVP's preference load reaches 1,050 mw before 1980, the commitment involves continuation of supply to these cities only at the level in effect at the time CVP reaches its overall limit.

Justice has alleged that the effect of the 1,050 mw capacity restriction is to reserve to PG&E the load growth of all those preference customers that are presently served by CVP. Justice also asserts that this restriction is not necessitated by PG&E's obligation to provide back-up for CVP's hydro generation. An arrangement by which CVP would notify PG&E of any addition to CVP's system capacity and would either compensate PG&E for the extra back-up burden or secure the extra back-up elsewhere would accomplish the purported purpose in a less restrictive manner.[83]

The staff of the House Subcommittee on Natural Resources and Power of the Committee on Government Operations has also criticized the CVP-PG&E 1967 contract. This subcommittee expressed the view, before the contract was executed, that Interior had not applied the preference law principle as was required and as a result preference customers were not afforded the full protection they were entitled to under the circumstances.[84]

The 1967 arrangements between PG&E and USBR require that any power and energy imported by CVP from the Pacific Northwest or generated by it must either be used to serve immediate demands of bureau customers or must be sold to PG&E. Moreover, rather than permitting an outright sale, the agreement credits such sales as "deposits" in non-interest-bearing "bank" accounts from which the bureau can theoretically call for power at some future time. There are a series of such accounts; they are not planned to be withdrawn until the period between FY 1998 and FY 2005.*

Sales of capacity to PG&E for banking purposes are made at the same rates charged preference customers.† Capacity is to be repurchased by USBR at the price paid to CVP by PG&E plus a 14 percent service charge. Energy banked prior to the interchange contract may be repurchased at a fixed price; energy banked thereafter is to be repurchased at its original sales price to PG&E plus 14 percent. However, rates for this class of energy are also to be adjusted to reflect changes in costs of generators in new PG&E units. In the event that capacity provided for CVP pumping is not returned by the United States within five years of its delivery, an additional 15 percent service charge is imposed.

---

*In 1973, the bureau estimated that energy would flow into the bank account until FY 1977. After some withdrawals between FY 1977 and FY 1986, a small amount of energy deposits would be made through FY 1997. Then complete drawdown would occur by FY 2005. Capacity account withdrawals would occur between FY 1973 and FY 1985, and then deposits could resume.

†Power produced in the government's share of the coal-fired Centralia station in Washington is sold to PG&E at a rate below its cost, which rate is derived by averaging costs of coal-fired and hydro generation.

Limitations of CVP Marketing in 1967 Accords

The 1967 agreement maintained contractual limits on the quantity and geographic location of load CVP might serve. In negotiations, PG&E maintained (successfully) that the contractual load limits applied to all loads served by CVP regardless of their locations or PG&E's obligation to support a particular load.[85] All CVP output in excess of that required by its limited load was to be sold to PG&E under the banking arrangements. Moreover, PG&E proposed a contract provision that would prohibit CVP from repurchasing banked energy and capacity to serve customers outside of the agreed upon service area.[86]

CVP preference customers can purchase supplementary power from a supplier other than PG&E only when the total demand of CVP's preference customers exceeds the load level PG&E is required to support; PG&E's obligation to furnish support is extended only to preference customers in its service area. As in the earlier agreement, nonfirm CVP power can be sold only to customers having back-up resources. The load-carrying limitations in the 1967 agreement limit CVP's dependable capacity to an amount that gradually expanded to reach 1,050 mw. This limitation on CVP loads is recognized within the bureau as a marketing limit.[87] It is not a product of government engineering studies.

The contractual arrangements under which CVP operates result in a situation in which a project with an installed capacity of 1,437 mw has been able to obtain support capacity rising in steps to serve only 1,050 mw of load. Indeed, the actual amount of support capacity has not risen above 925 mw. The CVP load is supplied through the PG&E system which has exclusive sales authority to supply supplemental power to CVP customers.

CVP is credited with only up to 850 mw of project dependable capacity on the basis of nonextant studies. The PDC is, moreover, subject to downward revision for five years in the event of a drought, as is presently (1977) being experienced in California. All capacity, including 587 mw of spinning reserve received by PG&E in excess of PDC is not paid for by that company. CVP finds itself paying for unit power from the Centralia coal-fired station which it then sells to PG&E for less than it costs.[88]

CVP does not make an independent evaluation of its PDC; rather, it relies upon a PG&E study. When the evaluation study was sought by CVP preference customers, PG&E reported that the study was no longer available.

> Project Dependable Capacity specified in the 1967 PG&E-CVP contract was based on plant capacity factors first established in the 1951 Bureau-PG&E contract and later modified in 1955 and 1963 amendments to that contract. Load curve data from which these capacity factors were derived were not retained in our files. These capacity factors assign CVP system plants to the peak portion of the area load curve.[89]

If studies of any substance were done, the engineering formulation supporting assignment of CVP hydro units to the peak portion of the load curve has not been carried over into PG&E-CVP contracts.

The 1,483.7 mw of installed CVP capacity should become increasingly dependable as peak load grows in the PG&E service area. This results from the decreased amount of energy that is required to provide dependable hydro capacity during short duration load peaks.

In 1967, the parties agreed not to reevaluate the amount of energy required to be associated with dependable capacity at CVP units (the capacity factors). Rather, "in the 1967 contract, Article 11(b)(2), the parties agreed that the capacity factors used in the 1963 amendment to the 1951 contract would be continued in the new contract and applied to then existing Project plants in all future redeterminations. Therefore, the Project plants that existed in 1976 have fixed capacity factors, not subject to renegotiation under the contract. . . ."[90]

The large number of nondependable mw-months CVP has provided PG&E (178 mw-months in 1972 alone) and the large (400 mw) difference between PG&E rating of CVP project dependable capacity and the installed machine generation capability of CVP projects indicate that the amount of CVP capacity that could be sold on a firm short-term basis has been understated. This indication is buttressed by PG&E's inability to produce the derivations of CVP PDC.

The limits on CVP project dependable capacity are embodied in agreements having an unusually long life--40 years. This agreement fixes PDC solely on the basis of federal hydro resources. In so doing, it may exclude from PDC nonfederal hydro or steam-electric resources (Agreement, Article 14). CVP may not sell power on a temporary allocation to a new preference customer.[91] USBR will not reduce its sales to PG&E to increase sales to preference customers now served by PG&E; moreover, if able to purchase power from elsewhere, CVP could not raise its dependable capacity.[92]

USBR has not challenged contractual wheeling restrictions even though it has been informed that these restrictions may contravene antitrust law.[93]

Completing Market Division Arrangements

The conclusion of the 1967 CVP agreement still left open several avenues through which bulk power competition might enter northern California. One such avenue was presented in the terms of federal transmission line rights-of-way permits to SCE and PG&E that permitted other parties to employ excess capacity in these lines.

Additionally there remained the remote possibility that a Northwest entity might attempt independently to enter California markets (for example, by building a line or forcing allocations in intertie facilities or contractual arrangements). Both avenues were closed in an arrangement that foreclosed access to Northwest power and resulted in market sharing by Northwest utilities.

Rapid conclusion of the intertie agreements and congressional action left some of CALPP members' commitments in oral form. One verbal understanding was that if they would enter into a series of agreements to plan and operate nonfederal lines in California in harmony with operations projected under their intertie proposal, Interior would waive its usual requirements that permittees allow others to use excess line capacity for wheeling. One remaining agreement concerned the California proposal to make purchases from Northwest agencies on a generally pro rata basis in order to afford each Northwest agency a fair share of the sales to California.

This agreement, the Seven Party Agreement, was executed on January 14, 1969.[94] Under it the four Pacific Northwest utilities--PP&L, Portland General Electric Company, Puget Sound Power & Light Company, and the Washington Water Power Company, trading as a group--sell the three CALPP members, trading as a group, hydro energy at BPA rates and purchase excess steam-generated energy at seller's incremental cost plus 15 percent. The agreement posed CALPP and the private Northwest entities as exclusive trading partners. This exclusivity was successfully opposed by the BPA as regards sale to the south from the Pacific Northwest. No changes were made, however, to protect California preference customers.

Interior, at the apparent instigation of BPA, wanted direct contractual obligations between the major private companies in the Pacific Northwest group and the California group for sales and purchases to and from the Northwest on a fair and equitable basis. BPA was aware that California pool members were trying to insulate themselves through the Seven Party Agreement from application of the preference law sections of the Bonneville Project Act and the Regional Preference Law. These preference sections specify that any power added to the federal system is subject to the marketing priorities for BPA power which are, first, Northwest preference customers public power systems; second, other Northwest entities; third, nonregional preference customers; and finally, other nonregional entities. These preference requirements would result in California public agencies' obtaining the largest share of Northwest exportable energy.[95] However, BPA did not secure access to Pacific Northwest energy for California preference customers. Sales of nonfirm energy from the Northwest are credited pro rata to all Northwest entities.

The secretary of the interior asserted, after these amending agreements were entered into, that subordinating the Seven Party Agreement to the two agreements would result in California public power systems having access to by far the largest share of the exportable energy available in the Pacific Northwest; and CALPP members would have access to Northwest exportable energy only after the demands of all California preference customers were met.

The Interior Department asserted that since the largest share of Pacific Northwest nonfirm energy is generated by BPA, such energy, together with that received by BPA from both the investor-owned and publicly owned utilities under exchange agreements, is subject to the preference clause, and a substantial portion of the exportable energy delivered annually by PP&L and Portland General will go to Los Angeles, Glendale, Burbank, and Pasadena, and any other public agency in California that contracts with BPA. Preference customers in California will have access to by far the largest share of exportable energy available in the Pacific Northwest. The California companies will have access only to a portion of the exportable energy available from the Pacific Northwest investor-owned utilities. The California companies will have access to the remainder of the Pacific Northwest exportable energy only after the demands of the California preference customers have been satisfied in full.[96]

In actuality, California preference customers did not obtain effective access to the Northwest energy exports. The agreements did not specifically provide for access by California preference customers to intertie transmission, nor did they change existing arrangements among California Power Pool members and public power systems in California. CALPP members have not provided necessary coordination services and wheeling to other California systems. Without access to pooling or to wheeling, most California preference customers have no other artery to Northwest energy.

Even though PG&E is given access to Northwest exportable energy only after the needs of California preference customers have been met, the 1967 CVP-PG&E contract insures that preference customers will have minimal needs. Thus the agreements merely avoided dissension among Northwest entities.

During a roughly contemporaneous period, two other arrangements were made that tend to exclude competitive factors in California. These arrangements precluded access by southern California municipal systems. One was the Interior Department's decision to waive federal wheeling requirements on lines of SCE into southern California from the Southwest.

The second, described above, limited marketing by the one large new intrastate source of generation, the California DWR.

> DWR was permitted to obtain rights to use 300 mw of capacity of 500kv Pacific Northwest Intertie, but DWR's use of the line was limited to interstate transmission; PG&E has consistently refused to permit DWR to use the California segment of the Pacific Northwest Intertie for any intrastate transmission, a refusal based solely on PG&E's desire to exploit fully its monopoly on transmission facilities.[97]

Preemptive Features in CVP-PG&E Contract

Under its contract with PG&E, CVP transmission to customers is dependent upon cooperation by a competitor, which has interpreted the coordination agreement as preventing CVP from seeking to compete for loads.[98] Dominance of CVP by PG&E appears to be coupled with a go-along attitude on the part of USBR and the Interior Department. This attitude is manifested in conduct toward PG&E and toward preference customers. One illustration of this is the handling of a request by the city of Santa Clara, California.*

The city of Santa Clara, which had been purchasing power from PG&E for its municipal power system, sought to obtain an allotment of power from CVP in 1964. To avoid problems relating to CVP's

---

*The earlier history of USBR refusals to allocate power are documented in reports of the House Committee on Government Operations previously referred to.

having allotted all of its firm power, Santa Clara sought an allotment that would be withdrawable by CVP at such time as the allotted power was required by more senior allottees.

PG&E vigorously opposed such an allocation, advising the Interior Department that it would not be in accordance with the city's contract with PG&E and that any such allotment would be counter to USBR's policy of not disturbing contractual arrangements.[99] The company informed Santa Clara and the Interior Department that power transmission services would not be provided for this power under the CVP-PG&E contract, as PG&E was not obligated thereunder to wheel (transmit) bureau power where such wheeling would terminate or affect an existing contract between PG&E and a preference customer.[100] Santa Clara, questioning the lawfulness of the CVP-PG&E accords,[101] persisted in its efforts. It obtained a withdrawable allotment of CVP power in December 1965.[102]

A Justice Department evaluation of PG&E's conduct cited Santa Clara's efforts first to break its long-term, all-requirements contract with PG&E and then to force the company to wheel USBR power to the city as the only instance known to the department of successful resistance by a municipal system to PG&E control tactics. As a result of this experience, Santa Clara was concerned about provisions in the intertie contracts. Santa Clara had been informed that in view of load growth forecasts, its temporary allocation would be withdrawn by 1970, and possibly as early as 1968, to meet the needs of preference customers that had not required some portion of their allotments prior to that time.[103] When the city applied to the bureau for nonwithdrawable power it was told that its application would be considered when more firm power became available as a result of the intertie. Later, however, Santa Clara was told that the additional firm power would all be allocated to preference customers with existing nonwithdrawable allotments.[104]

Santa Clara protested this action and formulated a plan whereby the city would build its own generating facilities and meet its needs through a combination of its own generation and purchased CVP power.[105] The city also proposed to provide the bureau with the 400 mw of power the United States would be required to import under the intertie agreement.[106] This proposal drew a negative response from the Interior Department.[107]

After some years of acrimony after which Santa Clara received only a temporary allocation, a proceeding was commenced in the U.S. District Court. In this proceeding, the court has found that the Interior Department lacked even an official system for allocations and ordered that one be promulgated.[108] Such a system was deemed a prerequisite to a determination of Santa Clara's eligibility.

Santa Clara was not the only preference entity whose misgivings about the intertie and power supply problems seemed to elicit little protective response from the Interior Department.[109] USBR was aware that other municipalities also viewed the intertie contracts with concern.[110]

CVP's Failure to Renegotiate Rates

The rates charged by CVP to PG&E could have been, but were not, renegotiated in 1971, under the terms of the PG&E-CVP contract. CVP chose not to review rates with PG&E on April 1, 1971, as permitted under the 1967 contract. Such a review might have resulted in rate changes preventing the losses GAO records. Interior argues that the rate review was omitted because it believes all customers should be charged the CVP system rate rather than the cost associated with a particular source; because the cost of Centralia power was not accurately known in April, 1971; and because by banking the power at a low rate CVP would reduce the charge to be paid to PG&E when it buys the power back.[111]

The single-rate argument fails to distinguish among classes of services; for instance, PG&E, unlike preference customers, obtains unit capacity and energy from the bureau's entitlement in the Centralia coal-fired generating station while CVP preference customers have not been able to obtain this capacity. Preference customers are largely wholesale purchasers taking all or most of their requirements from CVP. The prospect of future savings through the banking arrangement is no reason to compel present utility customers to pay for losses, especially when savings are speculative. Both present and future CVP customers are likely to benefit if CVP were to collect interest on its bank account.

Interior reports that PG&E gave notice that when the 1976 review procedure for Centralia rates got under way, PG&E would ask for increases in other rates it charged.[112] The underlying motivation for CVP's failure to renegotiate rates appears to be its lack of bargaining power vis-a-vis PG&E. This stems from PG&E's control of bulk power resources, particularly the intertie.

PG&E's overall response to the various charges of subsidy is that the 1967 contract provides CVP and its preference customers with enormous benefits. Thus PG&E asserts that the power support provisions enable CVP to serve twice as much preference load as it could without such support; that CVP may repurchase banked output at rates below cost;[113] that PG&E buys only power surplus to CVP needs, providing CVP with an assured market for all its surplus; and that PG&E provides CVP with all the power it needs to meet its fluctuating needs, and also with reserves.

PG&E has provided benefits to CVP. It has done so only as part of arrangements wherein PG&E precludes CVP from obtaining access to such benefits, and perhaps greater benefits elsewhere;* and wherein the competitive growth of CVP and preference customers is severely limited. The surplus PG&E takes would not be surplus to CVP were it able to have more bulk power supply trading partners through use of more open wheeling arrangements, or federally owned transmission facilities.

The allocation of benefits of the coordination between CVP and PG&E runs disproportionately in PG&E's favor. Further, the understatement of values and failure to specifically identify the value of various benefits in CVP-PG&E arrangements also indicate the weak bargaining position and limited prospects of CVP. When systems of equal bargaining power develop coordination arrangements, each component of coordination is separated and distinctly evaluated in the arrangements.

The ability of PG&E to threaten to retaliate with rate charges in other areas, should one rate CVP charges it be changed, reflects PG&E's great power over bulk power marketing in northern California. PG&E is apparently able both to relate prices and terms of service for different services--that is, to tie some services to acceptance of others--and to tie the price and terms of service it will accept to the price and terms of service it will offer.

By using its power over bulk power supply facilities to obtain below cost or uncompensated services, PG&E is increasing the CVP costs which its competitors, the preference customers, must bear. PG&E is better able thereby to undersell them.

Price Pressure on PG&E's Competitors

Retail rates charged by systems competing with PG&E are subject to conflicting pressures: the necessity of maintaining competition with the company forces rates down, while the costs of capacity and energy may force them up.

In 1973, CVP began to seek rate increases, from preference customers only. The bases for these increases were sharply questioned by its preference customers, apparently with some reason. In November 1974, the Interior Department revised the rate increase proposal so as to implement the increases in two steps, and

---

*If there had been a federal, or a non-PG&E-controlled, or open access intertie arrangement, CVP might well have been able to obtain both reserves and other coordination elsewhere.

announced it would seek rate increases from PG&E in their 1976 contract renegotiation. At the behest of the preference customers, however, the U.S. Court of Appeals for the District of Columbia Circuit ordered a refund of the increase, relying on procedural deficiencies in Interior's actions.[114]

Quite novel justifications were advanced by Interior for the rate hikes. The increases, had they become effective, would have generated a $490 million surplus from preference customers over 35 years. Increasing costs to preference customers sufficiently to generate such a surplus, with no assurance that the customers would receive a return in the form of future electrical services, could obviously damage the preference customers' competitive position vis-a-vis PG&E.

The approaches and procedures followed by CVP seem to suggest an adversary relationship with its preference customers--not close, collective relations capable of jointly providing a competitive alternative to the dominant electric utility in the region.

GAO has concluded that, except for providing for about $78.4 million in deferred costs, power rates should not be increased to create a surplus, and that the concept of CVP's rate and repayment study is inconsistent with the criterion used by other federal power projects and congressional mandates respecting the proper concepts for a rate and repayment study.[115] USBR calculated that by abandoning one erroneous concept, it could scale increases down to 36 percent from the 51.6 percent proposed.[116]

Members of the Subcommittee on Conservation and Natural Resources of the House Committee on Government Operations during hearings on the rate proposal repeatedly asked Interior to produce the computer studies that supported the abnormal concepts used. Interior officials testified that they were not sure that one computer study had ever been made but that a second study probably had been made but had since been discarded.[117] Members also asked why the bureau, in its rate and repayment study, did not plan on withdrawing some of the bank account energy to reduce costs. A bureau official agreed that this would indeed reduce problems, but that that amount of energy would have to be bought elsewhere later. Members then asked why the bureau was trying to project potential savings into the distant future when dealing with the immediate problem of the proposed rate increase. A bureau official testified that he did not know the answer to that question.[118]

Congressman McCloskey, citing USBR documents, established that in 1971 (though allegedly not at the time of the hearing) the bureau entertained the purpose of obtaining funds, through rate increases, to construct new CVP units not yet authorized. The bureau documents revealed a meeting--held before either the renegotiation of PG&E-CVP rates or data on the cost of Centralia power were available--to discuss programmed CVP rate increases. Another document cited by Congressman McCloskey disclosed a conscious bureau decision to maintain surplus CVP revenues at about $350 million. Even though the surplus would foreseeably hinder customer acceptance of rate increases, it would aid in obtaining congressional approval for future CVP additions to plant. The memorandum stated that the apparent conflict could be eased by "referring to the 'surplus' 'reserve' or 'contingency' with an explanation that this amount is considered necessary as a reserve fund which would be maintained in accordance with prudent business procedures and management."[119] Congressman McCloskey stated that the practice smacked of fraud; Congressman Moss called it "an excellent example of gymnastics."[120]

NCPA alleged that it and its members were denied due process by the absence of hearing procedures on the rate proposals which would permit testing of conflicting presentations. Though repeatedly urged by Congressmen Reuss, Moss, and McCloskey, Interior refused to provide public hearings with a record and opportunity for presenting and cross-examining witnesses.[121] Interior responded that public hearings are not required by the law and that there was no need for the expense, delay, and rigidities that more formal proceedings, with such features as sworn testimony, cross-examination, rules of evidence, and written briefs, would entail.[122]

Congressman McCloskey stated during the 1974 hearings on the proposed rate increase, ". . . I don't find any record, in all the testimony the Bureau has presented, that an orderly procedure was pursued initially to ascertain the basis for rates and then to increase rates."[123] Congressman Moss stated at the end of the 1974 hearings that "I have never seen a poorer case for action by a department or agency of the Federal Government, and I have seen some bad ones in more than 20 years of service in this House."[124]

On February 14, 1975 District Judge Gesell set aside the 1974 CVP rate increases because of Interior's failure to accord due process.[125] The secretary subsequently published procedural regulations that do not provide for trial-type proceedings, cross-examination, discovery, or decision based on a record.[126]

Interior Wheeling Stipulations and Bureau Contracts

As prior discussion has indicated, smaller entities and new entrants require a complex of coordination arrangements, sometimes including wheeling of energy. As of October 1973, the Interior Department had never sought to have electric utilities having right-of-way permits perform wheeling service.[127] As of mid-1976 Interior had not provided wheeling for others on bureau lines in California.[128]

The bureau's position has been not to remove anticompetitive provisions from contracts until a problem develops, that is, until someone seeks a service and is denied it under that provision.[129] The overall approach followed by the Interior Department creates uncertainty for new entrants. This contributes to an environment discouraging new entry or expansion by entities other than the dominant power suppliers in California.

Hostility to Development of Geothermal Units--NCPA Efforts

Small utility systems in California are faced with a hostile environment in markets for transmission services and coordination when they seek to develop their own geothermal generation.

NCPA and its member city, Santa Clara, have sought to obtain geothermal resources from potential suppliers in the Geysers area which are not under exclusive contract to PG&E. NCPA has recently entered into an arrangement with lessees in this area to participate in the financing of a drilling program; Santa Clara has recently obtained a lease from the state of California for acreage in this area. These entities are clearly in earnest in their efforts to develop geothermal generation. However, as the Justice Department has alleged, NCPA's attempts to develop such generation have been hindered and frustrated by PG&E's repeated refusals to provide NCPA with transmission services, or reserves and standby services.

When NCPA first approached Interior, in its efforts to enter into geothermal generation, Interior gave a favorable response. Discussions in 1968 focused on geothermal resource development by NCPA, the USBR, Geothermal Resources International, Inc., or joint efforts of the three.[130]

NCPA's objective during these discussions was to develop energy sources to meet greater needs than could be met with bureau power, and to obtain a bureau commitment to participate actively in this effort. Suggested areas of bureau involvement included wheeling NCPA power over government lines, when finally constructed; purchasing NCPA-generated geothermal steam power; providing supplementary power to NCPA cities; providing pumping power; providing cooling water; and assuring a preference for NCPA with respect to potential nuclear power plants.[131]

PG&E responded on August 19, 1969 that the intertie contract did not commit PG&E to participate in any particular program with NCPA; that PG&E would not wheel power from NCPA's project to

CVP's system and thence to NCPA member systems; and that PG&E was unwilling in principle to wheel power from any hydro plant for which NCPA might recapture a license from PG&E in forthcoming relicensing proceedings.[132]

NCPA then developed an alternative plan to construct its own transmission lines to the CVP system and thence wheel over CVP's transmission system to its members. The feasibility of the alternative plan was shown in a joint study by NCPA and the USBR.[133] This plan, according to NCPA, could not be executed because of obstruction by PG&E which used its 1967 contract with CVP and an array of alleged technical problems on the interconnected system as a basis for preventing the plan's execution.[134]

USBR relied on Contract 2948A as grounds for refusing to use spare CVP transmission capacity to wheel NCPA power without PG&E's consent; for refusing to arrange PG&E wheeling of CVP power to NCPA cities not then receiving such power; and for refusing to negotiate for purchase of capacity and energy from NCPA for delivery to preference customers or for replacing or augmenting Centralia power.[135]

The Department of the Interior, pointing out that NCPA's proposal involved more than wheeling, asserted that allegations that the bureau refused to wheel NCPA power were inappropriate. The bureau agreed that PG&E's approval was not required under the contract for CVP to wheel NCPA power,[136] but asserted that the barriers actually lay in NCPA's reserve and reliability problems.[137] Interior was not able to assist NCPA with generation reserves, however, since all bureau reserves were under contract.[138]

USBR declined to replace Centralia power with NCPA power, stating that it was bound to purchase up to a specified amount of Centralia power through December 31, 1981.[139] The bureau was willing to negotiate for purchase of NCPA power after that date,[140] assuming NCPA power would be available and competitive.[141]

Purchase of NCPA power for use other than as a substitute for Centralia power was limited by the bureau's limited authority to provide bulk power to preference customers.[142] The bureau was authorized only to dispose of CVP-generated capacity and energy in excess of project requirements, the 400 mw of Northwest power, and any energy acquired to firm up project capacity.[143] Purchase of firming energy from NCPA would require amendments to the bureau's contract with PG&E, which provided that PG&E was to be the sole supplier of such energy.[144]

When NCPA has sought to raise a complaint before the FPC regarding PG&E wheeling restrictions, it has been met by administrative indifference and hostility. The FPC becomes involved because it regulates some of PG&E's wholesale rates and interconnection agreements. The commission has declined to consider allegations of anticompetitive matters in rate cases. Rather, antitrust issues are placed in a separate proceeding, which is supposed to run in parallel to the relevant rate case. The rate cases are terminated one after another while the antitrust-oriented proceedings go on at a snail's pace and never reach a decisive point. This has been the pattern where PG&E has sought to raise its rates to captive NCPA members,[145] where PG&E filed the agreement whereby PG&E controls all the output of the SMUD nuclear plant,[146] or where PG&E filed the agreement giving it control of the transmission from a power source in the Pacific Northwest.[147]

After PG&E objected, the Interior Department failed to take any steps to implement the USBR-NCPA joint transmission plan, even though the features of the 1967 PG&E-CVP contract, which purport to afford PG&E a veto over use of CVP transmission by others, create bottleneck conditions thwarting economic development by other systems.[148]

NCPA's attempt to engage in coordinated development of independent generation has been thwarted, as Justice has alleged, by CALPP's denial of access to NCPA.[149] In 1970, NCPA requested participation in the pool for the purpose of coordinating the development of generation and transmission resources. PG&E declined this request.[150]

NCPA has alleged that PG&E has used exclusive dealing provisions and preemptive tactics to severely limit NCPA's access to geothermal resources.[151] The facts here are that on May 11, 1970 PG&E executed a revised agreement for the sale of geothermal steam with Magma Power Company and Thermal Power Company, the entities from which it had been purchasing geothermal steam for several years, and an agreement with Union Oil Company for the sale of geothermal steam. The agreements stipulate that Union, Magma, and Thermal agree to develop steam exclusively for PG&E from any leases they then held, or might afterwards acquire in a 180-square-mile red-lined area (unless PG&E fails to build additional generation capacity if the steam developers find at least enough steam for 50 mw of power in a year). This 180-square-mile area contains all of the Geysers land that had previously been developed for geothermal steam production and much land surrounding those steam wells. PG&E's obligation provides that it may defer construction for six years if it does not need the additional capacity, if construction would be inconsistent with cooperation in the development of water resources, or if such facilities would not be competitive with the cost and comparative reliability of power from alternative sources, all as determined by PG&E. The contracts continue in effect for a period of 50 years beyond the date when PG&E places its last unit in commercial operation. PG&E later entered into similar contracts with Pacific Energy Corporation and with Signal Oil and Gas Company. These contracts cover additional wells and areas within the red-lined area.[152]

NCPA has alleged to the California PUC that the exclusive dealing provisions of the Magma, Thermal, and Union Oil contracts are inconsistent with the antitrust laws and the public interest. The commission held that it had no jurisdiction to determine the issues that NCPA presented.[153] The California Supreme Court reversed the commission and instructed it to consider those issues.[154] The commission on remand held that Union, Magma, and Thermal's combined holdings (of 85 percent of the wells and 54 percent of the land) in the red-lined area were not a monopoly and that for PG&E to violate antitrust law or policy there would have to be a monopoly of all power resources in northern and central California.[155] The California Supreme Court denied review.[156]

The California PUC's decision is questionable on grounds relating to both antitrust policy and administrative procedure. As regards administrative procedure, the PUC decision fails to discuss or make findings regarding the underlying facts in the Geysers area. It merely sets out its ipse dixit conclusions. On antitrust grounds, the use of exclusive purchasing provisions for a substantial portion of a unique new energy source is inherently suspect. The questions of why these terms were sought and whether less preemptive terms would suffice should be addressed if the public interest is to be protected. They were not asked by the PUC.

Although the PG&E red-line agreements have not totally excluded others from the Geysers area, NCPA and Santa Clara would first obtain lease rights in that area only years after the PUC decision. In the meantime, many wells developed for production, on red-lined acreage, have stood idle.

LIMITATIONS ON THE SACRAMENTO MUNICIPAL UTILITY DISTRICT

The Sacramento Municipal Utility District (SMUD) is a large public-power system in central California which has the capability and interest to develop a geothermal plant on its own. SMUD is so located as to be surrounded by PG&E service area and lines. To obtain coordination and the benefit of cheaper hydro power from the Northwest, SMUD requires access to the PG&E high voltage grid, including the Pacific Intertie. In order to build larger units having lower unit costs, SMUD must obtain coordination services such as reserves and load growth coordination.

The contracts SMUD has obtained for coordination services preempt any SMUD output excess to its needs, limit SMUD's ability to market energy that the CALPP companies wheel from the Pacific Northwest, and limit SMUD's flexibility in obtaining access to new energy sources (such as geothermal sources). Terms for SMUD's access to the intertie lines are set out as noted above in an agreement between SMUD and the California companies as a group (PG&E, SDG&E, and SCE). Under the contract, shipments south over this grid consist of firm capacity sold for periods of not less than a year, dump energy, and power available from Northwest utilities (acquired under Canadian entitlement exchange agreements).

The California companies agreed to wheel only such capacity and energy as is acquired by SMUD "for use within its boundaries and for sale or assignment as agreed upon by the parties hereto,"[157] from any entity in the Northwest; and they also will accept from SMUD capacity and energy for delivery to any entity in the Northwest through the intertie high voltage lines. The California companies agreed to wheel up to 400 mw of intertie capacity, between April 1, 1971 and March 31, 1976, and otherwise 200 mw of such capacity. SMUD is also to inform companies immediately of any contract with a Northwest entity for power or energy to be transported over the intertie. Additionally, up to 225 million kwh per year of dump energy may be wheeled.

Except in regard to Canadian entitlement power, changes in quantities transmitted are to be made once a year only, generally in 25mw increments. Transmission services are provided at a rate expressed in dollars per kw of firm power shipped, and a sum per kwh of dump power shipped.

In the event that SMUD purchases Northwest firm power or Canadian entitlement power in excess of needs, it may resell all or part of the power to the California pool companies if SMUD pays for transmission, if the power is intended for future use by Sacramento within its boundaries, if notice is given by SMUD five years in advance of the initial date of delivery under the contract (for transmission clearance), and if SMUD gives one-year advance notice in the event it wishes to change the amount of power it will resell. Once SMUD has notified the companies that it is invoking the resale provisions, it must thereafter get the consent of CALPP companies before amending its contract for purchasing Pacific Northwest power for changes effective during a period less than 61 months.

The companies' obligation to purchase Canadian entitlement power that SMUD seeks to resell (lay-off, in trade parlance) is contingent on there being no change in the agreement by means of which Pacific Northwest utilities obtain Canadian power that would be unfavorable to CALPP companies in regard to prices, terms, and conditions for the purchase of such power. The pool companies' obligation to purchase Northwest firm power from SMUD is also conditioned on there being no changes in quantities, times, prices, terms, or conditions of sale. The pool companies need purchase only if they determine that they would have purchased such power in the same quantities and at the same contract terms for use in their own service areas, and if the power is firm, with sufficient associated energy to enable the companies to use such power in their own service areas consistently with their other obligations.* The Canadian and Northwest power is purchased from SMUD at SMUD's cost.†

The contract provides that SMUD may, until March 31, 1976, purchase CEP from CALPP companies. Five years' notice is required for such purchases, unless SMUD is coordinating its load growth planning with PG&E, in which case only one year's notice is required.‡ The contract prevents SMUD from getting any greater quantity of Canadian energy (per kw of Canadian-attributed capacity available to it) than PG&E obtains for its own system. The price of such power to SMUD is the pool companies' cost.

After the SMUD intertie agreement, the California companies amended their intertie agreement. This contract's flexible provisions for scheduling intertie shipments are interesting in juxtaposition to the terms given SMUD. The California pool companies arranged for SDG&E to have access to intertie transmission in this period when public entities were not using their allotted capacity. This supplemental agreement covered a two-year period. During this period the companies would be acquiring approximately 275 mw of Canadian power and associated energy from the California DWR, and anticipated acquiring all of SMUD's CEP--approximately 10,200 kw of capacity in 1968-69 and 38,800 kw in 1969-70). During this period, the provisions of the California companies' Pacific Intertie agreement were amended. These amendments provided that SCE and SDG&E were to purchase, in proportion to their contract shares, CEP which SMUD and the state might sell to the companies, together with associated transmission capacity. PG&E became obligated to resell to the state a portion of the amount of CEP to be acquired by

---

*Northwest power is to be sold in increments of not less than 5,000 kw, and transmission losses are to be absorbed by SMUD to the California-Oregon border.

†SMUD must purchase in 5,000kw lots and change only once a year.

‡The energy involved in SMUD resales counted against the maximum transmission capacity SMUD acquired.

SCE and SDG&E, and PG&E also became obligated to sell SMUD the amount of power SMUD wished to sell companies from its assignment of CEP in excess of the CEP power acquired by SCE and SDG&E under this agreement.

SDG&E was permitted both to obtain assured capacity, and to make interruptible or short-term use of intertie capacity not used by PG&E. Also it could use the line in emergencies. SDG&E obtained from PG&E, in the second year of the contract period, nine mw of assured intertie capacity (measured at the California-Oregon border) which it might use to acquire peaking capacity from the BPA or the Northwest entities, provided SDG&E had first used its other intertie capacity for CEP and firm power from north of California. The nine mw of transmission capacity would be available to PG&E for delivery of Northwest dump energy to its system unless the transmission was needed by SDG&E to supply load in an emergency when SDG&E would otherwise be deficient as to spinning reserve. In that event, SDG&E would later return energy in an amount equal to the dump kwh foregone by PG&E. While PG&E was to give SDG&E advance notice of its intention to schedule dump energy deliveries in this nine mw of transmission capacity, no notice period is specified. If PG&E had not scheduled dump energy, SDG&E might schedule capacity in return for its energy. It could only use the nine mw to receive dump energy from north of California. SDG&E was not required to give long prior notice, and was permitted to share fully in line capacity that happened to be disused when SDG&E could employ the line to purchase cheaper Northwest energy. The flexibility afforded SDG&E but not SMUD, for short-term or dump, or emergency transactions, could be valuable for a power generating entity. For example, such flexibility would make it easier for SMUD to coordinate, or obtain complementary services, from an entity other than PG&E.

1975 Amendments to SMUD-PG&E Contract--Geothermal Units

SMUD's tentative plans to install up to 400 mw of geothermal generation prior to the date of its next nuclear unit are reflected in the 1975 change in the SMUD agreement with PG&E. As noted above, SMUD agreed to sell any of its geothermal capacity and energy not used to meet SMUD's load; sales would be made at prices that could be below cost. Surplus capacity must be sold to PG&E at a price midway between SMUD's cost and the cost PG&E would have incurred in generating an equal amount of capacity and energy from

PG&E 's geothermal generation, so long as this is not in excess of PG&E's cost. Capacity purchased by PG&E each month will be considered to come first from geothermal and oil-fired generation, and secondarily from the first nuclear unit. Such purchases are to continue until SMUD installs its next large thermal resource, at which time any surplus power available for sale to PG&E will be deemed to come from SMUD nuclear resources. PG&E is not to pay more for geothermal power and energy than it would have cost PG&E to generate power using a dry system unit, even if SMUD constructs a hot water unit. If PG&E does not purchase all the energy associated with capacity purchased by it, the remaining energy is to be priced at an amount to be agreed upon, and PG&E shall be entitled subsequently to acquire equivalent amounts of energy at a surplus energy price until a sum for equivalent firm energy is liquidated.

PG&E also has preempted the surplus output of SMUD's proposed 150 mw of combustion turbine generation. PG&E will not only have the right to dispatch these turbines (at capacity factors up to 10 percent each month) but will also be entitled to purchase the turbines' output at the same rate as it pays for energy from SMUD's nuclear units,* a sum less than the costs of oil. If PG&E dispatches the turbines at greater capacity factors, it is to pay SMUD's cost for all energy generated in excess of that associated with the 10 percent capacity factor. The theory behind this turbine rate is that

> for capacity factors of up to ten percent, it is reasonable to assume that the capacity and energy from Sacramento's combustion turbines are being used to serve Sacramento's load. This has the effect of making more base-load nuclear power available for sale to Pacific. Therefore, if the turbines are operated at a capacity factor of ten percent or less, Pacific will purchase only surplus nuclear capacity and energy.[158]

It also appears reasonable that a system controlling its own resources would not operate its combustion turbines at times when it had excess nuclear capacity, unless it were making a sale to another system at a price covering its incremental costs for the sale plus a profit. Nuclear energy (kwh) costs far less than does

---

*Capacity factor is determined by dividing the output of a unit by the product of its rated capability and the hours in the subject period.

combustion turbine generation. PG&E's statement amounts to an assumption that the area peak load (to meet which, combustion turbines would be operated and for which PG&E dispatches the SMUD turbines) coincides with SMUD's peak. This may or may not be the case, but in any event it can be determined as a matter of fact and need not be assumed.

Since the 1970 contract, costs for both SMUD and PG&E have risen substantially above the fixed ceiling prices set for sales of nuclear output to PG&E.* This placed SMUD in a poor bargaining position, especially as it lacked an alternative coordination market for its planned new units.

In regard to nuclear sales, it is stated in the 1975 amendment filed with the FPC that, "as Sacramento's costs for capacity and energy from Rancho Seco are below Pacific's projected cost at Diablo Canyon, these amendments are in keeping with the underlying reasoning of the 1970 contract."[159] SMUD proposes that its second nuclear unit be 1,100 mw. The amended contract provides that after SMUD's second nuclear unit becomes commercially operable, capacity sales to PG&E from both SMUD nuclear units will be at a price midway between SMUD's average cost for its nuclear units and PG&E's cost for its ownership share of the second SMUD nuclear unit; surplus firm energy will be sold to PG&E at cost. Thus sales from the nuclear units, which will be separately dispatched, will not be at their separate incremental costs.

Nuclear energy not associated with the sale of firm capacity would be sold to PG&E at a price midway between SMUD's incremental cost for nuclear energy and PG&E's cost for using its fossil-fueled resources. Unlike the normal practice of determining purchaser's decremental cost as of the time of sale of economy energy, the amended contract prices PG&E's decremental cost by use of monthly average fuel costs at PG&E plants, excluding units burning refinery by-products. This price is stated as likely to be higher than the previous contract price of cost plus 10 percent. PG&E is not required to purchase this excess energy.

The original agreement with PG&E came into being when, subsequent to the intertie agreement, SMUD in 1970 wished to construct two nuclear generation units and found that to do so, it must enter into coordination or integration contracts with PG&E. This arrangement envisages a crude form of load growth coordination between

---

*Under the 1970 contract, SMUD sold energy to PG&E at cost plus 10 percent, but not in excess of 1.9 mills per kwh. Similarly, capacity was sold to PG&E at Sacramento's cost with a ceiling of $2.17 per kw per month.

PG&E and SMUD, PG&E dispatching of SMUD units in connection with its own revenue sharing, and coordination of unit maintenance.

Coordination of load resources planning was essential to SMUD because it could economically only put in nuclear units whose capacity would be substantially larger than SMUD could absorb at the time of construction. When SMUD loads grew larger than its capacity resources, it would have to defer building a second large new unit until it could employ a substantial share of that unit's capacity. Accordingly, SMUD sought to dispose of a portion of its first nuclear unit and to obtain energy and capacity from others in periods between nuclear units. SMUD also required reserves to back up nuclear units. The resulting agreement reflects and fosters PG&E's control of bulk power marketing in central California.

In return for PG&E's commitment to share reserves* and to supply SMUD with up to two million kwh a year to cover SMUD needs between nuclear units, SMUD was required to permit PG&E to dispatch SMUD generating units, control SMUD's acquisition of Northwest dump power,† and to receive only limited credit for the firm capacity SMUD supplies SMUD-PG&E joint loads.‡ The contract placed the burdens of inflation in construction costs on SMUD by requiring sales of nuclear capacity and energy to PG&E to be made under fixed rate ceilings. After its first nuclear unit is in operation, SMUD is to sell capacity in excess of SMUD's imputed needs to PG&E at a price which is the lesser of $2.17 per kw per month or cost; excess energy is to be sold to PG&E at the lesser of SMUD's cost plus 10 percent or 1.9 mills per kwh. While SMUD must provide PG&E with six years' notice of SMUD plans to add capacity, PG&E apparently is under no reciprocal obligation to arrange its planning with SMUD so as to fit new SMUD capacity into PG&E's mix of generating resources. PG&E's consent is required before SMUD can increase the planned size of its nuclear unit.

PG&E is authorized to schedule and dispatch SMUD hydroelectric capacity as an integral mix of resources serving area load in

---

*The contract provides for integrated operation of PG&E and SMUD generation and transmission resources. Reserves shared include reserves for unexpected load growth, scheduled and unscheduled unit outages, and construction delays.

†Under the contract, SMUD may only purchase dump hydro energy when PG&E also is scheduled to acquire such energy; SMUD relinquished its right to import dump energy under the intertie arrangements.

‡SMUD will receive credit only for supplying capacity to PG&E to repay capacity provided SMUD (between nuclear units) if PG&E has not otherwise provided for capacity to serve area load.

the same manner as it schedules its own units. However, SMUD is not credited with the contribution to dependable capacity made when SMUD units are fitted into an areawide load curve. Rather, SMUD is credited only with the capacity its units would have under SMUD's isolated load pattern.

Although credit for capacity is essential if SMUD is to amortize nuclear unit costs, SMUD capacity credits are restricted in that SMUD's nuclear units are to be dispatched by PG&E after preference is given to PG&E geothermal and minimum fuel consumption requirements, and to PG&E long-term power purchase arrangements. This last limitation further reflects the lack of reciprocal planning obligations on the part of PG&E.

Subsequent to the 1970 agreement, SMUD has experienced severe operating problems with its first nuclear unit.* The second unit has now been indefinitely deferred and substitute capacity is required. To provide for this capacity, SMUD has had to renegotiate its integration agreement with PG&E. This renegotiation resulted in a 1975 agreement.[160] The 1975 amendments allow SMUD higher prices for output from its nuclear unit, and permit SMUD to add nonnuclear generation including geothermal units. In addition to the output sales restrictions previously mentioned, the 1975 amendments price output to PG&E at rates that may be less than production costs.

Capacity purchased by PG&E each month will be considered to come first from geothermal and oil generation, and secondarily from the first SMUD nuclear unit irrespective of actual dispatching. Such purchases are to continue until SMUD installs its next large imputed thermal (oil, coal, or nuclear) resource, at which time any surplus power available for sale to PG&E will be deemed to come from SMUD nuclear resources. Geothermal and oil-fuel generation are likely to have lower capacity costs than present nuclear units. The capacity cost of an already built nuclear unit is likely to be lower than the cost of a new large nuclear unit. Thus, using imputed capacity sources as opposed to actual dispatched capacity enables PG&E to obtain capacity at less than cost. As heretofore noted,† PG&E will likewise be acquiring energy from SMUD at prices below SMUD's cost.

---

*Rancho Seco No. 1.

†If PG&E does not purchase the energy associated with capacity purchased by it, the remaining energy is to be priced at an amount to be agreed upon, and PG&E shall be entitled subsequently to acquire equivalent amounts of energy at a surplus energy price until a sum for equivalent firm energy is liquidated.

As SMUD seeks to add other units, and coordination, its expansion will be subject to agreement with PG&E. The coordination agreement between PG&E and SMUD appears to be limited to specified levels of additional SMUD capacity. Under the coordinating agreement, SMUD may add generation only when it forecasts that in seven years it would otherwise have deficient resources (of not less than 2 million kw per month) to meet its own requirements. SMUD appears to be precluded from acting jointly with others to supply smaller deficiencies, or from expanding its scope of operation at the wholesale power supply level.* Various contract terms are not consistent with both parties' obtaining the full benefits of coordination, and hinder SMUD capacity expansions. For instance, 100 percent maintenance reserves are required in the first year a SMUD unit is in operation.

Restrictive Contracts and Predatory Tactics

In addition to reducing the ability of CVP and SMUD to function as sources of bulk power supply to small utilities, the California pool members appear to have fostered a hostile environment for small utilities by their direct dealings with the smaller, retail-oriented utilities. Justice has alleged that PG&E's power supply contracts, by virtue of their all-requirements provisions, and their timing of renewals, have generally precluded attempts by small utilities to avail themselves of alternative sources of bulk power where such sources have become available.[161]

Alamada, Healdsburg, Lodi, Lompoc, and Ukiah, all municipal customers of PG&E, have been under five to seven year all-requirements contracts since 1955. PG&E for many years resisted these cities' requests for shorter contract terms and insisted upon terms precluding access to alternative sources of supply. Although PG&E has now reduced terms to two years or less, Justice has alleged that there is no certainty that PG&E will not again resort to long-term, all-requirements contracts to maintain its monopoly position; and that PG&E continues to refuse to offer all-requirements customers a resale contract with provisions that would allow the customers to provide for future load increases from an alternative source of supply.[162]

---

*The amended contract states that it is anticipated that others besides PG&E and SMUD will be co-owners of the proposed second SMUD nuclear unit which appears to have a specified size of 1,100 mw.

The preemptive nature of CALPP dealings with potential competition is seen in their 1966 contract with the California DWR to provide transmission services over the Pacific Intertie, and to provide power from pool members. This contract runs for 17 years, and specifies the sources from which DWR may obtain power. DWR must give pool members five to six years' notice and negotiate terms and conditions with them for use of generation from any additional facilities of its own or from any other sources.

The DWR has set out, in a brief filed with NRC, its view of action by PG&E and other major utilities in the California area in restricting competitive alternatives.[163] The course of action allegedly has included:

PG&E contracting with DWR on terms preventing DWR from competing with PG&E for bulk power supply, including provision for terminating the arrangement should DWR create its own thermal bulk power supply facility.

PG&E forcing DWR capacity excess to DWR immediate needs to be sold to only one bidder, PG&E, by refusing to wheel the power to others.

California pool members agreeing on a pooling arrangement that preempts their capacity and energy for each other, excludes use of such by nonpool members, and denies equal access to the pool by nonmembers.

PG&E using wheeling restrictions on entities other than DWR (CVP and SMUD) to limit their markets and competitive scope in a manner analogous to its conduct with DWR.

California pool members preempting access to Pacific Northwest power, by means of the Seven Party Agreement, and thus elimininating potentially competitive use of Pacific Northwest power by others in California.

PG&E's refusal to allow use of the intertie facilities for interstate transmission.

PG&E's resistance to DWR and other utility attempts to gain ownership interests in major new additions to California and California-Pacific Northwest transmission facilities.

The Justice Department has in various proceedings made assertions closely paralleling those of DWR.

PG&E has contracted to purchase the entire hydroelectric output of seven California county agencies and districts that lack transmission and distribution facilities. PG&E maintains that exclusivity is necessary in order to provide a long-term financial commitment which these agencies require in order to finance their hydro projects, that no other utility has indicated interest in purchasing such power, and that where PG&E has rejected a proposal the project has not been built. Justice maintains that both conduct by PG&E which inhibited attempts by others to develop nonhydro capacity, and refusals by PG&E to wheel power for others are relevant to the question of why no other utility has indicated an interest in purchasing this power.[164]

CONDUCT IN SOUTHERN CALIFORNIA, LIMITING DEVELOPMENT OF ALTERNATIVE POWER SUPPLIES

In southern California, as in the northern part of the state, there have been problems of access to high voltage transmission. Although LADWP has some high voltage transmission, and is interconnected with the direct current leg of the Pacific Intertie, SCE has the bulk of the power generation and high voltage transmission facilities in southern California.

In the past, SCE has followed a course of refusing to coordinate with small systems in its area, requiring long-term, all-requirements contracts and refusing to wheel power from other bulk power suppliers. In addition, SCE had a territorial allocation agreement with an irrigation district in its area, sought to acquire small systems, and imposed a price squeeze on its wholesale customers. SCE has all-requirements contracts with six municipalities (Anaheim, Azusa, Banning, Colton, Riverside, and Vernon) and one rural electric cooperative (Anza Electric Cooperative) for which it is the only available source of bulk power supply and high voltage transmission.

Justice alleged that SCE has pursued a policy of acquiring the systems of its competitors and customers. SCE attempted to block efforts of its all-requirements wholesale customers to receive bulk power from their own or other alternative sources.[165]

Between 1961 and 1971, SCE acquired four electric utility systems, two of which were all-requirements wholesale customers, and made offers to or indicated interest in purchasing the systems of two additional all-requirements wholesale customers.

SCE has denied requests to wheel federal power to Anza. Moreover, by means of renewing in 1967, for a 25-year term, a restrictive contract with the Imperial Irrigation District whose service area borders Anza, SCE obtained agreement, since rescinded, from Imperial not to sell or wheel power to any SCE wholesale customer, including Anza. SCE has also denied requests to wheel federal power to the city of Colton, another all-requirements customer. In 1964, SCE declined to agree to wheel a block of power from the intertie to Riverside and Anaheim. SCE's all-requirements contracts with these two cities precluded later attempts to obtain Northwest power.[166]

A pattern of refusals to wheel, and long-term, all-requirements contracts, effectively preventing wholesale customers from obtaining alternative sources of bulk power, makes such local systems vulnerable to predatory practices aimed at eventual acquisition.

Justice has alleged that SCE imposed a price squeeze on its municipal wholesale customers in 1957; it raised their rates above SCE's rates to large industrial customers, thus placing the municipals at a severe disadvantage in competition with SCE to attract large industrial loads to their service areas.[167] Justice Department allegations of price squeeze conduct are supplemented by reports from other sources. The mayor pro tem of Riverside has recently alleged that SCE is currently imposing a severe price squeeze on Riverside, Anaheim, and other cities. Riverside is at present an all-requirements customer of SCE and receives electricity at 66,000 volts over its own transmission lines which it delivers to its distribution substations, thus saving SCE the expense of facilities necessary to step down the voltage and, consequently, eliminating one possible justification for incurring higher rates than SCE's industrial customers.

Riverside found itself paying SCE 21.4 percent more for wholesale electricity than an industrial customer would under SCE's rate schedules. As a result, Riverside and other cities are presently at a severe disadvantage in competing with other communities served by SCE in obtaining new industries, new payrolls, and a broader tax base. Riverside recently lost a major industrial customer, Alcan Aluminum, that located nearby and is now taking electricity from SCE at a rate cheaper than Riverside could buy it.[168]

In July of 1963, SCE executed all-requirements contracts with wholesale customers which, among other things, prohibit the purchaser from operating any generating facilities in parallel with those of SCE. Consequently, any generation developed by a municipal would have to operate in isolation; it could not be integrated into the electric network supplied by SCE. Officials of the city of Riverside, a SCE wholesale customer, discussed possible development of peak-shaving generation with SCE during the 1963-64 period, but SCE remained firm in its stand that it would not allow its resale customers to develop their own generation.[169]

In April of 1969, the City of Anaheim, another SCE wholesale customer, indicated to SCE that it was interested in participating in the Navajo-Four Corners coal-fired generation project to provide a portion of its future bulk power supply. SCE responded by offering Anaheim a new ten-year, all-requirements contract at a lower rate, provided that Anaheim accept the offer within a short time frame, which would not permit Anaheim fully to consider participation in the Four Corners project. Anaheim made a counterproposal that it join with SCE in the construction and operation of future generating facilities. On May 29, 1969, SCE responded that such a proposal was beyond the scope of the subject under discussion.[170]

On February 2, 1971, Anaheim and Riverside requested participation in San Onofre Units 2 and 3 from SDG&E, and SCE refused the request because it was too late to alter the sizing of the units, but SCE indicated a willingness to discuss the matter. In March and April of 1971, SCE held five meetings with the cities at which it emphasized that although it would consider any specific proposal, it would be extremely difficult for the cities to make a feasible proposal. SCE set forth general criteria, on April 19, 1971, which any proposal by Anaheim and Riverside would have to meet to be acceptable to SCE.[171] The Justice Department has pointed out that SCE's position would require the transaction to accord significant benefits to SCE, including offsetting the loss of a full-requirements wholesale customer. This makes it just about impossible for a customer to submit a proposal that would satisfy SCE. The Justice Department has further alleged that reserve requirements to which CALPP members have adhered in dealing among themselves and with the small utilities, seem to allow little opportunity for a municipal wholesale customer to submit a workable offer of power supply coordination, though there is no technical necessity for such reserve arrangements.[172]

For 15 years, SCE acted to prevent cities in southern California from obtaining access to a potentially substitutable new power supply, the Colorado River Storage Project (CRSP), or from steam-electric generation in Arizona and New Mexico. SCE acted, at times in conjunction with other utilities and with the Interior Department, to reduce the amount of capacity available to preference customers. It did this by refusing to coordinate with the federal government, thus limiting CRSP capacity, by refusing to wheel, and by acquiring government power to the exclusion of others. The history of these actions is as follows.

The Interior Department was confronted, in the late 1950s and early 1960s, with the problems of how to obtain coordination with steam-electric generation for its Colorado River hydro projects, and how to obtain pumping power for CAP.

Because construction of coal-fired, steam-electric power plants in the New Mexico-Southern Utah-Arizona area requires permits from the Interior Department regarding transmission rights-of-way and, usually, water rights, Interior was not without bargaining capability in seeking coordination.[173] The utility systems of the Southwest looked to large coal-fired units for the future base-load generation required to meet their rapidly rising loads. An organization, Western Energy Supply and Transmission Associates, was created to permit joint government-utility planning for the construction of a six-plant, 12,000mw scheme.* Under the provisions of the Colorado River Basin Project Act of 1968,[174] the Interior Department became a direct financial participant in one plant, the Navajo project, with the stated purpose of obtaining pumping power. USBR obtained a 561mw entitlement, discussed hereafter.

The Navajo project consists of three coal-fired units of 770mw dependable capacity each, and several 500 kv transmission facilities. The project is owned as follows:

Participants	Percent	Number of kw
SRP, manager	21.7	501,270
LADWP	21.2	489,720
APS	14.0	323,400
NPC	11.3	261,030
TG&E	7.5	173,250
USBR	24.3	561,330
Total	100.0	2,310,000

The hydro plants of the CRSP must be firmed up through coordination with steam-electric systems if their dependable capacity is to be fully realized. When the joint ventures owning large new coal-fired stations in the southwest came to Interior for permits required for these units and attendant high voltage transmission lines, these permits were provided, without the inclusion of provisions for coordination with CRSP.

Joint venturers in these plants are both private utilities and public power entities; they control the major power generation and transmission facilities in the area. In return for their commitment to a future coordination agreement with CRSP, the owners of these plants sought and obtained an exemption from wheeling requirements normally found in federal transmission right-of-way permits, for the high voltage transmission facilities associated with the two plants, including high voltage lines into California.[175]

Later, in 1968 and 1969, the Interior Department sought to obtain coordination agreements for integrating CRSP hydro together with southwestern electric capacity. It was unsuccessful, and

*Subsequently, one plant--the Kaiparowits project--was indefinitely deferred.

Interior instead agreed to sell surplus federal hydropower to non-federal systems. According to the Interior Department, this refusal to consummate coordination arrangements rested on the private utilities' desire to avoid having federally owned energy flow to their existing customers.

Assistant Commissioner of Reclamation N. B. Bennett, Jr. reported to the secretary of the interior that as regarded a power coordination agreement, ". . . the sticky problem is still that the utilities will not commit themselves to supply energy to support federal capacity to pirate their customers."[176] Assistant Commissioner Bennett also informed the secretary that "freedom from the right-of-way regulations" and "even limited territorial integrity" were the two major features desired by participants in coordination discussions.[177]

During 1967 and 1968 negotiations for coordination, SCE took a vigorous stand in opposition to coordination and wheeling.[178] In this it was joined by the other participating systems. On February 24, 1969, Mr. David Barry of SCE wrote to a member of a Navajo-Four Corners Task Force (No. 10) that before SCE would agree to participate on the Navajo project, the government must waive its wheeling stipulations. Following collapse of efforts to obtain system coordination agreements, there still is no general arrangement for wheeling of power generation in the western and southwestern states.

SCE has taken further action to prevent smaller California systems from obtaining independent access to power generation east of California. For example, SCE was a member of a consortium to build the large Navajo coal-fired plant in Arizona, until the operators of the plant agreed to admit the city of Anaheim into the ownership consortium. Facing the prospect of being required to wheel Anaheim-owned power to Anaheim, SCE withdrew from the project. By letter of August 23, 1969 SCE refused to wheel any power to Anaheim. Anaheim then withdrew from the Navajo project --lacking any means of obtaining access to the plant's output. After withdrawing from ownership participation in the Navajo plant, SCE effected purchase arrangements with USBR which gave SCE control of most of the energy from the Navajo plant moving into southern California.

THE BUREAU OF RECLAMATION (USBR)

USBR participated in the Navajo project in order to obtain energy to pump water from Lake Havasu, on the Colorado River, to the Phoenix-Tucson area and points south. As it now appears this energy may not be needed for water pumping before 1985, and might be considered as potentially available for sale to government preference customers. However, the Interior Department currently is selling its 425 mw of Navajo capacity as follows: to Southern California Edison, 336 mw; to LADWP, 75 mw; and to Nevada Power Co., 14 mw.

The Interior Department never developed criteria for marketing its excess Navajo power and sold (laid off) that power without making any offer of sale to preference customers (except SRP and LADWP, which had entered into its own arrangements with SCE). This procedure has been held to be in violation of the preference laws.[179] USBR will market its Navajo lay-off power only to entities providing their own reserves and transmission. In the absence of either an overall system coordination agreement in the western-southwestern states area, or requirements for wheeling to California, the practical effect of this provision has been to prevent preference entities from obtaining this power.

POTENTIAL EFFECT OF RECENT ANTITRUST ACTION ON BULK POWER COMPETITION IN CALIFORNIA

Antitrust proceedings in conjunction with nuclear plant applications have yielded agreements between Justice, municipal systems, and SCE, and between Justice and PG&E, respectively, which are intended to be included as conditions to NRC licenses. The SCE license condition is final, while PG&E's proceeding is still pending (May 1977) at the instance of municipal intervenors.

The conditions agreed to in general require that the utilities offer smaller systems an opportunity to participate in new nuclear generation plants. They also entail promises to afford coordination and wheeling services, and partial requirement sales contracts. These conditions appear to hold out prospects for improved possibilities for entry and expansion of new smaller bulk power supply facilities, and other competitive elements in the electric utility systems in California (though not in other states).

However, it is still uncertain to what extent conditions will improve. The agreements to enter into future agreements with other electric utilities are rather general, and their efficacy remains to be demonstrated. They contain clauses that may allow a substantial amount of avoidance. Among these clauses are ones leaving undisturbed the limitations on, respectively, CVP, the power pool agreements, and other past arrangements. The companies signing the agreement have dominant positions allowing a great deal of freedom in planning and implementing their courses of activity: they still have opportunities for exclusionary conduct. The conditions involved for SCE, in paraphrase are as follows:

First, SCE agreed to permit participation in new nuclear units by any entities within or contiguous to SCE's service area which at that time do not have access to alternative, comparably priced sources of bulk power supply. Second, SCE agreed to permit interconnection and coordination of reserves by means of agreements for the sale and purchase of emergency bulk power with any entity within or contiguous to SCE's service area. Third, SCE agreed to sell bulk power to or purchase bulk power from any entity within or contiguous to SCE's service area. Fourth, SCE agreed to transmit bulk power between or among entities with which it is interconnected over its transmission facilities within its service area and facilitate such transmission over facilities outside its service areas.

In the agreement reached on June 27, 1974 by Justice and SCE, a set of procompetitive license conditions are qualified by the following statement:

> . . . SCE should not be obligated to enter into such an arrangement if (1) to do so would violate, or incapacitate it from performing any lawfully existing contracts it has with another party or (2) there is contemporaneously available to it a mutually exclusive competing or alternative arrangement with another party which affords it greater benefits.[180]

Interpretation of these conditions is clouded by the qualification that SCE will permit participation in new nuclear units (not San Onofre units 2 and 3), sell emergency power, and wheel if the requesting party makes it an offer that is more desirable to SCE than alternative uses of its facilities. The conditions do not call for admission of others to CALPP. In a letter indicating agreement to these conditions, SCE states that "Edison does not intend to become a common carrier by reason of these conditions."[181]

SCE has entered into agreements with all of its large resale customers which call for, among other things, integrated operations and coordinated planning of customer and SCE resources, partial requirements service, transmission service, and participation by such customers in certain future SCE units.[182] Those agreements leave critical areas for future agreement or resolution before they can be implemented.

On August 2, 1972, the Justice Department provided antitrust advice on the application of PG&E to construct the Mendocino nuclear plant. The letter concluded that an antitrust hearing should be held because conduct of PG&E had created a situation inconsistent with the antitrust laws, which situation would be maintained by PG&E through operation of the Mendocino units. No hearing was held, as the application was withdrawn.

Justice recommended at that time that license conditions should require PG&E to grant access to Mendocino to the members of NCPA, to eliminate provisions in its contract with CVP that restrict importation, wheeling, and marketing of energy, and to eliminate provisions in its contract with SMUD limiting the generation SMUD may plan and construct.[183]

Justice commenced an investigation under the antitrust laws into whether PG&E was monopolizing the relevant markets for electric power, natural gas, and geothermal steam, and whether it was a party to certain contractual restraints of trade. Prior to completion of this study, PG&E indicated a desire to facilitate licensing of the San Joaquin nuclear project in which it and seven other utilities were participating. Justice agreed to accept certain conditions in the San Joaquin license in lieu of an antitrust hearing. Under the conditions, PG&E would grant ownership access to San Joaquin to any distribution utility in its service area, would wheel participants' shares of San Joaquin power, would provide reserve support to participants' shares of San Joaquin power, and would purchase surplus San Joaquin energy from participants.[184]

On May 5, 1976, after completion of an antitrust investigation of PG&E, Justice issued its letter of advice on the application of PG&E to construct Unit 1 of the Stanislaus nuclear project. Justice advised that it had reached agreement with PG&E on a statement of commitments which Justice believed would obviate the antitrust problems posed by PG&E's activities, and would remedy the situation inconsistent with the antitrust laws which it believed had previously existed. The statement of commitments was requested to be included as conditions to the Stanislaus license in lieu of an antitrust hearing.

The Stanislaus conditions are quite extensive. PG&E agreed to interconnect with utilities in or adjacent to its retail service area without imposing limitations on use or resale of capacity and energy sold or exchanged, or on further interconnections except for reliability considerations; to sell full- or partial-requirements firm power, emergency power, reserves, and to coordinate maintenance schedules with any utility to which it is interconnected; to wheel power for those to whom it is interconnected; and to provide access to Stanislaus or to any other nuclear unit for which PG&E applies for a construction permit during the next 20 years.[185]

Certain further qualifications and omissions, however, should be noted. The Stanislaus conditions do not require PG&E to seek admission of new members to CALPP, nor do they specifically require PG&E to void previous contracts that have anticompetitive provisions, such as the SMUD or CVP contracts discussed above. The conditions do not require PG&E to wheel power or energy from PG&E hydro plants with expiring FPC licenses which may be licensed to others, or to wheel power or energy to any PG&E retail customers, or to wheel power and energy related to the intertie if such action would impair PG&E's own use of the intertie. The intertie is now heavily loaded. Without coordinated planning the intertie is unlikely to be developed so as to be accessible to others.*

PG&E has agreed to sell short-term, limited, and long-term capacity and energy, and economy energy to any generating utility in or adjacent to its service area under any rate schedule or agreement it files with the FPC. It has not agreed to provide such service to distribution utilities--the bulk of its wholesale customers. The conditions do not preclude PG&E application of a price squeeze on a distribution utility. Moreover, nothing in the conditions prevents PG&E from attempting to acquire another utility.

PG&E's control of access to alternative power supply sources in central California is unimpaired. All provisions are subject to "Good Utility Practice" stipulations. These stipulations allow PG&E to require as a condition of interconnection or transaction that the other party use practices, methods, and equipment, including levels of reserves and provisions for contingencies, that are commonly used by PG&E, provided that such stipulations are prudent from the safety, reliability, conservation, and environmental standpoints. They are very complex, involving technical considerations on which there are no industrywide standards. PG&E could, if it so desired, frustrate attempts by small utilities to obtain benefits under the license conditions by shrewd application of the "Good Utility Practice" stipulations. PG&E found technical problems on an interconnected system to frustrate NCPA's attempt to construct its own transmission lines to the CVP system. It appears that the hostile environment for small utilities in California may not be completely eliminated by the license conditions.

A final qualification is that nothing in the license conditions affects PG&E's preemption of geothermal resources by use of redlined areas as described above or its preemption of hydro sites. Even if the license conditions made it easier for a distribution utility

---

*Systems installing base-load geothermal generation would find the intermediate and peaking load-carrying ability of hydro projects appealing. Good hydro sites in California are in general available only in relicensing proceedings.

to obtain the reserves, coordination, and wheeling services required to develop an independent generating capability, the utility still would have difficulties developing geothermal generation if PG&E preempts resources. Other parties to the Stanislaus proceedings before NRC have not agreed to the conditions of the PG&E-Justice accord, and the proceedings continue.

OTHER WESTERN STATES

There is hostile environment for small utilities in certain other parts of the western United States as well. The situation regarding the Central Arizona Project (CAP) and the exclusion of California preference customers from access to Navajo unit output have been described. Preference customers east of California were similarly excluded when Interior privately sold power from its entitlement in the Navajo project.

Arizona Power Pooling Association (APPA), a nonprofit Arizona corporation comprising Arizona Electric Power Cooperative, Electrical District Number Two of Pinal County, and the city of Mesa, in a suit joined by the Arizona Power Authority, a state agency, Intermountain Consumer Power Association, and Bountiful, Utah,[186] have alleged that Interior has violated a duty imposed by federal reclamation laws requiring preference to be given to entities such as APPA in the sale of federally owned electric power. The suit seeks to compel Interior to negotiate with plaintiff for the purchase and sale of Navajo power. On September 24, 1975, the U.S. Court of Appeals for the Ninth Circuit reversed the judgment of the district court (which had held that Interior's decision with respect to which entities were to be allowed an opportunity to purchase power was not judicially reviewable, and that Congress had approved any possible violation of the preference provision by accepting Interior's plan for the CAP as submitted, and appropriating funds for its implementation).

A key point in the Court of Appeals judgment was the undisputed refusal by Interior to offer APPA the opportunity to purchase the power prior to offering it to the private utility companies. In fact, the preference customers had sought, and had been refused, the chance to purchase the interim power.[187] Subsequent to the judgment, Interior has still not proffered this power to preference customers (May 1977).

Utah Power and Light

Robert Gordon, corporate secretary and attorney for Utah Power and Light (UP&L), recently testified against proposed federal legislation that would mandate joint use of bulk power facilities, and access to transmission, and would authorize the FPC to prohibit unfair methods of competition.[188] He argued that the incentive for voluntary interconnection is already present without additional legislation, and that legislative action mandating access to generation and transmission would unfairly disregard the rights of existing customers which pay the full cost for these facilities and which should be entitled to their full benefits as their future load growth absorbs current surplus capacity.[189]

Prompted by this testimony, the manager of Bountiful City Light and Power, a Utah municipal, described situations involving UP&L's "voluntary" behavior. Two days after UP&L announced plans to construct the Emery County coal-fired generating plant, in November 1975, Bountiful asked UP&L if Bountiful might join it (with a 5 percent interest) in construction of the project. The next day Bountiful appeared as a witness before the Utah Public Service Commission in a hearing on UP&L's application to proceed with the construction (which was already well underway). Bountiful supported UP&L's application and filed a petition with the commission for an order requiring joint participation. On March 9, 1976, UP&L responded to Bountiful that all UP&L units up to and including its 1980 unit were fully committed, but that a modification or change could possibly occur and that UP&L would be happy to negotiate with regard to power from an unspecified 1982 unit. The short notice by UP&L to regulatory authorities has made it almost impossible for a small utility to succeed in gaining participation in a unit.[190]

Early in January of 1976, Bountiful attempted to purchase surplus energy for the summer of 1976 from UP&L. There was a certain sense of urgency in the attempt because the alternative was to purchase energy from USBR's CRSP and the deadline for applying for summer energy was January 23, 1976. On January 26, 1976, UP&L notified Bountiful that it regretted that time had run out before UP&L could decide whether to sell surplus energy or not. (In the meantime, UP&L continues to sell surplus power and energy to many utilities outside the State of Utah.)

When on February 5, 1976, Bountiful indicated to BPA that it would like to purchase 18 million kwh annually during the next two years on an if-and-when-available basis. The BPA responded on February 13, 1976, that it might have some surplus energy available but that transmission available to BPA on the Montana Power Company's facilities is fully committed to USBR and that Bountiful would have to make its own arrangements for transmission, perhaps with UP&L.

Still seeking to obtain this BPA energy on March 10, 1976, Bountiful tried to get the bureau to wheel 8 million kwh of if-and-when-available energy from BPA to CRSP reservoirs where it would be stored as water. The water could be used for generating energy to be delivered to Bountiful when needed. On March 26, 1976, the bureau responded that this type of service was not then available.

Bountiful requested, on May 10, 1976, that UP&L wheel the BPA energy to Bountiful. UP&L replied that they were willing to negotiate. Unfortunately, the reply came on July 28, 1976, after the date established by the bureau by which Bountiful had to commit itself to purchase energy during the 1976-77 winter from CRSP. The terms offered by UP&L for wheeling are 3.2 mills per kwh, maintenance of reserves which UP&L would sell at a rate above the pool rate, and installation of under-frequency relays more extensive than those employed by UP&L. The 3.2 mill rate is quite high compared to UP&L's 0.6 mill rate for wheeling CRSP power.[191]

The need for small utilities to approach UP&L to obtain transmission service needed to receive energy from USBR is a consequence of the bureau's refusal to sell economy energy to small utilities while it presently sells such energy to UP&L and other investor-owned utilities, and Interior's failure to build federal transmission facilities after Congress appropriated funds in the early 1960s.

Bountiful is presently trying to obtain geothermal energy. It has bid on geothermal lands, but lost out to Phillips Petroleum. Bountiful approached Phillips about purchasing heat produced at seven wells drilled at Milford, which apparently could sustain 65 mw for 30 years. Phillips, however, appears to be holding out for higher prices. Prime Utah geothermal areas are leased to oil companies; other Utah areas are too risky for Bountiful to enter.[192]

It appears that the environment for small utilities in Utah is little different from the environment in California. The manager of Bountiful City Light and Power testified that UP&L has been giving Bountiful the run-around on requests for participation in new units, and that valuable time and resources have been wasted as a consequence; that UP&L gives the same run-around on requests for transmission service; and that USBR has become as difficult to work with as UP&L.

Municipal and cooperative systems in Utah are members of the Intermountain Consumers Power Association (ICPA), a nonprofit corporation. Most of these systems purchase power from UP&L lines. ICPA's arrangements with UP&L, described in the discussion of pools, exemplify the satellization of wholesale customers that substantially inhibits these firms from helping meet their needs through self-generation. Long notice periods for introduction of new sources, lack of available tariffs for reserves coordination and short-term sales, and lack of an alternative transmission system all work to preclude economic pooling by ICPA members inter se or with others.

Pacific Northwest

Unlike California and Utah, the Pacific Northwest offers a more favorable environment for small utilities. In this area, the small systems appear to be able to obtain transmission and coordinating services through the Bonneville power grid. BPA dominates regional transmission with over 12,000 miles of transmission facilities (about 80 percent of the region's bulk power transmission capacity). BPA markets power to 149 customers, mainly preference customers.

In this region, planning for existing and future thermal and hydro resources is coordinated by a joint planning council. Membership includes 5 investor-owned utilities, 110 publicly owned agencies, BPA, and Washington Public Power Supply System (WPPSS), a municipal corporation and a joint operating agency of the state of Washington which engages in bulk power supply and is made up of 18 operating public utility districts in Washington and the cities of Richland, Seattle, and Tacoma. There is a high degree of coordination and cooperation among utilities involving generation and transmission. Small utilities appear to have no difficulties in obtaining power and energy, wheeling service, or participation in large new units. For example, 29 municipalities, 28 public utility districts, and 47 cooperatives will share in the output of a 1,250mw nuclear unit being constructed by WPPSS.[193]

BPA, with its extensive transmission system, can prevent geographical isolation of small utilities within the service areas of large investor-owned utilities. Because BPA operates its own transmission system, it is not at a strategic disadvantage in bargaining with investor-owned utilities, as is the case for USBR.

In the BPA area, utilities such as the Raft River Cooperative with its geothermal program or the utility system of Eugene, Oregon, with its program for using waste forest products for fuel, can build generating units with confidence that they will be able to market output. The Raft River Cooperative was able to get BPA's assurance that up to 150 mw of power could be transmitted west from its geothermal area at a very early stage in Raft River's planning. The Eugene, Oregon, system has been able to sell 50 mw of power to Burbank, California (over the d.c. leg of the Pacific Intertie).

To date, very high temperature geothermal resources have not been located in the BPA region. The development of less promising geothermal resources, such as are sought in the Raft River

Valley, may be expected to go forward in the Pacific Northwest more rapidly than in the Southwest--when the capacity cost of such power is less than that for alternative sources, which are also more readily accessible to small northwestern systems.

A SUMMATION OF HOW GEOTHERMAL ENERGY DEVELOPMENT IS AFFECTED BY THE CONDITIONS IN BULK POWER MARKETS

Geothermal energy appears to be a resource that could be developed through installation of a number of 50mw to 150mw generating units. Up to 10,000 mw of such capacity appear to be attainable for less than the cost of more conventional coal-fired or nuclear power stations. Lead times on geothermal units should be several years shorter than the decade required for other base-load generation.

The limited size of geothermal units coupled with the regulatory process that must be gone through before a unit may be installed--which is comparable to that for a much larger unit--makes geothermal power more attractive to smaller utility systems than to larger ones. Units of 50-100 mw can provide a small system with a way to enter into power generation and provide such a system with a means of hedging against prices of alternative fuels and delays in the development of coal-fired or nuclear units. Large systems do not obtain a meaningful hedge from 50-100 mw of capacity. The efforts of their management are necessarily directed toward 800mw to 1,200mw units. Small systems are, accordingly, the market geothermal energy must principally look to for rapid development.

This market in small systems exists in part because small systems find new or expanded self-generation projects economically more promising than a wholesale power market largely shaped by anticompetitive practices. Development of geothermal resources for power generation would not, of course, be undesirable or superfluous even if conditions in the wholesale power market were more openly competitive. Rather, market conditions that small power systems now face make development of moderate-sized units a near necessity rather than simply a desirable adjunct to existing generation resources.

Anticompetitive conditions exist not only in the wholesale firm power market (in which all-requirements distribution customers of large systems are the purchasers) but also in such areas as transmission and coordination services (including load growth coordination). These conditions severely hinder smaller systems in their efforts to supply some of their own types of generating units.

Industrial firms, confronted with rapidly rising energy costs, face the same problems of securing access to transmission and coordination.

Conditions in the regulation of electric utilities leave wholesale customer utilities and industrial firms with limited, time-consuming remedies for maintaining a competitive yardstick in bulk power supply. In the meanwhile they are subject to the FPC's shunting aside of issues relating to competition, and its slow review of rate increases that permits one unapproved rate increase to be made effective on top of another.[194]

Even where large utilities have accepted reactor license conditions which require that they engage in transactions with small utilities, small utilities have been subject to what they assert are exorbitant charges for service, excessive uncertainty, and delay in negotiations for services. USBR, as part of the hostile environment, has been reluctant to provide such services even where restrictive contracts with large investor-owned utilities do not expressly forbid it. Consequently, self-generation may be viewed as one of the few means of survival for small systems. Also, self-generation has advantages in cost and in unit size relative to other forms of generation which small systems require at a time when shortages of capacity are widely foreseen.

The difficulties confronting small utilities or industrial firms desiring to operate a geothermal unit are not limited to those arising from the grip fastened by large utilities upon transmission and coordinating services. Small systems and industrial firms will require access to the limited number of high quality geothermal sites to develop this resource at a level of risk they can bear. This risk is attractive only if it holds out the promise of a lower cost energy source.

In seeking such an energy source, would-be users are confronted by the present speculative pattern of resource leasing and development.

NOTES

1. David W. Penn, James B. Delaney, and T. Crawford Honeycutt, Coordination, Competition, and Regulation in the Electric Power Industry (NUREG-75/061) (Springfield, Va.: National Technical Information Service, 1975), pp. 28-35.

2. U.S. Federal Power Commission, The 1970 National Power Survey (Washington, D.C.: U.S. Government Printing Office, 1971), p. I-17-4. Frequency control is described in the FPC's opinion in Florida Power & Light Company, 37 FPC 544 564-66 (1967).

3. Firm power is defined as a power supply considered to be continuously available to serve a particular load or demand of a particular size at a particular location. Users of electric power desire and expect such power to be continuously available. Initial decision, Alabama Power Company, Nuclear Reg. Commission Docket 50-348A (April 8, 1977), pp. 46-47. Mimeographed.

4. 1970 FPC Power Survey, II-1-58.

5. FPC, Glossary of Important Power and Rate Terms, Abbreviations, and Units of Measurement (Washington, D.C.: U.S. Government Printing Office, 1965).

6. See testimony of Bruce C. Netschert, Abraham Gerber, and Irwin M. Stelzer, in U.S. Congress, Senate, Committee on the Judiciary, Competitive Aspects of the Energy Industry, Hearings before the Subcommittee on Antitrust and Monopoly, S. Res. 334, 91st Cong., 2 Sess., 1970, pt. 1, pp. 215-27.

7. Bruce C. Netschert, Abraham Gerber, and Irwin Stelzer, "Competition in the Energy Markets: An Economic Analysis," in Economics of Energy: Readings on Environment, Resources and Markets, ed. Leslie E. Grayson (Princeton, N.J.: Darwin Press, 1975), pp. 54-55.

8. Leonard W. Weiss, "The Possibilities for Competition in the Electric Power Industry," in Electric Power Reform: The Alternatives for Michigan, ed. William H. Shaker and Wilbert Steffy (Ann Arbor, Mich.: University of Michigan, 1976), p. 211.

9. Booz, Allen & Hamilton, Inc., A Study of the Eastern Industrial Coal Market, contract report (1967); Dow Chemical, Energy Industrial Center Study, Draft, p. 26 (1976).

10. Walter J. Primeaux, Jr., "A Reexamination of the Monopoly Market Structure for Electric Utilities," in Promoting Competition in Regulated Markets, ed. Almarin Phillips (Washington, D.C.: The Brookings Institution, 1975), pp. 175-200.

11. Both the substantive and procedural aspects of electric rate regulation by the FPC allow ample opportunity for price squeezes. Rates for sales to other power systems often involve sales in interstate commerce. Federal Power Commission v. Florida Power & Light., 404 US 453, 30 L.ed 2d 600 (1972); Federal Power Commission v. Southern California Edison Co., 376 US 205, 11 L.ed 2d, 683 (1974); and United States v. California PUC, 345 US 295 (1953). Retail rates are filed with state commissions. Under tariffs filed with the FPC, it can suspend rates for up to five months, after which the rates go into effect subject to refund. 16 USC 824 (d). Rate proceedings often take far longer than five months and a second rate increase is often filed while a prior one is pending or just decided. When decided, rates allowed are usually less than those filed. As a result of this situation, wholesale customers find

themselves having to pay high rates, and having to pass such high rates along to their customers. See GAO Report, Management Improvements Needed in Federal Power Commission's Processing of Electric Rate Increases Cases, EMD-76-9 (September 7, 1976); cf. F.P.C. v. Conway Corp., 426 US 271 (1976) (FPC required to consider price squeeze allegations in rate cases). In its Order No. 563 issued March 24, 1977 (FPC Docket No. RM 76-29) the commission prescribed its policy concerning "filings of comparative rate" information on cases of alleged price squeezes. Under the policy, customers intervening in a rate case must establish a prima facie case of price discrimination and its anticompetitive effects. This case must include a showing that the customer competes in the same market as the filing utility, and that retail rates are lower than proposed wholesale rates for comparable service. When price squeeze allegations are made, intervenors are only given 30 days from the filing of the supplier's response to their data request to prepare and file their case-in-chief. As stated in the order "the Federal Power Act does not permit any reparation for damages suffered by the complainants due to any price differentials in superseded rates." The FPC indicated that discovery should be limited to present, not past anticompetitive effects. It is not clear whether the FPC's policy and statement seek to preclude consideration of situations in which a course of conduct involving price squeezes has excluded a customer from a market.

12. See Norman Perch, "Get the Standby Power You Need," Power Magazine (May 1970), pp. 23-24; Federal Council on Science and Technology, Total Energy Systems, Urban Energy Systems, Residential Energy Consumption (October 1972), p. 124; files of American Gas Association on Group to Advance Total Energy program; telephone interview with Frank Morse, Southern California Gas Company, May 1976.

13. H. Murray Echols, "Problems of Total Energy, Part III," Actual Specifying Engineer (January 1971). Echols wrote a three-part series of articles on various problems of total energy systems. For a view opposite to Echols on many points, see Paul Chamberlin, "In Defense of Total Energy," Actual Specifying Engineer (April 1971) (noting that several systems have been purchased by electric utilities at a profit to sellers and that several had been retired when lower electric rates were offered). Only limited amounts of natural gas for back-up units are available. Southern California Gas Company will not back up another primary source of energy beyond a limit of 1,000 cubic feet per hour (about 100 kwh).

14. U.S. Congress, House, Interstate and Foreign Commerce Committee, statement before Subcommittee on Energy and Power, April 6, 1976.

15. Candlewood Mall Shopping Center v. Utah Power and Light Company, 440 F.2d 36 (CA 10, 1971). A utility may be more amenable to using an industry's waste heat for utility generation than using the waste heat for industrial electricity generation. The Los Angeles Department of Water and Power has announced that it is considering a 60 mw generating station utilizing heat produced during a petroleum coke calcinating process. Public Power Weekly Newsletter (September 13, 1976), p. 7.

16. See, for example, David W. Penn, James B. Delaney, and T. Crawford Honeycutt, op. cit., pp. 23-25.

17. See, for example, Crown Zellerbach Corporation v. F.T.C., 296 F.2d 800 (1961) (clustering); and a series of commercial banking cases: United States v. Phillipsburg National Bank, 399 US 350 (1970); and United States v. Connecticut National Bank, 418 US 656 (1974). United States v. Grinnell Corp., 384 US 563 (1966) is authority for combining a number of serviced products into a single market where the combination reflected commercial realities.

18. In United States v. E.I. du Pont de Nemours & Co., 351 US 377 (1956) the court stated, "Determination of the competitive market for commodities depends on how different from one another are the offered commodities in character or use, how far buyers will go to substitute one commodity for another," and a "market is composed of products that have reasonable interchangeability for the purposes for which they are produced--price, use, and qualities considered."

19. Standard Oil Co. v. United States, 337 US 239, 299 (1949).

20. Cf. Brown Shoe Co. v. United States, 370 US 294, 325-26 (1962).

21. Lorain Journal v. United States, 342 US 143; Eastman Kodak Co. v. Southern Photo Materials Co., 273 US 359 (1927); United States v. Aluminum Co. of America, F.21 (CA 2, 1945).

22. Otter Tail Power Co. v. United States, 410 US 366, 380 (1973); Associated Press v. United States, 326 US 1 (1945); United States v. Terminal Railroad Association, 224 US 383 (1912); Gamco, Inc. v. Providence Fruit & Produce Building, Inc. 194 F.2d 484 (CA 1,1952), cert. denied, 344 US 817 (1952).

23. Associated Press, supra.

24. Hughes NRC 207, p. 40. Initial Decision in The Toledo Edison Company, NRC Docket 50-346 A et al., p. 42 (January 6, 1977).

25. Cf. David W. Penn, Coordination Competition, and Regulation in the Electric Utility Industry, NRC., p. 24 (NUREG-75/061, 1975). An atomic safety and licensing board in a recent initial decision, found there to be separate product markets for regional power exchanges (pooling), and bulk power services (individual

contracts). The Toledo Edison Company, NRC Docket No. 50-346A (Initial Decision issued January 6, 1977), pp. 47-51. These groups can be better categorized as submarkets for coordinating services with individual transactions being interchangeable with some aspects of pools.

26. Leonard W. Weiss, op. cit., p. 210.

27. William Lindsay, Pricing Intrasystem Power Transfers in the United States, unpublished paper available from Federal Power Commission, Washington, D.C.

28. The anticompetitive practices in the electric utility industry have long been known. This is evidenced by congressional recognition of problems in the Federal Water Power Act of 1920 which seeks comprehensive development, hydraulic coordination, and to restrict manipulation of power sites; and the Public Utility Act of 1935. Under the Federal Power Act, one part of this latter enactment, the FPC is supposed to consider the anticompetitive impacts of rate proposals, financings, and mergers. Under the Public Utility Holding Company Act, the other part of the 1935 enactment, the Securities and Exchange Commission was instructed to restructure holding companies along competitive lines and to prevent mergers and acquisitions unless the mergers did not lessen competition, and did provide positive benefits to the public. The legislation providing for licensing of the construction and operation of nuclear power plants provides for antitrust review by the attorney general and the Nuclear Regulatory Commission. (Section 105 of the Atomic Energy Act, 42 U.S.C. 2011.)

David W. Penn of the NRC reported in a paper entitled "The NRC's Antitrust License Conditions and the Structure of the Electric Utility Industry" (March 10, 1976), that 69 large systems had been subjected to review and that license conditions had been required for 23 nuclear applications. Later, similar work published as D. W. Penn, J. B. Delaney, and T. C. Honeycutt, "The U.S. Nuclear Regulatory Commission's Antitrust Review of Nuclear Power Plants: The Conditioning of Licenses," U.S. Nuclear Regulatory Commission, Antitrust and Indemnity Group, Nuclear Reactor Regulation, Washington, D.C. (May 1976), (NR-AIG-001).

Regulatory and Sherman Act cases have dealt with market allocations, Pennsylvania Water & Power Co. v. Consol. Gas, Elec. Lt. & Power Co., 184 F.2d 552 (CA4, 1950); United States v. Florida Power Corp., 5 Trade Reg. Rep. 11971 Trade Cases para.73637 (M.D. Fla., July 19, 1971) (consent decree), Georgia Power Co. v. FPC, 373 F.2d 485 (CA 5, 1967) (upholding FPC rejection of wholesale tariff's restrictions on resale), refusals to trade (and refusals to grant access to bottleneck facilities); Shrewsbury Municipal Light Dept. v. New England Power Co. v.

FPC., 349 F.2d 258 (CA1, 1965); Otter Tail Power Co. v. United States, 410 US 366 (1973), and political interference, Otter Tail, supra, and cf., Gulf States Utilities Co. v. Kauper, D.C.M.D.La., Civil Action 71-102, Antitrust & Trade Reg. Reporter No. 583, p. A-11 (Oct. 10, 1972) (granting discovery).

Courts have attempted on numerous occasions to instruct the FPC as to its duty to consider and foster competition in bulk power markets. See, for example, FPC v. Conway Corp., 426 US 271, 48 L.Ed. 2d 626 (1976) (price squeeze); Gulf States Utilities v. FPC, 411 US 747 (1973) (financing). Courts have upheld FPC action requiring coordination of operating reserves, Gainesville Utilities Dept. v. Fla. Power Corp., 402 US 515 (1971); and have required the SEC to consider exclusionary practices when it passes upon joint generation ventures. Municipal Elec. Assoc. of Mass. v. SEC, 413 F.2d 1052 (CADC 1969). In price squeeze cases, court action under the antitrust laws may supplement FPC jurisdiction. City of Mishawaka, Ind. v. Indiana & Michigan Electric Co., 1975-1 Trade Cases, para. 60318 (N.D. Ind., 1975).

29. Alfred E. Kahn, The Economics of Regulation, vol. 2 (New York: Wiley, 1971), pp. 105-6.

30. David W. Penn, James B. Delaney, and T. Crawford Honeycutt, op. cit., p. 23.

31. A series of discussions concerning public utility productivity are found in Public Utility Productivity: Management and Measurement, a symposium published by the New York State Department of Public Service (1975). Papers given discuss the problems of measuring productivity.

32. Described in Power Replacement Corporation v. Air Preheater Company, Inc., 356 F. Supp. 872 (1973).

33. The disputes between fuel suppliers were accompanied by allegations of unfair promotional practices. See U.S. Congress, House, Select Committee on Small Business, Report on Promotional Practices by Public Utilities and Their Effect Upon Small Business, H. Report No. 1984, 90th Cong., 2 Sess. (1968).

34. The SEC employed a case by case procedure to determine if the statutory requirement of divestiture should not be followed, in order to avoid such a loss of economies as would make the separated systems nonviable. See generally, SEC v. New England Electric System, 384 US 176 (166); and 390 US 207 (1968); and Philadelphia Co. v. SEC, 177 F.2d 720 (CADC, 1949).

35. Union Electric Company, HCAR 18368 April 10, 1974; and Northern States Power, 36 SEC 1 (1975); and the Initial Decision in Delmarva Power & Light Company, Ad. Power File 3-3640 (June 26, 1974).

36. Testimony of John Landon, Exhibit 227, Delmarva Power & Light Co., SEC Admin. Proc. 3-3640 (1973).

37. FPC News Release No. 23024, "Power Production Generation Current Data for 1970 to 1975" (March 25, 1977). Industrial generation totaled 2,118 mw in the Western System states at the end of 1976.

38. This resistance is reflected in coordination afforded the Intermountain Consumer Power Association; and in U.S. Congress, Senate, Interior Committee, Hearing on Oversight of Electric Power Contract, Montana, 91st Cong., 2 Sess. (1970). Its continued existence was also reported by interviews.

39. Transmission capacity was crudely computed from circuit miles of transmission lines data reported in Federal Power Commission, Statistics of Publicly Owned Electric Utilities in the United States, 1972, line 54 of statements for individual utilities; and in Federal Power Commission, Statistics of Privately Owned Electric Utilities in the United States, 1972, Section VII, line 66.

40. California Power Pool Agreement.

41. Pacific Power & Light Co., FPC Rate Schedule 105, dated June 29, 1972, filed July 3, 1972.

42. Contract between California Suppliers and the State of California for the Sale, Exchange, and Transmission of Electric Capacity and Energy for the Operation of State Water Project Dumping Plants, November 18, 1966 (DWR Contract No. 855503).

43. Department of Justice, Letter of Advice concerning PG&E Mendocino Power Plant (August 2, 1972), p. 8. This restriction was incorporated in Article 20 of a superseding contract of September 6, 1966, but expired in April of 1971.

44. DJ Letter of Advice, Mendocino, op. cit., pp. 9-10. In litigation pertaining to the FPC's acceptance of the 1970 contract as a tariff, the FPC was sustained in its contentions that it lacked the authority to compel load growth coordination or to mandate unit sizes and ownership such as to permit participation by others. NCPA v. FPC, US App. D.C., 514 F.2d 184 (CADA 1975), affirming, Pacific Gas and Electric Company, 45 FPC 1157 and 48 FPC 1103.

45. PG&E Rate Schedule 45, FPC Docket ER 76-296.

46. Contract 175r-2650, executed April 5, 1951.

47. Contract 175-3428, executed October 1, 1951.

48. The rate set, one mill per kwh, Contract 175r-2650 at Section 10(a), exceeded the 0.75-mill rate originally sought by USBR. GAO later asserted that the one-mill rate was too high and failed to fairly reflect both the load transmitted and the distance covered. U.S. Congress, House, Committee on Interior and Insular Affairs, Subcommittee on Irrigation and Reclamation, GAO Audit Reports on Central Valley Project for Fiscal Years 1953, 1954,

1955, and 1956, excerpted in Hearings on H.R. 6997, 7407, and 10005, 85th Cong., 2 Sess. (1958), p. 231. (Herafter cited as 1958 Hearings.) The bureau responded to this charge by arguing that the rate was a firm one, freely negotiated, and that a single level transmission charge on a systemwide basis had been found most satisfactory in circumstances such as those contemplated by the bureau negotiators. Ibid., pp. 231-32.

49. USBR memorandum from E. K. Davis to Regional Counsel's files, Region 2, January 31, 1951. During negotiations, USBR's representatives stated that the bureau did not feel bound to provide power to customers outside of the area that might be economically served by the bureau's own facilities. Ibid., August 18, 1950.

50. Ibid., January 24, 1951.

51. Contract 175r-2650, Sec. 9(c).

52. Ibid. This last provision represented a retreat by USBR from a position taken at a January 24, 1951 negotiating session. At that time the bureau maintained that a restriction on PG&E's wheeling obligation such as one concerning service to new preference customers would be unacceptable. USBR memorandum from E. K. Davis, op. cit., January 31, 1951.

53. Ibid., February 2, 1951.

54. Ibid.

55. Ibid., September 7, 1950.

56. 1958 Hearings, p. 232. USBR responded to the GAO, asserting that integration benefits were reflected in the 300mw PDC, since that PDC exceeded the PDC of the CVP operation in isolation. The bureau further asserted that PG&E's 518 mw reflected better than normal stream flows. Ibid.

57. During the contract negotiations, USBR expressed its desire to firm up its hydro capacity to perhaps an amount in excess of the 450mw nameplate capacity. It wanted to do this either by energy purchases from PG&E, to be sold back to PG&E when the CVP had excess hydro energy, or by building a government steam plant. Neither contingency was provided for. USBR memorandum from E. K. Davis to Regional Counsel's files, Region 2, July 17, 1950. The company's initial negotiating position made clear its opposition to any agreement which contemplated construction of a federal steam-electric plant. USBR memorandum from E. K. Davis, January 5, 1950.

58. Contract 175r-3428, Sec. 10(b).

59. 1958 Hearings, p. 233.

60. Ibid., p. 234.

61. U.S. Congress, House, Committee on Government Operations, Sale and Transmission of Power (Bureau of Reclamation,

Central Valley Project, California), Hearings on H.R. 2221, 86th Cong., 2 Sess. (1960), pp. 4-28. (Hereafter cited as 1960 Hearings.)

62. Ibid., pp. 15-17; see Contract 175r-3428, Section 10(b) and Article 6 of Amendment 4 thereto.

63. 1960 Hearings. The City of Redding was denied an allocation of CVP power in 1952. In 1956, the bureau somehow found itself able to allocate 7,250 kw to the city of Roseville after a congressional investigation. U.S. Congress, House, Committee on Government Operations, Availability of Power to Public Preference Customers from Central Valley Project (Roseville, California), H. R. Rept. No. 218, 85th Cong., 1 Sess. (1957), cited in 1960 Rept. supra, note 5, p. 13.

64. Amendment 4, Contract 175r-3428.

65. Ibid., p. 29.

66. Article 12(a) in Supplement No. 5 to Contract 175r-2650; Article 8 in Amendment No. 4 to Contract 275r-3428, cited in 1960 Hearings, pp. 30, 32.

67. Ibid., p. 32.

68. 1960 Hearings, p. 28. The report stated that USBR should sell PDC to customers on a withdrawable basis if PDC exceeded other customers' elements. Ibid., p. 23.

69. Ibid., pp. 21-22.

70. Ibid.

71. Ibid., pp. 34, 39-46.

72. U.S. Department of the Interior, Report of the Secretary to the House and Senate Appropriation Committees on the Pacific Northwest--Pacific Southwest Intertie, June 1964, Sec. 11, p. 1.

73. Letter from Stewart Udall to Congressman John E. Moss, USBR-BPA attachment, June 25, 1964.

74. Congressman John E. Moss questioned the proposal's failure to reduce the wheeling charge from its present level or to make any change in the wheeling boundary. The National Rural Electric Cooperative Association (letter from Clyde Ellis, general manager, to Senator Warren Magnuson, July 6, 1964) cited GAO's view that the wheeling charge was possibly excessive. Interior in turn asserted that the charge was reasonable when viewed as the average cost of servicing all customers, and that the current wheeling area provided maximum benefits for the majority of preference customers while service to isolated loads outside the boundary could result in increased costs for all customers. A compromise involving a broader wheeling area coupled with different wheeling charges related to actual costs of transmission in different zones was not considered.

Congressman Moss questioned the proposal's restrictions on the government's ability to change a preference customer's contract rate of delivery. Interior responded that this was necessitated by bookkeeping, scheduling, and planning problems. Ibid. Congressman Moss also questioned whether the amortization plan in the proposal was less beneficial than that of the federal yardstick plan. Interior replied that the proposal was in reality of greater benefit to the government when viewed over a 75-year period (50-year amortization, 25-year postamortization) yielding $240 million more in benefits than the yardstick plan. Ibid. However, these benefits were based on proposals and assumed completion of contracts and additional commitments from systems not then parties to the negotiations. Interior claimed that preference customers would receive 55 percent (1,750 mw) of the proposed line capacity. (Letter from Charles Luce to Senator Warren Magnuson, July 2, 1964.) However, only 47 percent (1,500 mw) would really be available for preference customers. Even this amount was subject to California exercising options which were to expire in 1967. Otherwise, the preference customers would only be entitled to 33 percent (1,050 mw) of the capacity. (Letter from Clyde Ellis to Senator Warren Magnuson, op. cit.)

75. Letter from Clyde Ellis to Senator Warren Magnuson, op. cit.

76. Letter of Kenneth Holum to Senator Thomas H. Kuchel, October 11, 1966.

77. The California Companies Pacific Intertie Agreement, August 25, 1966.

78. See Interior Department, Remaining Major Issues with USBR Regarding Contracts Relating to the Northwest Intertie (April 6, 1966).

79. Contract 14-06-200-2948A, executed July 31, 1967.

80. DJ Letter of Advice, Mendocino, op. cit.

81. Ibid., pp. 6-7.

82. Ibid., p. 11. This article precludes construction of a publicly owned steam-electric plant such as the proposed Delta steam plant, which is the only federal steam plant ever authorized by Congress outside the TVA system.

83. Ibid., p. 8.

84. Staff of House Subcommittee on Natural Resources and Power of the House Committee on Government Operations, Staff Memorandum Regarding Proposed Contracts Between the Bureau of Reclamation and Pacific Gas and Electric Company, Relating to the Pacific Coast Electric Intertie, 90th Cong., 1 Sess., March 8, 1967, pp. 1-4.

85. USBR memorandum from C. H. Kadie to commissioner of reclamation, August 19, 1966.

86. Ibid.

87. USBR memorandum from Loy Kirkpatrick to Ray Coulter, September 1, 1966.

88. GAO, Report on California's Central Valley Project--Proposed Rate Increase, G.A.O. Rept. B-125042, November 19, 1973, pp. 10-11. PG&E, unlike preference customers, purchases unit power. CVP preference customers have not been able to obtain allocations of Centralia power from CVP.

89. Letter of December 17, 1973, to Assistant Regional Director, J. Robert Hammond, USBR, from J. D. Worthington, PG&E.

90. Ibid.

91. Memorandum to files from assistant solicitor, Power, Department of the Interior (January 11, 1972).

92. Letter of October 2, 1972, to Mr. Gillmor, NCPA, from Assistant Secretary of the Interior, James R. Smith. This letter said that any supplemental service provided by CVP to NCPA in conjunction with NCPA self-generation efforts would first require tripartite discussion with PG&E, and tripartite agreement regarding reserves and reliability.

93. Memorandum to assistant secretary, Water and Power Resources, from assistant solicitor, Power, Department of the Interior, September 22, 1972. That USBR contracts with utilities are not immune to antitrust scrutiny was made clear in Otter Tail Power Co. v. United States, 410 US 366, 378-379 (1973).

94. Pacific Power & Light Co., FPC Rate Schedule 105, dated June 29, 1972, filed July 3, 1972.

95. Statement of Rogers Morton before the FPC, March 12, 1973, p. 28. At BPA's insistence, the seven utilities agreed to afford all Northwest utilities access to the intertie for exporting nonfirm energy, and to share with them all nonfirm energy imported from California.

96. This explanation, which had generally appeared before, appears at page 30 of the statement of the secretary of the interior in Pacific Power & Light Company FPC Docket No. E-7796 March 12, 1973 (Proceeding on Seven Party Agreement filing). The letter of January 14, 1969 transmitting the Seven Party Agreement to BPA simply spoke, in vague terms, of providing for sales to preference customers.

97. State of California DWR, Petition to Intervene and Request for Hearing, In re Pacific Gas and Electric Company, NRC Docket No. P-56A-A (October 15, 1976).

98. Cf. Memorandum to files of associate solicitor, Water and Power Resources, November 4, 1965.

99. Letter of September 8, 1965 from Robert Gerdes to Stewart Udall.

100. Letter from Frederick T. Searls to Donald Von Raesfeld (Santa Clara's city manager), of September 20, 1965.

101. Letter of September 10, 1965 from Donald Von Raesfeld to Frederick T. Searls.

102. U.S. Congress, House, Government Operations Committee, Staff Memorandum 60 (July 14, 1967); and see 37 Fed. Reg. 1624 (1972).

103. U.S. Congress, House, Government Operations Committee, Staff Memorandum 66 (July 24, 1967).

104. Ibid.

105. Letter from Donald Von Raesfeld to Stewart Udall, June 7, 1967.

106. Ibid.

107. Memorandum from Deputy Solicitor to Assistant Secretary for Water and Power Development, Edward Weinberg, July 14, 1967.

108. City of Santa Clara v. Kleppe, N.D. Ca. Civ. No. C-76-1574-SC.

109. The city of San Francisco was another preference customer with plans to develop additional capacity, in this case at existing plants, switching the Hetch Hetchy plant to peaking operation. The city sought a banking and exchange arrangement with PG&E and municipal agencies. The Raker Act dictated the terms of this agreement, however, requiring that such a proposal provide for maximum development of the resources; make benefits of low cost public power available to the area's public; call for the city to retain full control of the marketing of the projects output; and set up arrangements with PG&E and others on an exchange basis rather than a sale-and-subsequent-purchase-of-energy basis. Memorandum from Richard Pelz, acting assistant solicitor, Power, to Harry Hogan, Department of the Interior, December 6, 1966.

110. Letter from Donald Von Raesfeld, Santa Clara city manager and president of the Northern California Municipal Electric Association, to Stewart Udall, July 28, 1967.

111. GAO, Report on CVP Proposed Rate Increase, p. 12.

112. U.S. Congress, House, Committee on Government Operations, Hearings on Power Rate Increases, Bureau of Reclamation, Central Valley Project, California, 93rd Cong., 2 Sess., 1974, pp. 452-53. Hereinafter referred to as 1974 Hearings.

113. This assertion neglects the time value of money. It is based on hypothetical suppositions regarding the price of returned capacity and energy. Ibid., p. 270.

114. Secretary of the Interior v. NCPA (US CA, DC. No. 75-1572).

115. GAO, Report on CVP Proposed Rate Increase, pp. 23, 32.

116. Ibid., p. 31.

117. 1974 Hearings, pp. 491-93.

118. Ibid., pp. 917-18.

119. Ibid., p. 929.

120. Ibid., pp. 927-29.

121. Letters to Secretary Morton dated February 2, 1973, June 7, 1973, and July 5, 1973.

122. Letter to Chairman Reuss from Assistant Secretary Horton dated June 21, 1973.

123. 1974 Hearings, pp. 918-19.

124. Ibid., p. 929.

125. NCPA v. Morton, 396 F. Supp. 1187 (US DC), aff'd Secretary of the Interior v. NCPA (USCA, DC, No. 75-1572.

126. 40 Fed. Reg. 34431, August 15, 1975.

127. Memorandum of October 15, 1973 to under secretary of the interior from solicitor, "Draft on Wheeling Stipulation." The stipulation had primarily been useful in bargaining for transmission for USBR.

128. Letter of May 7, 1976 to James F. Trout from M. A. Catino, assistant regional director, USBR.

129. See Memorandum of March 16, 1973 for acting associate solicitor, Water and Power Resources, to assistant commissioner of reclamation from assistant solicitor, Power, Department of the Interior.

130. Minutes of NCPA meeting attended by NCPA officers and representatives and Interior Department personnel, including Kenneth Holum, assistant secretary for water and power, August 13, 1968.

131. Ibid. NCPA representatives were in agreement at that point that an atomic generating facility could be built in cooperation with Interior and PG&E.

132. DJ Letter of Advice, Mendocino, pp. 10-12.

133. NCPA Geothermal Power Project, Joint Transmission Study, April 1972.

134. 1974 Hearings, p. 320.

135. Letter from Gary G. Gillmor, chairman, NCPA, to James R. Smith, assistant secretary for water and power, Department of the Interior (DI), September 7, 1972.

136. Memorandum from Richard Pelz, assistant solicitor, Power, to assistant secretary, Water and Power Resources, DI, September 22, 1972.

137. Letter from James R. Smith, assistant secretary for water and power, DI, to Gary Gillmor, chairman, NCPA, October 6, 1972.

138. Memorandum from William Wilson, acting deputy assistant secretary, Water and Power Resources, DI, to Jared Carter, deputy under secretary, DI, January 26, 1973.

139. Letter from James Smith to Gary Gillmor.

140. Ibid.

141. Memorandum from William Wilson to Jared Carter.

142. PG&E's approval was required under the interchange contract but the company could not refuse consent unreasonably, or for reasons unrelated to engineering considerations, or to achieve monopolistic ends. Memorandum from Richard Pelz, assistant solicitor, Power, DI, to assistant secretary, Water and Power Resources, DI, September 22, 1972.

143. Ibid.

144. Ibid.

145. FPC Docket E-7777, Order of September 28, 1973.

146. FPC Docket E-7597, Orders of June 8, 1971 and May 15, 1973.

147. FPC Docket E-7796.

148. Cf. Otter Tail Power Company v. United States, 410 US 366.

149. DJ Letter of Advice, Mendocino, p. 13.

150. Ibid.

151. 1974 Hearings, op. cit., pp. 314-22.

152. Ibid.

153. 71 Cal. P.U.C. 543, Decision 77918, November 10, 1970.

154. NCPA v. PUC (1971) 5 Cal. 3d 370, 96 Cal. Rptr. 18, 486 P.2d 1218, 91, PUR 3d 246.

155. Decisions 78402 and 78403, November 23, 1971, 72 Cal. PUC 704 and 718.

156. S.F. No. 22879, June 7, 1972.

157. Contract between California Companies and Sacramento Municipal Utility District for Extra High Voltage Transmission and Exchange Source, Art 9.

158. Letter of PG&E accompanying filing of this contract with FPC.

159. Letter to Kenneth F. Plumb, secretary of the Federal Power Commission, from W. M. Gallavan, PG&E, of November 21, 1975, FPC Docket No. ER 76-296.

160. Filed by PG&E with the FPC in November 1975 as PG&E Rate Schedule 45, FPC Docket No. ER 76-296.

161. DJ Letter of Advice, Mendocino, p. 4.

162. Ibid., pp. 4-5.

163. Petition of State of California DWR to intervene and request a hearing; Pacific Gas and Electric Company, N.R.C. Docket No. P-564-A (October 15, 1976).

164. Ibid., p. 9.

165. Department of Justice, Letter of Advice concerning SCE and SDG&E, San Onofre Nuclear Generating Station Units 2 and 3, AEC Docket Nos. 50-361-A and 50-361-B, p. 38. (Hereafter cited as DJ Letter of Advice, San Onofre.)

166. Ibid., p. 3.

167. Ibid., pp. 3-4.

168. U.S. Congress, Senate, Commerce Committee, testimony of Mayor Eric Haley, *Hearings on Electric Utility Rate Reform and Regulatory Improvement,* S. 1666, S. 2208, S. 2502, and S. 2747, S. 3011, S. 3310, and S. 3311, 94th Cong., 2 Sess., 1976, pp. 345-53. (Hereafter cited as *1976 Hearings*.)

169. Ibid.

170. Ibid., p. 5.

171. Ibid., p. 6.

172. Ibid., p. 11.

173. The federal government owns, or holds in trust for Indians, the major portion of land in the Southwest. The United States also owns 45 percent of the coal. Department of the Interior, Southwest Energy Study, Summary Report (November 1972).

174. 43 USC 1501 et seq.

175. By an exchange of letters. The first letter is dated April 7, 1966 and is from the Secretary of the Interior to SCE Vice President Jack K. Horton; the second is a letter of April 11, 1966 from Mr. Horton to the secretary. Navajo Plant owners are: SCE, LADWP, NPC, SCE, Public Service Co. of New Mexico, SRP, TG&E, and El Paso Electric Co.

176. Memorandum of July 29, 1968.

177. Memorandum of October 21, 1968; and see letter of March 11, 1967 of Howard P. Allen of SCE to the secretary of the interior.

178. Also see Memorandum of January 7, 1969 from Bob Bennett to Secretary Udall.

179. *Arizona Power Pooling Association v. Morton*, 527 F.2d 721, modified on rehearing, 527 F.2d 728 (CA 9, 1975). It preempted an otherwise competitive source of power from reaching Southern California.

180. Agreement of June 27, 1974 between the Justice Department and SCE.

181. Letter to assistant attorney general, Antitrust Division, from William R. Gould, SCE, dated June 6, 1974.

182. Ibid.

183. DJ Letter of Advice, Mendocino, p. 16.

184. Department of Justice, Letter of Advice, San Joaquin Nuclear Project, NRC Docket No. P-499-A, November 24, 1975. Justice noted that PG&E was only a 23 percent owner of the project and that it would soon issue advice on the Stanislaus Nuclear Project, solely owned by PG&E.

185. Department of Justice, Letter of Advice, Stanislaus Nuclear Project, Unit No. 1, NRC Docket No. P-564-A, May 5, 1976.

186. APPA v. Morton, US Court of Appeals, 9th Circuit, Slip Opinion Nos. 74-1167, 74-1168, and 74-1173. Amici curiae briefs were filed on appellant's and intervenors' behalf by Anaheim, Riverside, and Banning.

187. Ibid., pp. 9-10.

188. Robert Gordon, Testimony, 1976 Hearings, p. 364.

189. Ibid.

190. Ibid., pp. 374-83.

191. Telephone conversation with Barry Hutchings, Manager, Bountiful City Light and Power, August 16, 1976.

192. Ibid.

193. WPPSS Unit No. 4. Sec. Dept. of Justice, Letter of Advice, WPPSS Unit 4, NRC Docket No. 50-513A, February 13, 1975.

194. William W. Lindsay of the FPC staff, testified that during the last six months of 1975, 79 wholesale rate cases were received, 29 cases were disposed of, and 229 cases remained pending. Some cases have remained pending for over 30 months. 1976 Hearings, pp. 775-90.

CHAPTER

4

THE RAW FUELS INDUSTRY AND GEOTHERMAL ENERGY

To obtain geothermal energy at a price related to its cost of production, rather than at a higher price tied to that of alternative fuels--that is, to obtain geothermal energy in a manner that makes it attractive--a utility or other user generally must be prepared to become a fuel company. The user cannot expect to be able to go to the petroleum companies and their affiliates and service companies--the group dominating lease holding--and obtain prices set on a basis other than the price of other fuels.

The raw fuel industry has little if any incentive to reduce utility fuel costs. It profits handsomely when these costs rise; it is so structured as to make it unlikely that a large fuel company will undersell the market and smaller petroleum firms are not likely to do so either.

Large petroleum companies are in an excellent position to warehouse prime domestic energy sites; as will be discussed, they have an incentive to avoid competition among the divisions of their enterprises. Smaller oil companies, as will also be discussed, are unlikely to undersell because they are often dependent upon larger firms for leases and financing; because they are likely to view themselves as service companies for larger oil companies, providing exploration services and some risk capital; and because they sometimes have the cash flow to speculate on geothermal leases, while the share of the market they see in the event of a price decline may not be sufficient to motivate them to cut prices to obtain such added sales. A price decline would also tend to depress the speculative value of their leases.

The independent enterprise specializing in geothermal work may be willing to undersell the market, or to enter into a joint venture with a co-venturer who seeks cheaper energy, to obtain a positive cash flow or perhaps even to "get the ball rolling." However, even such an independent geothermal developer may be constrained from entering into such transactions. It may have its prospects tied up in a joint venture contract with a large oil company, or perhaps a large utility, or may look forward to such an arrangement. Moreover, if an independent enterprise does act to sell its output for less, it foregoes the glimmer of out-sized profits that may have attracted some of its capital, although this loss may appear to be only temporary and may be offset by the prospect of some cash sales.

The independent enterprise will generally require a financing partner in its project. That partner may have to be an energy user, if it is not to be an oil company or an oil-associated financial institution. The partner could be a financial institution secured by federal loan guaranty, but such guaranties do not appear to be large enough to cover both the delineation and development of a field. Lenders secured by a guaranty and associated with large petroleum companies may constrain a debtor's marketing.

The rate at which geothermal development goes forward appears to reflect the leasing and warehousing practices of larger oil companies. The slow development and concern with opportunity costs that appear to characterize their activity are to be anticipated from the nature of utility fuel markets and their control therein.

To understand the effects of large oil company ownership of other fuels, which will sometimes be referred to as cross-fuels ownership, one must begin with a look at utility fuel markets.

WESTERN ELECTRIC UTILITY INDUSTRY AS A FUEL MARKET

The electric utility industry consumes over 28 percent of U.S. energy production.[1] Utilities obtain about three-fourths of their generation from coal, oil, and gas, consuming about 72.6 percent of domestic coal consumption* and 88.8 percent of residual fuel oil demand (1975).[2]

Electric utilities constitute a distinct fuel market because of their aggregate size,† the size of their fuel purchase contracts, and

---

*This assumes utilities burn only bituminous coal and lignite. Inclusion of the small quantity of anthracite consumed would yield no material difference.

†Electric utilities alone use civilian nuclear plants. They use unit trains for coal, and are almost the sole present large market for steam coal, though in the future there will probably be other large coal sales for gasification units. The future of coal gasification

the importance of fuel in overall generating costs. Distinct facilities and sources are, of course, devoted to the supply of this market. Only a limited number of firms compete to supply it (except for spot sales) and competition is limited to a few points in time; for example, when a new unit is being planned, or perhaps when a long-term agreement is lapsing. The utility fuel market has been growing with rising demand. Much of this growth is directed toward large generating units. For the nation as a whole, planned new capacity is in the size ranges indicated in Table 4.1.

TABLE 4.1

Fossil and Nuclear Steam Unit Size Ranges for Proposed New Capacity

Unit Size Range (mw)	Fossil Steam Capacity* mw	Fossil Steam Capacity* Percent	Nuclear Capacity mw	Nuclear Capacity Percent
399 or less	25,953	15.61	630	0.41
400-599	60,266	36.25	482	0.31
600-799	56,033	33.70	2,018	1.81
800 and over	23,989	14.43	151,018	97.46
Total	166,241	99.90	154,938	99.99

*Excluding combined cycle units.

Source: Federal Power Commission, Staff Report on Proposed Capacity Additions, December 9, 1975.

Although larger steam-electric generating units have clear economies of scale, a number of small fossil units are planned for the 1975-84 period. The distribution of new units is as follows: 399 mw or less, 132 (36.27 percent); 400-599 mw, 123 (33.79 percent); 600-799 mw, 82 (22.53 percent); over 800 mw, 27 (7.4 percent)--for a total of 364 units.[3] The large number of small units evidences a lack of pooling and/or transmission services. Fossil steam units comprise 43.2 percent and nuclear steam units 40.26

---

is clouded, as unit outputs may be limited by water requirements and output cost in excess of alternative supplies. Utilities use low sulfur residual fuel oil in quantities overshadowing other markets. Utilities have distinct contractual supply arrangements.

percent of the 384,788 mw of capacity scheduled nationally for completion in the 1975-84 period. Geothermal units (1,172 mw) constitute 0.3 percent of this total.*

Growth estimates for electric generating capability in the later 1970s and early 1980s have been reduced, declining from a projected average annual growth rate of 7.7 percent in 1974 to a projected rate of 5.67 percent in 1976.[4] Similarly, ten-year energy growth rate projections were reduced to 6.15 percent from 6.73 percent in 1975.

Long-term supply contracts increasingly dominate utility fuel purchases. A limited number of purchasers take a very significant share of sales on a national basis. The 20 largest utility coal purchasers acquired 54.6 percent of utility coal purchases, paying 55.5 percent of coal fuel bills. These systems included PP&L and Arizona Public Service Company. The 20 largest utility purchasers of oil bought 67.8 percent of utility oil purchases, paying 69.0 percent of oil fuel costs.[5] These systems included SCE (first in oil deliveries received), PG&E (ranked ninth), LADWP (ranked eleventh), and SDG&E (ranked fifteenth). It is interesting to note that the 20 largest utility purchasers of coal and oil did not appear to receive volumetric price discounts. Instead, they paid a greater share of fuel costs than is proportional to their energy receipts. This disparity may indicate that their extremely large contracts must be negotiated in an even more concentrated market than would smaller purchasers--there being fewer single suppliers capable of filling such a large order. If this is so, it may in turn indicate an excessive predilection for "one-stop shopping" by these large utilities.

Natural gas purchasing was still more concentrated. The 20 largest electric utility buyers of natural gas received 74.8 percent of the utility deliveries of gas, paying 70.5 percent of gas costs. Among these utilities PG&E ranks fourth, SCE fourteenth, and Public Service of Colorado nineteenth.

Prices

Coal, gas, and residual fuel oil† prices are particularly relevant costs of energy alternatives to geothermal power for base-load

---

*In addition to 11 projected geothermal units (all in the West), 468 mw (390 in the West) of fuel cell capacity are forecast, of which 180 mw (all in the West) will be equipped to burn refuse and 251 mw (all in the West) to employ waste heat.

†Utilities consume residual fuel oil (for example, No. 6 oil or Bunker C oil) in their boiler furnaces, while they use lighter oils for peaking gas turbines.

generation. These are the fuel prices that utilities may seek to escape by adopting innovative energy sources. They are the prices the energy companies may seek to match in selling geothermal energy.*

Table 4.2 and Figures 4.1 and 4.2 show the percentages of energy consumption accounted for by oil, gas, and coal for the 1973-75 period in the Pacific and mountain regions and the United States as a whole; and prices and price trends in terms of cents per million Btu.

TABLE 4.2

Utility Fuel Usage and Prices

	Usage (percent of total Btu in region)			Prices (cents per million Btu)		
Region	1973	1974	1975	1973	1974	1975
Pacific						
Oil	49.1	53.0	57.8	94.0	200.9	249.6
Gas	45.6	38.7	34.2	42.1	58.8	104.6
Coal	5.3	8.3	8.0	38.9	36.9	56.5
Mountain						
Oil	9.2	7.6	7.5	101.0	185.2	210.1
Gas	28.7	24.1	18.5	38.7	51.8	73.0
Coal	62.2	68.3	74.0	23.1	26.0	31.8
United States						
Oil	23.0	20.6	18.9	80.3	192.2	202.0
Gas	21.5	22.3	19.8	33.8	48.1	75.4
Coal	55.4	57.1	61.3	40.5	71.0	81.4

Source: Federal Power Commission, "Annual Summary of Cost and Quality of Steam-Electric Plant Fuels, 1973."

*Geothermal pricing based on substitute fuel prices (perhaps in excess of geothermal production costs) might yield a rate of return on energy company investment commensurate with that found in the production of crude oil. Such prices would maintain the stability of fuel markets.

FIGURE 4.1

Average Annual Prices of Fossil Fuels Delivered to Electric Utility Plants (25 mw or greater)

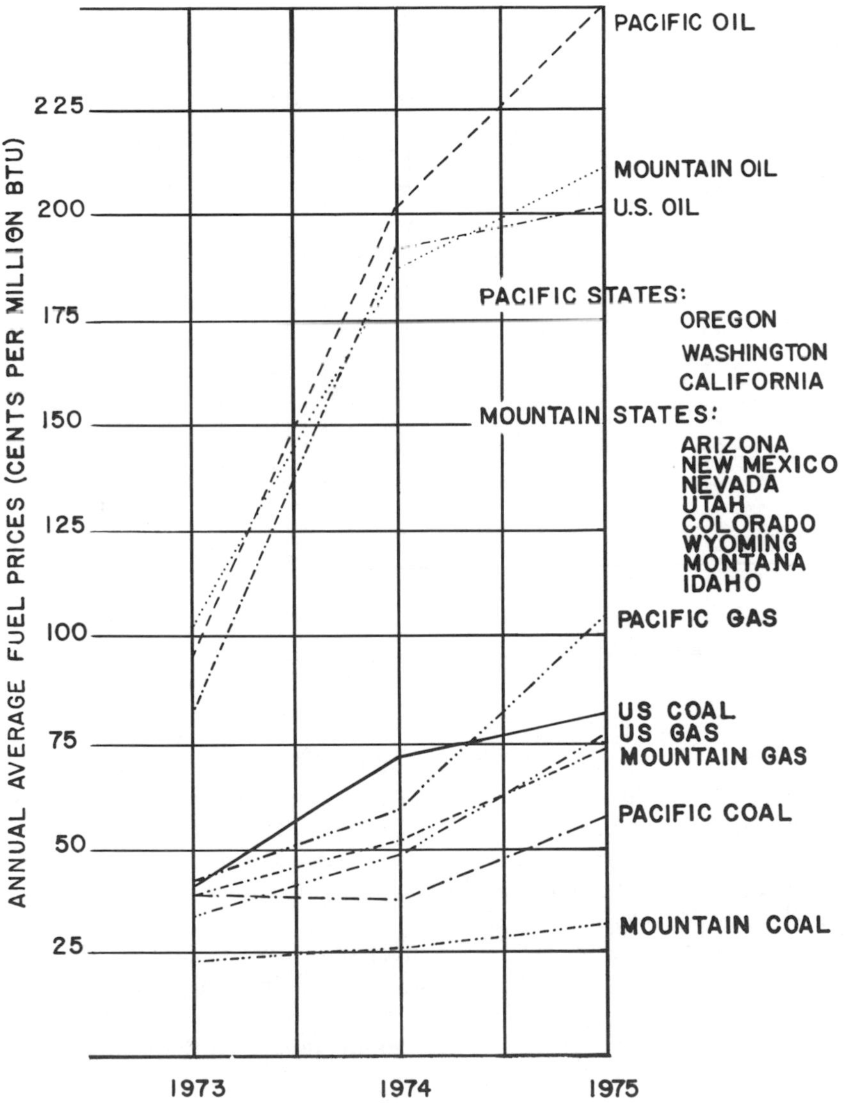

Source: Federal Power Commission, National Power Survey, p. 111-3-179.

FIGURE 4.2

Usage of Fossil Fuel at Electric Utility Plants

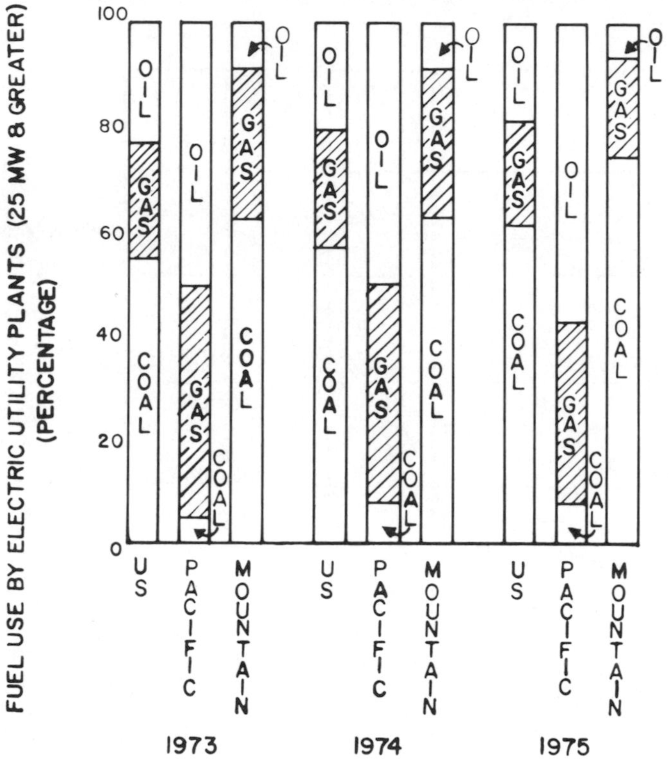

As the foregoing illustrate, on the West Coast gas and oil prices are far higher than national averages. Residual oil prices rose sharply between 1973 and 1975, while coal prices rose at a lesser rate. Western utilities, faced with declining gas deliveries, substantially increased their oil consumption, while also increasing their use of coal.

Contract rather than spot purchases are the dominant mode for acquiring fuel oil in California.* In the mountain states, coal is the dominant energy source, particularly in Montana, Utah, and Wyoming. Coal usage has grown tremendously in the southwestern

*The dominance of contract purchasing was reflected in December 1975, when 7,762,500 barrels of residual fuel oil were purchased under contract by West Coast utilities, and only 59,000 barrels were purchased on the spot market.

area of the mountain states--Arizona, Nevada, and New Mexico. The price of coal in the mountain states has been far below national average levels, while oil and gas prices have followed national trends.

Factors Affecting Price and Usage

As other fuel prices escalate, they could increase the market area open to geothermal energy. However, this will be the case only to the extent that utilities are fuel-cost-conscious.

Utility incentives to control fuel costs appear to be lessened by the practice of passing rising fuel costs through to customers under fuel adjustment clauses in utility tariffs.[6] Fuel adjustment clauses are not uniform among utilities and may even differ in a single utility's tariffs for different services. The pass-through of increasing fuel costs by automatic fuel adjustment clauses has usually been accomplished without regulatory review. These adjustment clauses can permit evasion of regulation of utility profits. If a fuel adjustment clause is drafted so as to permit inclusion of nonfuel charges or of transactions arrived at other than by arms-length bargaining, or if the clause is otherwise allowed to be an avenue for profit, utilities have an incentive to seek out energy sources whose prices or patterns of sale will give rise to such profits.[7]

There is evidence that certain California utilities have used fuel adjustment clauses to pad their customers' bills.[8] Auditors from the FPC reported non-fuel-cost overcharges in fuel clause adjustments of several million dollars by SCE (in a 14-month period ending June 1975). The audit report states that SCE passed through a variety of ineligible costs including excessive charges for operation of SCE's fuel oil pipeline system; interest charges for oil purchased through a financing trust; fuel exploration charges attributed to SCE subsidiary Mono Power, which produced no fuel; outdated costs linked to pipeline operations; and tank maintenance charges.[9] The FPC audit reports that the sums attempted to be rolled in for SCE's Mono Power subsidiary included oil, gas, and geothermal exploration advances of investment money for SCE to Mono Power.

A utility owning a raw fuel source can sell the fuel to itself or to others. Fuel sales among corporate affiliates are not required to be at cost-based prices, nor are fuel subsidiaries required to follow a uniform system of accounts or report regularly to utility commissions. Obtaining fuel from an affiliate can at times lead to high fuel prices, and vertical integration also can channel, to a degree, into utility fuel affiliates the financial assistance utilities sometimes offer fuel companies.

Advance payments and options payments to fuel producers and long-term contracts with broad escalation and force majeure clauses shift fuel industry risks first to the utility and then in turn to the utility customer. Advance payments to a utility's fuel affiliate (unlike advance payments to nonaffiliates) can lead to future profits for the parent, through their inclusion in the utility rate base. This increases the incentive to integrate into raw fuels and channel financial assistance to affiliates.

A utility owning its fuel source is most unlikely to share the profits of that affiliate with ratepayers (even though undeveloped fuel resource lands may be in the rate base). The profits of the fuel affiliate may not be treated as utility income available to offset the revenue requirement to be recovered through rates. Even in intracorporate sales, the price of coal is often passed on through fuel adjustment clauses, with additional profits possibly arising from increased working capital allowances. A comparison by Whitfield A. Russell and Associates[10] of coal costs from captive (that is, utility-controlled) and noncaptive mines shows that in 1973 captive mine costs were below those of noncaptive mines in 7 out of 10 cases. In 1974 the situation was reversed: captive mine costs were the higher in 6 of 10 cases. In this sample the cost of captive-mine coal rose along with that of coal from companies not owned by utilities; the percentage increase in prices for all major captive-mine sales in 1973-74 was roughly equal to that for noncaptive contract sales. Utilities' captive-mine production was as follows:

	1973	1974
Millions of tons	33.5	34.6
Average price per ton	$6.77	$9.32
Total FOB plant costs (millions of dollars)	227.0	322.7

Unregulated captive-mine profits would seem likely to yield a higher return on investment than utility operations. As utility operations must be provided for in any event, further escalation in coal prices would create an additional incentive for utilities to enter the fuel business.*

Insofar as utilities are not pushed (by regulation or competition) toward lower cost energy sources, they will have less incentive

---

*Some western utilities (for example, UP&L, PP&L, or TMPC) own huge coal reserves from which they will be selling fuel to others as well. And some utilities (for example, PG&E and SCE) have entered into uranium ventures.

to determine whether geothermal energy could reduce their costs. Unregulated fuel adjustment clauses in an essentially noncompetitive environment could have this effect. In addition, attractive vertical integration prospects in coal or uranium will compete for attention and capital with geothermal development opportunities.

Large Long-Term Contracts

The utilities need for new capacity is matched by a concern on their part for assurance as to fuel supply. They have attempted to achieve this assurance largely through the use of long-term supply contracts; they have used subsidiary firms to perform fuel supply functions to a smaller extent. Large-quantity, long-term utility fuel contracts affect possibilities for entry into the utility fuel supply business. As the requirements for capital and time to assemble the resources needed to support such contracts increase, the number of entities capable of participating in the market decreases.

Long-term contracts may be entered into with an intent to tie up supply or purchase markets. The contracts do involve situations which match up large, costly facilities which can best be used continuously; this suggests other than preemptive intent. On the other hand, the contracts typically have relatively broad exculpatory provisions and seem highly protective to the seller in terms of price assurance. The agreements tend to shift risks of supply cost increases and supply interruptions to buyers with detailed and comprehensive price escalation clauses, sometimes including price escalation provisions for costs not otherwise identifiable. Residual fuel oil suppliers have even been able to cause the price to utilities for residual fuel oil to include all cost increases in crude oil, even though residual fuel oil may account for only about 20 percent of the output of refined crude.

Shifting financial risk, via contract provisions or advance payments programs, enhances the seller's economic power and frees the seller's money to procure further resources. Moreover, the buyer's general lack of alternatives to long-term contracts raises questions about supply markets.

Joint Endeavors

Uranium and coal resources are sometimes developed by joint endeavors involving utility and raw fuel companies. These endeavors take forms that include joint ventures, contract mining of utility-owned reserves, and utility participation in providing or guaranteeing the debt portion of an energy project's financing. Joint endeavors may tend to cause utilities to become solicitous of the needs of their joint endeavor partners and to be less prone to seek alternative avenues of supply. The utility might be motivated, in its energy supply planning, to protect a party whose debt it has guaranteed.

Joint endeavors among utility and raw fuel companies require mutual agreement and harmony among the participants. Development schedules must be mutually agreed upon, and must be framed to accommodate any strong desire of one party to withhold present supply in anticipation of future price rises, or in consideration of a party's alternative investment opportunities that promise a higher return on investment. Joint endeavors appear to affect the extent to which utilities may encourage producer competition by credibly threatening to seek alternative suppliers, including utility-affiliated firms.

The community of interest among electric utilities and their fuel suppliers, which long-term contracts and joint enterprises tend to create, is augmented by the disincentive on the part of the utility, resulting from fuel adjustment clauses, to bargain for lower fuel costs. In the light of this community of interest, a utility may be reluctant to enter into energy ventures that might be viewed as disruptive of present utility fuel supply arrangements or industry structures.

Western Region

Utility fuel markets appear to fall into regional patterns. The West Coast is a distinct area separated from others by population and transportation distribution factors, and since there is indigenous production of raw fuel. The West Coast region differs from others in, among other things, the mix of utility fuel consumed, and a greater tendency toward large, long-term fuel contracts.

The terrain of the Rocky Mountains has largely separated western and eastern utility and fuel markets. For transportation cost reasons, mines in Utah or the Four Corners area of the Southwest (the junction of Arizona, Colorado, Utah, and New Mexico) will largely serve western utilities, while mines in northeastern Wyoming, southeastern Montana and the Dakotas will serve the plains states and the Middle West. *

---

*Geothermal developers are restricted in sales to utilities within limited transmission distance of their wells.

Trade arrangements for West Coast fuel supply have involved a relatively few large coal mines, refineries, and long-term contracts. This trade pattern tends to limit this market to large firms capable of entering into such arrangements.

Traditionally, California utilities have consumed, in addition to hydro-generation, large quantities of natural gas, brought in from other states and Canada. As the better hydro sites have been developed, fossil fuels have come to play an increasing role in power generation. Declining gas supply has been replaced by low-sulfur residual fuel oil, and to a quite limited extent by coal and nuclear power. This is unlike the situation in other regions where gas (not moving interstate) and coal have been the dominant fuels. Air pollution control regulations on the West Coast result in utility use of low-sulfur residual fuel oil for plants there, and in the locating of coal-fired plants in the mountain states.

In the western states of Washington, Oregon, California, and Arizona,* electric utilities received 60.6 percent (85,734,200 barrels) of the 141,410,000 barrels of residual fuel oil marketed.† Residual fuel oil is consumed in industrial and utility boiler furnaces. Lighter, more expensive petroleum products are consumed in turbine units.

If only because nuclear units will not be put into production in sufficient numbers to meet western power demand, new coal and geothermal generation will be employed, using coal and geothermal resources from within the western region. Western coal production is almost entirely used for generating electricity. In the future, western coal is expected to go to electric utilities as coal, or be converted to gas at the mine mouth.[11]

In 1975, coal mined in Arizona, Nevada, New Mexico, and Washington was delivered to plants in those states. Ninety-three percent of the coal mined in Colorado remained in that state; almost all of the balance went to Iowa and Nebraska. Fifty-eight percent of the coal mined in Utah remained there, while 17 percent left the region for eastern destinations. The northern plains area, eastern Montana, eastern Wyoming, and North Dakota, is a source of considerable amounts of coal for eastern utilities. North Dakota coal that is not consumed locally is shipped east (about one-third of production). All but 6 percent of Montana production is shipped east, while 56 percent of Wyoming shipments move east of Wyoming and Colorado.

---

*These states plus Alaska and Hawaii comprise Petroleum Administration for Defense (PAD) District V.

†Residual fuel oil is a heavy petroleum product produced after lighter products, such as gasoline and distillate oil, are extracted from crude oil.

Increased coal usage is forecast by the Western States Coordinating Council (WSCC). Tables 4.3, 4.4, and 4.5 list the new capacity planned for the West Coast. They show that an increasing number of units in states such as Utah and New Mexico are planned to produce energy for the West Coast consumer. Pending completion of coal-fuel and nuclear units, fuel oil use by utilities on the West Coast has risen, reflecting the declining availability of natural gas, and delays in development of nuclear power.

TABLE 4.3

Present Fuel Usage and Future Capacity Plans, West Systems Coordinating Council

	Production of Electricity in 1975 by Energy Source (percent of mwh)	Planned New Capacity for 1976-85 as Fueled (percent of mwh)
Coal	17.44	33.30
Gas	11.56	--
Oil	15.01	9.27
Hydro	52.51	20.39
Nuclear	2.58	33.35
Other	0.90	2.59 (geothermal)
		0.38 (fuel cells)
Undefined	--	0.41 (waste heat)

Source: Western Systems Coordinating Council.

TABLE 4.4

Present Fuel Usage and Future Capacity Plans, California

	Production in 1975 (percent of mwh)	Planned New Capacity[a] for 1975-84 (percent of mwh)
Coal	--	30.67
Gas	22.04	0.30
Oil	38.53	16.73[b]
Hydro	31.98	8.42
Nuclear	4.84	37.43
Other	2.60	3.56[c]
Undefined	--	--

[a]Includes some Nevada capacity.
[b]10.28 percent, combined cycle; 6.45 percent, combustion turbine.
[c]Geothermal.
Source: Western Systems Coordinating Council.

TABLE 4.5

Production of Electricity in 1975, by Energy Source

	Mountain Region[a]		Pacific Region[b]		WSCC Region	
	Thousands of mwh	Percent of mwh	Thousands of mwh	Percent of mwh	Thousands of mwh	Percent of mwh
Hydro	32,527	29.37	158,744	62.63	191,271	52.51
Coal	57,418	51.84	6,093	2.40	63,511	17.44
Oil	6,324	5.71	48,362	19.08	54,686	15.01
Gas	14,475	13.07	27,636	10.90	42.111	11.56
Nuclear	--	--	9,380	3.70	9,380	2.58
Other	14	0.01	3,266	1.29	3,280	0.90
WSCC	--	--	--	--	36,238	--
Total	110,757	--	253,481	--	--	--

[a]States of Arizona, Colorado, Idaho, Montana, Nevada, New Mexico, Utah, and Wyoming.

[b]States of Washington, Oregon, and California.

Source: Western Systems Coordinating Council.

Individual Fuels

More detailed descriptions are given below of the market conditions affecting each of the major energy sources for electric utilities in the western United States: residual fuel oil, natural gas, coal, and nuclear energy.

All of the residual fuel oil (resid) delivered to West Coast steam-electric plants in 1975 went to California destinations (80,108,800 barrels). In the mountain states residual fuel oil was burned (7,747,600 barrels) extensively only in Arizona, Nevada, and Colorado. Much of the resid consumed is replacing curtailed deliveries of natural gas.

The three large private utilities in California, SCE, PG&E, and SDG&E, as previously noted, rank among the top utility fuel oil consumers. They consume far more resid than other users in the West. Residual fuel oil used on the West Coast comes from refineries there.

PAD District V (Alaska, Washington, Oregon, Hawaii, California, and Arizona) 1955 refinery production of residual fuel oil totalled 136,044,000 barrels, and domestic waterborne movement from the Gulf Coast added 626,000 barrels (including No. 4 oil). Imports of residual fuel oil were 10,645,000 barrels with all but 1,000 of these barrels going to California. Exports from PAD V totalled 4,758,000 barrels and refinery stocks of residuals in that district rose 3,021,000 barrels.

Two firms--Tesoro and Natomas--are particularly oriented to western residual markets. Distinct transportation facilities are used to serve this market, as is also the case with natural gas. In the future, petroleum production from the North Slope and from offshore California will bolster the unique market conditions that exist there, where 55 percent of refinery crude oil runs are still indigenous oil. Domestic California crude oil production tends to be heavy and to have a high sulfur content. This crude oil generally must come to market through an interrelated grid of private pipelines owned by major firms, and requires desulfurization treatment in special units.

The unique fuels situation on the West Coast will be reinforced when deliveries of oil from Alaska's North Slope begin. Combined with new production from offshore areas, the Alaskan crude oil will result in a relative abundance of crude supply on the West Coast.*

The rapid decline in availability of natural gas as a boiler fuel has eliminated gas as a fuel choice for new units. Utilities in California and the Southwest present a tremendous replacement market. Because of boiler design limitations, much of the gas supply lost must be replaced by oil, if substantial reductions in capacity ratings are to be avoided when new capacity is developed.

If the growth in oil usage is to be mitigated, other fuels must be used for base-load power. Coal, geothermal, and nuclear fuels, the alternative fuels for base-load units, are almost exclusively consumed by utilities.

The supply market for residual fuel oil for utilities is concentrated in the hands of a limited number of large integrated petroleum companies. The effects of horizontal integration by these firms into dominant positions in the holding of geothermal resources is best assessed in the light of their position in the petroleum industry--which they dominate at a succession of levels.

---

*Recent long-term contracts for the purchase and sale of residual fuel oil between SCE and SOCAL peg prices to those for Saudi Arabian crude oil, and may deprive SCE of the benefits of changed conditions resulting from North Slope production. SOCAL reports building its two large California refinery expansions to use Arabian, not Alaskan type crude, although this assertion has been questioned. Supplies of Alaskan oil available to the West Coast may be reduced if Alaskan oil is shipped abroad in conjunction with oil exchange agreements.

PURCHASING OF RESIDUAL FUEL OIL BY CALIFORNIA UTILITIES

The three members of CALPP and LADWP have substantial amounts of fossil fuel capacity that will consume increasing amounts of residual fuel oil as gas supplies decrease. These systems are substantially dependent upon oil-fuel capacity.

Southern California Edison Company (SCE)

SCE purchased more fuel oil than any other utility in 1975, producing 77.9 percent of its kwh supplies in steam-electric units. Seventy-seven percent of SCE generating capacity (12,191 mw) is dependent on gas and oil fuel, 13 percent on coal, and 3 percent on nuclear fuel. SCE projects 5,900 mw of additional capacity for commercial operation through 1985, 37 percent designed to use gas and oil, 20 percent coal,* 42 percent nuclear fuel, and the remaining 1 percent to be hydroelectric. Information regarding the sources of energy and the unit costs of fuels in the past five years is set forth in Table 4.6.

TABLE 4.6

SCE Energy Sources and Costs of Fuels, 1971-75

	Percentage of Energy by Source					Average Cost (cents per million Btu)*				
	1971	1972	1973	1974	1975	1971	1972	1973	1974	1975
Oil	23	27	45	38	46	61	78	97	199	267
Natural gas	41	39	21	17	13	34	37	39	58	87
Coal	10	16	15	15	14	18	21	23	27	31
Nuclear	5	4	3	5	5	19	19	17	11	17
All fuels	79	86	84	75	78	39	46	66	119	176
Hydroelectric	9	7	9	10	8	--	--	--	--	--
Purchased and interchanged	--	--	--	--	--	--	--	--	--	--
Power	12	7	7	15	15	--	--	--	--	--

*The company's average fuel costs expressed in cents per kwh for the year 1975 were 2.638 for oil; .886 for natural gas; .330 for coal; and .183 for nuclear. For the year 1975, the net cost of purchased and interchanged power (primarily hydroelectric) was .750 cents per kwh.

Source: Southern California Edison, Annual Report, 1975.

---

*The projected coal capacity included the Kaiparowits plant, which appears to be suspended.

SCE received about one-half of its fuel oil requirements from one supplier in 1974. Its 1975 annual report indicates that SCE was negotiating a definitive contract with its principal supplier (SOCAL) that reportedly would obligate it to purchase fuel oil, contemplated to be primarily of Middle East origin, in an amount estimated to average 64 percent of its anticipated fuel oil requirements over a ten-year period commending in 1976. SCE anticipated that prices under this contract would be based on Saudi Arabian oil prices, with various adjustments for other factors similar to those under its existing contracts.

SCE is also reported to have entered into a ten-year contract to purchase approximately one-fourth of its fuel-oil requirements from Pertamina.[12]

> This marks the second large long-term fuel-oil contract negotiated by Southern California Edison this year. About 60% of SCE's fuel-oil needs will be filled by Standard Oil of California. SCE Vice President William Seaman says that prices paid to Pertamina will be lower than the Company average price ($15.20/bbl) paid for other fuel-oil but that those savings might be offset by the Standard Oil of California contract prices, which are still being renegotiated.[13]

SCE's 1975 annual report states that the ten-year contract with Pertamina is expected to provide 19 percent of its fuel needs for that period. In a financial report for the first quarter of 1976, SCE states that it has a long-term commitment through a letter of intent with SOCAL for a minimum sum of $6.4 billion, and total long-term fuel purchase commitments of $7.3 billion.[14]

Pacific Gas and Electric (PG&E) Company

Gas and oil-fueled units supplied 57.2 percent of PG&E's system net dependable capacity, as of July 1975, and 39.8 percent of the system energy generation in the 12 months ending June 30, 1975.[15] PG&E's expansion program for the ten years ahead envisages a 50 percent increase in capacity owned. Much of this capacity is nuclear. The system will still have a great deal of gas and oil combustion, and continued use of oil-fueled units. In 1974, PG&E used approximately 11 million barrels of oil, or about one-sixth of the total oil used by California utilities that year, and in 1975, PG&E oil deliveries totalled 13,447,000 barrels. With declining natural gas availability, normal weather, and a normal water flow year, PG&E requirements for 1976 were projected to be approximately 30 million barrels.*

Fuel costs to the PG&E system have risen dramatically in recent years.[16] The cost of fuel oil rose from $0.49 per million Btu in 1971 to $2.09 per million Btu for the 12 months ending June 30, 1975. During this time, nuclear costs, after declining in 1971 and 1972, remained flat at $0.20 a million Btu, while the price of natural gas more than doubled between 1971 and 1975.†

Table 4.7 shows PG&E's fuel oil purchases in the second quarter of 1974 and the first quarter of 1975, the latest obtainable period. As can be seen, ARCO, Phillips Petroleum, and Perta Oil Marketing (oil from Indonesia, acquired from Pertamina, the state company, and to a limited extent from Peru), supply together over 72.45 percent of PG&E residual fuel oil. General Energy Company (a broker) provided over 12 percent; this oil was rather high-priced.

PG&E purchases its fuel largely under long-term contracts,‡ for example, with ARCO and Union Oil. Under one agreement PG&E causes crude oil to be delivered (for PG&E's account) to the Pacific Resources refinery in Hawaii, and Pacific Resources in turn provides PG&E with fuel oil. There is evidence that oil prices charged PG&E are kept in line with those charged SCE.

Mr. R. P. Benton, manager of PG&E's procurement activity, testified in a proceeding before the California PUC that PG&E has a contract with ARCO, its largest supplier, that runs through 1981, for 750,000 barrels per month.[17] Reduced fuel oil requirements in late 1974 combined with a widening gap between the price paid by PG&E to ARCO and the prices charged by ARCO to SCE and to LADWP were stated by the witness to have provided PG&E with a basis upon which it (in March, 1975) temporarily negotiated postpone-

---

*Before discussing PG&E's oil acquisition program, it should be noted that PG&E entered into agreement with the Island Creek Coal Company, a subsidiary of Occidental Petroleum, for the purchase of a coal property near Price, Utah, if the property is found to have deposits of at least 150 million tons of coal.

†A comparison of capital costs for nuclear and geothermal shows that the planned nuclear power units, at the Diablo Canyon plant, have estimated costs of construction of $465 per kw (or at 80 percent load factor about 12 mills per kwh). The highest fixed charge for a Geysers unit ($200 per kw) compares favorably in this regard (using an 18 percent annual fixed charge rate, only 5.1 mills per kwh).

‡Some long-term fuel contracts of California utilities are summarized and discussed in the Appendixes.

ment of any increase in price beyond $12.88 per barrel (which became effective April 1, 1974). ARCO agreed to defer increases to PG&E pending negotiations between it and SCE. Later, Benton indicated, ARCO informed PG&E that if PG&E would not pay a high price from the point at which it was first intended to become effective, ARCO would terminate the contract and make no deliveries. PG&E acquiesced in paying the higher price. (SCE decided not to enter into a contract with ARCO. A contract for the sale of fuel oil to the LADWP is reported to have been negotiated to have a price of approximately $15 a barrel.)

TABLE 4.7

PG&E Fuel Oil Purchases
(Second Quarter 1974 and First Quarter 1975)

Supplier	Number of Barrels	Percent
ARCO	4,132,716	44.96
Union Oil	438,326	4.78
Phillips Petroleum	1,324,868	14.41
Pacific Resources	432,644	4.71
Perta Oil	1,202,615	13.08
Singapore Petroleum	326,522	3.55
Utility Petroleum and Refining	38,207	0.42
Western Refining	150,000	1.63
General Energy Corporation	1,145,796	12.47
Total	9,191,694	100.01

Source: California Public Utilities Commission.

PG&E has an agreement with Union Oil Company of California providing for approximately 180,000 barrels per month through 1980. As with the ARCO agreement, Union Oil obtained an upward renegotiation of the price of its oil supplied PG&E. Union initiated negotiations in June 1974, seeking to raise prices from $8.54 to $12.00 per barrel and six months later an additional two dollars per barrel. After Union stated it would cancel the contract unless its requests for higher prices were acceded to* the Union contract was extended

*FEA regulations permitted higher prices than Union was offering. PG&E asserted to the California PUC that it felt compelled

for six months to be renegotiated in the second half of 1975.[18] A sharp price increase resulted.

Phillips Petroleum Company supplied residual fuel to PG&E on a month-to-month basis at prices quoted each month ($13.50 at the refinery in March 1975).

PG&E has chartered a fleet of oil tankers, and has become a substantial importer of residual fuel oil.[19] A large portion of PG&E's residual fuel oil supply is imported directly by the company from Indonesia. Some of this residual oil is purchased from Perta Oil Marketing at a price that has gone down from $15.45 to $13 a barrel. Another portion of this oil is obtained in return for crude oil supplied to Pacific Resources in Honolulu.

San Diego Gas and Electric Company (SDG&E)

All of SDG&E's 2,070 mw of net dependable capacity as of December 31, 1974, consumed oil or gas. At the beginning of 1975, SDG&E planned, by 1983, to add 456 mw of nuclear capacity, and 992 mw of fossil-fuel generation (including 700 mw of coal-fuel units). It has plans for a nuclear plant with an initial capacity of 900 mw in 1988, rising to 1,140 mw, and has sought permission to build a 292mw fossil-fueled plant.

Under a contract with Tesoro Petroleum, SDG&E purchased, in the three years ended September 30, 1975, 3,516,000, 3,371,000, and 3,815,000 barrels of low-sulfur resid. These purchases were made under a contract expiring June 1977, which provides for a base price plus adjustments, reflecting Tesoro's raw material costs.[20] The oil supplied by Tesoro apparently comes from Tesoro's Kenai, Alaska, refinery.

SD&E has contracted with the HIRI refinery in Hawaii for HIRI to expand its refinery capacity and provide SDG&E with 10 million barrels of oil per year starting in 1977. This is a cost-plus, fixed-price contract in which the refinery's profit will be renegotiated from time to time.

In late 1974, Union Oil entered into a contract to provide several million barrels of oil to SDG&E at $12 per barrel. Union, in 1975, was seeking a $2.76 increase in this price. SDG&E has a subsidiary, New Albion, in the fuel business.

---

to renegotiate and, in addition, asserted that the price of $12.26 was not unreasonable under the circumstances.

Los Angeles Department of Water and Power (LADWP)

LADWP, while not a member of CALPP, is a major utility system in the state. In the year ending June 30, 1975, LADWP obtained 89 percent of the energy it generated and 72.3 percent of the energy it disposed of, from steam-electric units burning oil and gas.[21] Its planned construction through 1988 consists of a pumped storage unit (1,325 mw), participation shares of 2,027 mw in coal plants, and 1,846 mw in nuclear.* On February 27, 1975, LADWP said that it expected fuel oil purchases would increase by 95 percent for the 1975-76 fiscal year. This reflects declining use of natural gas.

In the 1974-75 fiscal year, a long-term contract provided one-third of LADWP's fuel oil, at a cost of about $5.35 a barrel. When the contract expired July 1, 1975, it was expected that almost all oil suppliers would sell at current market prices--approximately $15 a barrel.†

California Utility Fuel Purchasing Power

The large utility consumers of residual fuel oil in California have not been able to employ buying power to hold resid prices down. While the national average of delivered oil prices of resid rose 5.5 percent between 1974 and 1975 (from 190.6 cents per thousand Btu to 201.1 cents per thousand Btu) California utility fuel prices rose 25.5 percent (to 250.1 cents per thousand Btu from 199.3 cents for contract prices). Higher West Coast price increases may reflect the lower sulfur content of resid consumed on the West Coast, and West Coast utilities, in this period, were increasing their resid consumption to replace natural gas. However, gas delivery declines were foreseeable, and sulfur content limitations on fuel use did not change. Accordingly, lack (or nonuse) of West Coast utility buying power appears to be manifested. Long-term contract arrangements apparently have not prevented rapid price escalation.‡ Whether the limitation of alternatives involved in use of such arrangements contributed to a weak bargaining position is not clear.

---

*LADWP is building a hydroelectric project (the Castaic Project) and is participating in the Navajo Project near Page, Arizona. These projects will tend to alleviate its relative dependency on oil.

†LADWP acquired resid from SDG&E in 1975 when SDG&E found that it had purchased resid in excess of its own needs.

‡Spot sales were about 1 percent of the total deliveries.

Mountain States

Mountain state utilities (including the members of the Intercompany Pool) consume increasing quantities of petroleum. In the Pacific Northwest, PGE consumes petroleum distillates in its combustion turbines (824 mw, of which 439 mw are to be converted to add 160 mw of combined cycle capacity).[22]

In the Southwest, TG&E uses gas and oil for 727 of its 1,116 mw of capacity. In 1972, coal accounted for 19 percent and oil accounted for 12 percent of its total Btu consumed, with gas accounting for the remaining 69 percent. By 1975, coal's contribution to Btu consumption had risen to 51 percent and oil had risen to 33 percent, while gas had fallen to 16 percent. During this same period, the cost per million Btu of coal had risen from 15 cents to 25 cents, oil had risen from 81 cents to 164 cents, and gas had risen from 39 cents to 58 cents. Most of the company's fuel oil is purchased from Southern Union Oil Products Company under a ten-year contract initiated in August 1974. The price under this contract is tied to the cost of crude oil to the refinery that supplies Southern Union. During 1973, 1974, and 1975, the utility received 58 percent, 47 percent, and 37 percent, respectively, of the natural gas it would have used for power generation, if available, from El Paso Natural Gas Co.[23]

APS has also had a sharp rise in its cost of fuel. Of its 2,260.6 mw of capacity, 1,092.7 mw are fueled by oil and gas. In 1976, APS planned to add 225 mw of combined cycle capacity. In 1971, coal accounted for 76.1 percent of generation and cost 16.15 cents per million Btu; gas accounted for 23.5 percent and cost 38.45 cents; and oil accounted for 0.4 percent and cost 46.63 cents. By September 30, 1975, coal accounted for 79.3 percent of generation and cost 25.77 cents per million Btu; gas accounted for 7.7 percent and cost 74.15 cents; and oil accounted for 13.0 percent and cost 211.88 cents. APS receives all of its natural gas from El Paso Natural Gas Co. and had residual oil contracts for approximately half of its projected 1976 oil consumption, the balance being acquired on short-term bases, as and when storage capacity and transportation permit.[24]

Public Service Company of Colorado (PSCC) has increased its consumption of oil from 1 percent of its energy input in 1970 to 4 percent in 1975. During that period, the price per million Btu rose from 38.8 cents to 233.4 cents. PSCC's subsidiary, Fuel Resources Development Co., is engaged in the exploration, development, and production of gas and oil in Colorado, Wyoming, Utah, and Montana, and entered into a contract in October 1973 with an oil refinery and an exploration company, under which the refinery agreed to make scheduled deliveries of 207 million gallons of No. 2 fuel oil and

56 million gallons of residual fuel oil in exchange for the funding of a crude oil exploration program. PSCC has converted one plant from gas to residual fuel oil.[25]

REFINERY SOURCES OF RESIDUAL FUEL OIL TO THE WEST COAST

The feedstock employed by the West Coast refineries that supply utilities providing about 89 percent of the fuel oil in 1975 comes from domestic and overseas sources of oil. West Coast refinery yields of resid are significantly higher than national norms.[26] For PAD District V in 1975, of the approximately 705.5 million barrels of new supply of crude oil, 310.8 million barrels per day (44.1 percent) come from foreign fields.* Of the runs to refinery stills in California--562.5 million barrels during this period--372.5 million barrels per day (66.2 percent) were domestic and 191.2 million barrels per day (34 percent) were foreign oil.[27] 314.6 million barrels were of California origin.

There are a number of planned refinery expansions in California so that locally burned resid is likely to continue to be locally produced. Feedstock for West Coast refineries will come from the North Slope via the TAPS pipeline system, and from offshore California fields. It appears likely that the percentage share of West Coast refinery feedstock held by imports will decline at least in the short term unless North Slope oil is exported in substantial quantity, or used to displace California production.

Ownership of the potential large new sources of crude, and of related new transport and refining facilities, is generally even more concentrated in major oil companies than existing production. The large petroleum companies that have long-term fuel supply contracts with West Coast utilities are vertically and diagonally integrated among themselves, and control, individually and in joint agreements, the means of transportation and production of petroleum and natural gas. Their extensive interconnections, in a market with a small number of suppliers, promote parallel action, and give them significant market power. These same firms also own substantial interests in coal, uranium, and (for some) geothermal energy.

---

*In 1974, West Coast refinery inputs totalled 682,045,000 barrels, 400,971,000 barrels of domestic crude oil and 281,074,000 barrels of foreign crude.

Domestic Refining on the West Coast (PAD V)

Crude oil refining capacity in District V is characterized, in a recent FTC staff report, as moderately highly concentrated[28] in spite of some decline in concentration during the decade ending in 1975. At the beginning of 1975, the four largest refiners accounted for 54.1 percent of District V's refining capacity and the eight largest for 76.5 percent. These capacity shares represent apparent declines, over the preceding ten years, of 6.7 percent for the big four and 10.7 percent for the big eight.[29]

Concentration in ownership of PAD V refining is shown in Table 4.8.

TABLE 4.8

Largest Refining Companies in PAD V
(January 1, 1976)

Company	Operating Capacity: Crude Oil Distillation, barrels per calendar day
SOCAL	526,500
Shell	268,400
ARCO	261,000
Union	219,000
Mobil	195,000
Texaco	128,000
Phillips Petroleum*	98,000
Exxon	88,000

*Subsequently sold to Toscopetro pursuant to antitrust decree requiring divestiture.

Note: Total of top four companies, 1,274,900 (55.25 percent of PAD V capacity); total of top eight companies, 1,783,900 (77.4 percent of PAD V capacity); total PAD V capacity, 2,304,642.

Source: U.S. Bureau of Mines, Mineral Industry Surveys, Annual Survey of Petroleum Refineries in the United States and Puerto Rico.

Throughout the past decade, SOCAL has been the dominant refiner in District 5 with market shares of branded gasoline sales in excess of 20 percent in every period.[30] Its share has gradually declined over time while the shares of the other top four refiners have remained virtually unchanged.

Concentration in refining may be higher than ownership figures indicate, because some of the small refineries in California may have little freedom as competitors. For instance, the small refinery that was supplying the city of Burbank is completely dependent upon ARCO for petroleum and pipelining. Indeed, Burbank had to institute a proceeding before FEA in order to require ARCO to continue supplying this refinery so that the refinery could continue to supply the needs of Burbank's municipal power plant.*

Future shifts in relative strength may be expected as Phillips Petroleum complies with the Supreme Court's order to divest itself of the Tidewater Oil Company refinery it acquired from Getty Oil. Phillips has agreed to sell the facility to Toscopetro.[31] Exxon's market share may be expected to increase in view of its shares of California offshore resources and North Slope production, coupled with its indicated plans to expand its California refinery capacity.

New refinery construction in California is wholly through expansion of existing facilities. Crude oil refinery facilities are being built by USA Petroleum Corporation, 15,000 barrels per day; Lunday-Thagard Oil Company, 3,278 barrels a day; and SOCAL, 350,000 barrels a day.[32] SOCAL's expansion is particularly interesting since it clearly reflects marketing expectations influenced by the Trans-Alaskan Pipeline. The SOCAL expansion is limited to distillation units, and does not include increased cracking capacity. This indicates an intention to shift the refinery product mix toward a greater relative output of fuel oil. Frank Parker, SOCAL's assistant to the vice-president for refining, has indicated[33] that SOCAL has a profit incentive to maximize the refineries' ability to produce more middle distillates and residual products in order to tap a utility and industry market that is expected to grow at about 6 percent annually because of dwindling natural gas supplies. SCE has recently entered into a substantial contract with SOCAL for the purchase and sale of residual fuel oil, with prices pegged on overseas crude oil prices.

SOCAL, of course, has substantial avails of Gulf Coast crude oil, deliverable to the Middle West by pipelines, and is therefore in a good position to acquire TAPS crude from BP on exchange, since

---

*Without data on through-put and exchange volumes, a complete picture of refinery control is not obtainable. However, some smaller independents have held crude oil purchase contracts with the majors which commit a substantial portion of the uncracked residuum back to the major for further refining.

BP has no West Coast refining capacity, but needs crude deliverable to Ohio. Exxon, also, could utilize SOCAL's Gulf Coast crude.

Alternate Refinery Sources of Residual Fuel Oil

Residual fuel oil cannot be pipelined for any long distance because of its very high viscosity. It must either be produced near its point of use or else be shipped by tanker or by barge. Thus a West Coast utility must choose between a District V refinery source or a distant, foreign refinery. The choice of a foreign source will be affected by fuel sulfur content restrictions. Table 4.9 presents West Coast sources of fuel oil.

TABLE 4.9

West Coast (PAD V) Imports of Low-Sulfur Residual Fuel Oil

	1974		1975	
Country of Origin	Amount (thousands of barrels)	Percent of Total	Amount (thousands of barrels)	Percent of Total
Bahamas	--	--	949	9.1
Netherlands	--	--	--	--
West Indies	3,272	21.4	2,930	28.1
Venezuela	4,416	28.9	1,110[a]	10.6
Peru	691	4.5	772	7.4
England	--	--	278	2.7
Belgium	--	--	273	2.7
Saudi Arabia	79	0.5	--	--
Singapore	20	0.1	483	4.6
Indonesia	4,910	32.1	2,084	20.0
Japan	--	--	1	--
Australia	--	--	1,556	14.9
Virgin Islands	1,596	10.4	--	--
Total	15,304	100.0	10,436	100.1[b]

[a]Also, 205,000 barrels of less than 1 percent sulfur content less oil imported.

[b]Due to rounding off.

Source: U.S. Bureau of Mines, Mineral Industry Surveys.

The Dutch West Indies sources are most probably the two refineries there, owned respectively by Shell and Exxon, which operate principally on Venezuelan crude oil. The Venezuelan residual may have come from Venezuelan refineries owned by the four major companies, Shell, SOCAL, Mobil, or Exxon.* It should be noted that two-thirds of the Venezuelan refining capacity is either owned or marketed by Exxon and Shell Oil, and that Gulf, Chevron, and Mobil also own substantial Venezuelan capacity. Indonesian residual fuel oil comes in part from Pertamina, the Indonesian national oil company. A number of West Coast refiners participate heavily in production done under contracts with the Indonesian government, from which Pertamina gets its product. They are SOCAL, historically affiliated with Texaco in the Caltex Joint Venture, Shell Oil, Union Oil Company, ARCO, Mobil, Exxon, Natomas, and Tesoro.[34]

In 1973, Indonesia produced 1,339,000 barrels per day of crude oil. American companies account for about 80 percent of the foreign investment in Indonesian oil. P. T. Caltex Pacific Indonesia (CPI), jointly owned by SOCAL and Texaco, is the biggest foreign producer. CPI production, all onshore and mostly on Sumatra, was 965,000 barrels per day in 1973. Nearly 410,000 barrels per day came from CPI's Minas field, the largest single producing field in the country. Other major U.S. companies are Union Oil (off eastern Borneo), ARCO (Java Sea), Natomas Company (Java Sea and on Java), Mobil Oil (north Sumatra), and P. T. Stanvac, a consortium of Mobil Oil and Esso (mainly onshore Sumatra). In addition, numerous American independent oil companies operate in Indonesia. Among the more important independents are Carver-Dodge Oil Company, Getty Oil Company, Genard Oil and Gas, Mapco Inc., North Central Oil, Southern Cross Ltd., and South Pacific Oil Company. Mapco, North Central, Southern Cross, and Trend Exploration (U.S.) are partners in drilling in Irian Jaya in a major new venture.[35] The Petroleum Economist reports Indonesian oil production as 1.37 million barrels per day in 1974 and 1.34 million barrels per day in 1975, with Caltex producing 0.9 and 0.83 million barrels per day in those years, respectively. Productive capacity in 1975 was estimated to be 1.5 to 1.6 million barrels per day.

> Production by the main producer Caltex, with fields onshore Sumatra, appears to have declined for the second year in succession, the

---

*There is also a small refinery owned by ARCO and one each owned by Phillips and Gulf, and a total of perhaps 12 enterprises in Venezuela that have refineries.

> slack being taken up by new producers which include Arco (Ardjuna field), Natomas (Cinta), Union Oil/Japex (Attaka), Huffo (Bakale), CFP (Bekapai), Pertamina, and Petromer Trend in Irian Jaya. . . .[36]

Indonesian production at mid-1976 is shown in Table 4.10.

TABLE 4.10

Indonesian Production of Crude Oil, Mid-1976

Company	Production (barrels per day)
Stanvac[a] and Caltex[b]	830,000
Union	130,000
ARCO	125,000
Natomas	105,000
Total	91,000
Petromer Trend	65,000
Asamera	14,500
Tesoro	8,000
Cities Service	3,500
AAR	1,150
Pertamina	90,000
Total	1,500,000

[a]Exxon and Mobil.

[b]SOCAL and Texaco.

Note: The four largest producers account for 72.3 percent of total production; Caltex and Stanvac, for 55.33 percent.

Source: Petroleum Economist 63 (September 1976): 359.

The residual fuel oil imported from Australia probably was produced at one of that country's 11 refineries. Of the 750,935 barrels per calendar day of crude input capacity listed for Australia,[37] 262,000 barrels per day (35 percent) are owned by Caltex subsidiaries, 153,335 barrels per calendar day (20.4 percent) by BP subsidiaries, 147,000 barrels per calendar day (19.6 percent) by Shell Oil subsidiaries, and 143,600 barrels per calendar day (19.1 percent) by a subsidiary owned jointly by Mobil (65 percent) and Exxon

(35 percent). Of the other two refineries, one (25,000 barrels per calendar day) is owned by a subsidiary of Standard Oil of Indiana, and one (20,000 barrels per calendar day) by the Total Oil system--part of the French National CFP company.[38]

Peruvian refining is almost entirely done by Petroperu, using crude from that firm and Belco Petroleum (on the Pacific Coast). Petroperu and Belco are the most likely export sources. Occidental Petroleum has Peruvian crude production shipped out via the Amazon, pending completion of a pipeline west.

Singapore refineries are owned by Exxon, Mobil, Shell, and BP.

Petroleum Refining in PAD IV

Because of geographic constraints, the mountain states of Utah, Wyoming, Montana, and Colorado have a number of refineries smaller than those found in Coastal regions. These refineries are not so situated as to be likely to supply residual oil to the West Coast. Heretofore area utilities have not consumed much resid.

PAD IV refineries use little imported (Canadian) feedstock, and produce far less resid per barrel of crude oil than is the case on the West Coast. The refineries in PAD IV do produce the No. 2 oil used for diesel generation in small isolated power systems, and peaking gas turbines. Refinery ownership is rather concentrated for the region; smaller local markets can, of course, be even more concentrated.[39]

TRANSPORTATION OF CRUDE OIL TO REFINERIES PRODUCING RESIDUAL FUEL OIL

Crude oil moves through pipelines or by water to refineries. Other modes of transportation are far more expensive. The Joint Committee on Public Domain of the California State Legislature conducted an inquiry in 1974 into the crude oil pipelines in California and control of crude oil markets, and it reported that "major oil companies control the California crude oil market by their ownership and control of the crude oil pipelines. This control artificially depresses prices paid to independent producers and restricts the supply of crude oil to the independent refiners."[40]

Transportation can be a significant percentage of total costs at a refinery.[41]

> The pipeline is the cheapest way to move oil over land. The cost of moving oil by pipeline

is considerably less than by truck and rail.

The following is a general cost relationship; short haul rates are higher:

	Cents per ton mile
Barges	0.15 to 0.60
Pipelines	0.17 to 0.60
Tank trucks	3.00 to 5.00
Rail tank cars	2.00 to 7.00[42]

Because of the cost advantage, pipelines have come to occupy a major place in the transportation system. Nationwide, they handle over 75 percent of the total crude oil tonnage. The cost advantage is important because crude oil is relatively heavy in relation to unit value, and transportation can be a significant percentage of the total production cost.[43]

The sources of the 706,848,000 barrels of crude oil input to PAD V refineries in 1975 were intrastate pipelines with 285,563,000 barrels or 40 percent of inputs; tankers and barges from overseas with 250,632,000 barrels, or 35 percent of inputs; pipeline carriage of 60,203,000 barrels of foreign crude or 9 percent of inputs; interstate tankers and barges, 46,799,000 barrels or 7 percent of inputs; interstate pipelines, 6,886,000 barrels or less than 1 percent of inputs; interstate tank cars and trucks, 8,626,000 or over 1 percent input barrels; intrastate tankers, 36,807,000 barrels or about 5 percent of inputs; and intrastate tank cars and trucks, 12,005,000 barrels or about 2 percent of inputs.[44] Foreign tankers and intrastate pipelines were responsible for approximately equal portions of the supply with over 60 percent of domestic oil moving through intrastate lines.

The FTC staff report on the western petroleum industry, and the California Report[45] discuss problems of nonmajors in securing pipeline access to markets and supplies, to demonstrate that pipeline access is interrelated with the pricing of crude oil and the ability to acquire government lease tracts. The degree of interconnection is such that even major firms cannot meet their refinery requirements without using pipelines owned by others. Pipelines control field prices for crude.

Three companies post prices paid producers in California,[46] which vary only slightly inter se.* Currently, the price differential

---

*Three major firms, Exxon, ARCO, and SOCAL, are partners in the Santa Ynez joint production venture (HAS) which could provide up to 10 percent of California's output by 1977.

for oils of different gravity is disproportionate. While the prices of high gravity (relatively light) crude oils in California are roughly the same as or higher than prices of high gravity crudes in the rest of the United States, large gravity price differentials (relatively large price differentials for increases in crude density), posted in California have resulted in prices for low gravity (relatively heavy) crude below prices for such oil elsewhere in the country.[47]

The inland production areas of California, such as the San Joaquin Valley and inland Los Angeles Basin, are not served by waterways, and pipelines are the only economically feasible method for inland shipping or for interurban transport.[48]

The crude oil pipeline system is dominated by majors. Except for three lines,* the system is made up of private carriers rather than common carriers. In this private system, carriage of oil produced by others than the owner of the line is arranged through exchange, or purchase and resale. Crude oil is acquired from laterals by the firms owning trunk lines and resold to the owners of laterals at the trunk line's destination point.

This practice of exchanges is the basis of an integrated transportation system. It is illustrated by reference to a Mobil pipeline linking the San Joaquin Valley and the Los Angeles Basin. The Mobil pipeline connects 34 lines owned by others in the basin. In this system no public tariffs are posted, and independent refineries are forced to purchase oil from majors' pipelines instead of from field producers.[49]

Periodic round-robin settlement and clearance of exchange balances among companies reflect the crude supply interrelationships of firms.† Control by a few firms of transport may tend to reduce opportunities for transport cost differentials, and for reflecting any favorable supply conditions in sales markets. To the extent exchanges are made by or under the influence of a few dominant firms, opportunities for independent market challenges tend to be reduced.

The FTC states[50] that the four largest companies, ARCO, SOCAL, Union, and Shell, control 70.1 percent of crude oil pipeline

---

*The Four Corners Pipeline, originally owned by SOCAL (25 percent), Shell (25 percent), Gulf (20 percent), ARCO (10 percent), Conoco (10 percent), and Superior (10 percent), now owned wholly by ARCO, which has announced its intention to reverse the line's flow; and two systems owned by SOCAL that terminate in private lines.

†In such settlements, Firm A owing a balance of deliveries to firm C arranges for firm B which owes deliveries to A to make delivery to firm C instead.

capacity in District 5; the eight largest companies (the others being Mobil, Getty, Texaco, and Gulf) control 98.9 percent. The FTC Report states that it is suspected that these figures reflect the situation of the past ten years or more. Most California pipelines are intrastate, are not common carriers, and are not regulated. While most of these lines apparently hold federal public land right-of-way permits, obligating them to perform as common carriers, there is no ready access to records of such permits for private enforcement of the permit requirement. Interior has never brought such actions, nor has it formally published the operating regulations required by the Trans-Alaska Pipeline Act of 1973 amending the Mineral Leasing Act.[51] The possible utilization of the pipelines to exclude or limit competition is discussed in the FTC Report.

Crude oil currently coming to California from Alaska--for example, from the Kenai Peninsula and Cook Inlet--goes to market through the Cook Inlet and the Kenai pipelines. The Cook Inlet Pipeline is jointly owned by ARCO, Marathon,[52] Mobil, and Union Oil. The Kenai Pipelinc Company is owned by ARCO and SOCAL in even shares.* The Trans-Alaska Pipeline is discussed later in the section regarding new crude oil supply sources.

On a nationwide basis, it also appears that control of the lines through which crude oil flows provides control over crude oil. Independent producers typically sell their product to the owner of the gathering system serving the field. Through ownership of extensive crude gathering systems, major petroleum companies appear able to control disposition of a large amount of independently produced crude.

Both the FTC staff and the Department of Justice have found little competition among gathering systems.

> . . . the field market for crude oil displays all of the indicia of a monopoly market. The physical pipeline connection and the basic division order may be the subject of competition initially, with occasionally real rivalry as to which purchaser-pipeline is to be chosen for the initial connection. But once installed, the expensive physical connections and the complex legal arrangements are long enduring, not easily responsive to supply-demand fluctuations or to price behavior.[53]

---

*In 1973, ARCO shipped 33.6 percent and SOCAL 31.52 percent of the through-put on the Kenai line.

Table 4.11 is a summary of independent crude sales. Commenting on these data, an FTC preliminary staff report states:

> The data indicate that the eight largest firms purchased an average of 42.5 percent of independently produced crude oil during the survey period (average of Table 1, Column 1). In most cases where the eight largest firms purchased crude from independents, they also owned the gathering lines used to transport it (Table 1). Other results of our survey indicate that once gathering lines are erected, there are few other buyers of crude from that particular field. Control of these gathering lines, in addition to ownership of approximately 52 percent of crude sources, gives the eight largest firms substantial control of domestically produced crude oil.[54]

TABLE 4.11

Independent Crude Producer Summary, 1967-71

	1967	1968	1969	1970	1971
Crude sales to eight largest majors as percent of total sales	40.2	41.5	42.0	44.6	44.4
Crude sales to other majors as percent of total sales	25.3	26.3	24.8	27.1	25.8
Percent of gathering lines operated by eight largest majors	39.3	40.7	37.6	42.0	36.0
Percent of gathering lines operated by other majors	5.2	21.7	3.8	24.9	24.8

Source: "Preliminary Federal Trade Commission Staff Report on Its Investigation of the Petroleum Industry" (1973), p. 6.

The report of the Senate Judiciary Committee on the Petroleum Industry Competition Act of 1976[55] refers to a series of statements by industrial personnel to the effect that control of pipelines is second only to ownership of proven and developed acreage in controlling crude oil supply.[56]

Like the gathering systems, the crude trunklines funnel the supply of crude oil to the major integrated firms. Table 4.12 indicates the extent to which the refining industry depends upon the majors' controlled pipelines for crude supply. In 1973 over 92 percent of the crude oil pipeline shipments reported to the Interstate Commerce Commission (ICC) originated in lines that were owned or almost completely controlled by the integrated majors.

TABLE 4.12

Shipments Originated in Major Oil Company Pipelines, 1973

	Number of Barrels (millions)	Percent
Crude oil lines		
8 largest majors	1,897.4	64.3
8 other majors	442.5	15.0
Joint venture lines*	385.3	13.1
Total, 16 majors	2,725.2	92.4
U.S. Total	2,949.9	100.0

*Lines owned or controlled by groups of two or more major companies.

Note: ICC reporting lines, except Lakehead, Trans-Mountain, and Portland pipeline companies. Volumes exclude shipments received from connecting carriers.

Source: Interstate Commerce Commission, Transportation Statistics in the United States for the Year Ended December 31, 1973, Pt. 6, Pipelines, 1974, hereinafter cited as Transportation Statistics 1973; U.S. Senate, Judiciary Committee, "Report on the Petroleum Industry Competition Act of 1976," Report no. 94-1005, 94th Cong., 2 Sess., p. 25.

Common carrier obligations imposed on interstate pipelines by the Hepburn Act of 1906 are readily avoided: there are no certification procedures and a pipeline company need not jointly plan routes, capacity, or operating cycles with nonowners. Likewise, the pipeline company need not and rarely does provide essential tankage. Devices available to a pipeline operator seeking to deny others access to its lines include requiring large minimum tenders, granting irregular shipment dates, and imposing unreasonable quality specifications.[57]

Under the Hepburn Act, which imposes common carrier obligations on oil pipelines, a single company-owned pipeline transporting only its own oil is a common carrier only for reporting purposes, and cannot be required to file rates or to transport the oil for others.[58] The attorney general recently concluded that special conditions to protect competition had to be put in licenses for operation of superports and their ancillary pipelines. "A cursory examination of the ICC's present powers, however, and an understanding of past ICC regulatory efforts in this area reveal that the ICC cannot effectuate any of these (necessary) competition rules . . . the ICC has no power over the form of corporate organization, etc."[59]

Oil pipeline rights of way may be granted by the secretary of the interior upon the express condition that such oil pipelines shall transport, or purchase without discrimination, oil produced in the pipeline in such proportionate amounts as the ICC may, after a full hearing due notice, determine to be reasonable.[60] This clause, providing for something less than common carrier operation, has been avoided by designating offshore lines as gathering lines, which in effect exempts them.[61]

Many large pipelines are jointly owned by several petroleum companies. Joint ownership may be corporate or held as undivided interests. The latter is arranged so that each co-owner files a separate tariff for its proportionate share of a system. A would-be outside shipper must therefore not only seek to coordinate his shipment with the pipeline operating cycles, but must also deal separately with co-owners for a portion of their time-share of the pipeline's shipping space. "To show the similarities and dissimilarities between joint venture pipelines and undivided interest pipeline systems," the following information will be useful.[62]

A joint venture pipeline, as it is known in the industry, is a corporate entity consisting of two or more owners, usually oil companies, which control the pipeline through ownership of the majority of the stock. The stock is usually not traded publicly. An undivided interest pipeline system is similar to a joint venture pipeline but in itself is not a corporate entity. The pipeline facilities (assets) are owned in an undivided interest by two or more pipeline or oil companies. This undivided interest does not separate or identify the physical assets by individual ownership. One company, a participant in the system, is usually selected and acts under an agreement as the agency operator of the pipeline facilities.

There follows a tabulation in which joint venture pipelines are compared in various aspects to undivided interest pipeline systems. Shown first are the similarities of both, followed by those in which they differ.

	Joint Venture	Undivided Interest System
Similarities:		
Participated in by two or more oil and/or pipeline companies	Yes	Yes
Shippers over the systems	Predominantly	Predominantly the participants or their affiliated or parent companies
Through-put agreement among participants	Yes	Yes
Certificate of necessity required by the ICC to construct or abandon pipeline facilities which may lead to the discrimination of small shippers	No	No
Dissimilarities:		
Incorporated	Yes	No
Stock issued	Yes*	No
Original construction and subsequent expansion of system	Planning and supervision performed by company personnel, actual construction done by outside contractor.	Designated operator designs line which plan is subject to approval of participants. Line is constructed by outside contractor and supervised by operator personnel.
Operation, maintenance, and accounting	Performed by its own personnel	Performed by designated operator
Payment of expenses, invoices, ad valorem taxes, and so forth	Paid by corporate entity	Paid by designated operator who collects each party's proportionate share.
Tariffs filed with ICC	Yes	Not by or for the system itself. Each participant files its own tariff as a separate common carrier, and collects its own revenue.
Annual Report Form P and quarterly reports filed with the ICC.	Yes	Not by or for the system itself. Each participant records its proportionate share of the system in its own reports.

*Unlike many corporations in which the stockholders are passive, the joint venture stockholders are active participants in the entire operation of the pipeline. In the TAPS system, the operator, Alyeska Pipeline Company, is a corporate venture owned jointly by the undivided interest participants in the line in shares roughly proportionate to their undivided interest.

Pipelines have significant economies of scale, so that access to large lines--for example, joint endeavor ownership--can have a significant competitive impact.

SOURCES OF CRUDE OIL

The crude oil needed for refinery inputs can come from a variety of sources. These sources, because of transportation mode configurations, vary among different regions.

On a national basis, domestic production comes increasingly from a narrow group of firms. The share of crude oil production of the four, eight, and twenty largest firms rose between 1955 and 1974 from 21.2 percent, 35.9 percent, and 55.7 percent, respectively, to 33.1 percent, 54 percent, and 76.9 percent.[63]

Tables 4.13 and 4.14 present sources of crude oil for western refineries.

TABLE 4.13

Crude Oil Input to Refineries in PAD V, 1975
(thousands of barrels)

State	Total Input	Intrastate	Interstate	Foreign
California	562,462	313,638	57,988	191,190
Washington	105,792	--	3,683	102,448*
Other states	38,594	20,736	640	17,197
Total	706,848			

*59,694,000 barrels to Washington refineries in 1975 were Canadian oil shipped by pipeline. These shipments are being rapidly curtailed by the Canadian government, and are likely in the future to be supplanted by Alaskan oil.

Source: U.S. Bureau of Mines, mineral industry surveys.

TABLE 4.14

Crude Oil Input to Refineries in PAD IV, 1975
(thousands of barrels)

State	Total Crude Input	Intrastate	Interstate	Foreign*
Colorado	16,204	4,498	11,539	162
Montana	42,589	9,766	19,817	12,656
Utah	42,797	17,736	25,082	--
Wyoming	53,805	47,338	3,100	3,567
Total	155,395			

*Canadian oil
Source: U.S. Bureau of Mines, mineral industry surveys.

Domestic Crude Oil from California

There are seven major producers in California, which produced 53.4 percent of the state's crude oil output in 1974, and a large group of independents. In the older California fields, 13.2 percent of the state output was by unit operation. The ten largest fields in the state accounted for some 61 percent of state production. They produced 561,692 barrels per day in 1974, down from 580,068 per day in 1973 (63 percent of the 1973 state total).* Crude oil production in California is, and probably will continue to be, moderately concentrated in ownership.

Production shares of the seven major companies operating in California and the independents are shown in Table 4.15. There has been some shifting from their 1960 positions, owing principally to a decline in ARCO's share, perhaps reflecting ARCO's heavy investment in the North Slope.

The exploitation of offshore resources provides a major portion of the domestic oil supply† and is expected to produce an even larger portion of future supplies.[64]

*The ten fields have 8,310 of the state's total of 38,688 wells. In 1974, 855 wells were completed in the ten fields, representing 56 percent of the 1,414 total completions statewide.

†In 1975 offshore production of crude petroleum was 501,270,000 barrels or 16.4 percent of total U.S. production (3,052,048 barrels). 1975 California offshore production was 79,096,000 barrels (15,435,000 from federal and 63,661,000 barrels from state lands) which was 24.2 percent of total California production.

TABLE 4.15

California Crude Production

Company	Percent of Total Number of Barrels per Day				1974 Production (barrels per day)	
	1960	1972	1973	1974	From Unitized Acreage	Total
ARCO	9.9	3.6	3.7	3.7	13,565	33,091
Mobil	6.3	5.6	5.5	5.3	5,605	46,594
Shell	8.9	7.6	8.5	8.5	2,154	74,751
SOCAL	14.0	14.0	13.4	13.2	16,626	116,971
Texaco	4.9	4.5	5.0	5.3	4,096	47,127
Getty	5.2	10.9	11.3	11.4	14,099*	100,363
Union	9.0	6.6	6.3	6.0	8,118	52,978
Total seven majors	58.2	52.8	53.7	53.4	64,263	471,879
Independents	41.8	47.2	46.3	46.6	52,040	411,734
State total (barrels per day)	830,736	947,519	919,839	883,613	116,303	883,613

*Tidewater.

Source: Conservation Committee of California Oil Producers, Annual Review of California Oil and Gas Production, 1974 (Los Angeles: Conservation Committee of California Oil Producers, 1975).

Alternate Sources: California Offshore

The offshore area of southern California in the federal and, to a lesser degree, the state domain is a potentially rich source of oil. The area off Santa Barbara County is particularly promising.

A review of the prior and recent leases and of drilling platform and mooring facilities indicates that major oil companies, alone and in joint ventures, control this resource. Through control of pipelines, moorings, and refineries, major oil companies maintain an advantageous lease-bidding position; they can best afford the high costs and long lead times associated with remote offshore exploration, which can be seven to ten years. Since the majors control transportation and refining, any other lessee would be selling them its production.

Ownership of oil tanker mooring facilities in the southern California coastal area has been reported by the Interior Department (1972) as shown in Table 4.16.

Operators of platforms for the production of oil and gas as listed by the Department of the Interior (1972) are shown in Table 4.17.

TABLE 4.16

Oil Tanker Mooring Facilities in Southern California Coastal Area

Operator	Function	Diameter (inches)	Length (feet)	Volume (barrels)
Phillips	Crude oil loading	10	2,000	200
Getty	Crude oil loading	12	4,800	680
Shell	Crude oil loading	16	2,890	700
Signal*	Crude oil loading	11	2,530	300
SOCAL	Crude oil and refined	20	2,550	970
	products loading and	10	2,500	240
	unloading (three-ship			
	facility)			
Union	Crude oil loading	10 and 20	9,120	2,050
Getty	Crude oil loading	18	4,300	1,350
SCE	Fuel oil unloading	24	4,480	2,520
SOCAL	Crude oil and refined	8	5,300	330
	products loading and	12	7,780	1,090
	unloading (three-ship	14	7,900	1,470
	facility)	16	2,100	530
		20	3,200	1,250
		26	7,780	5,100
Gulf	Crude oil unloading	24	5,950	3,320
SDG&E	Fuel oil unloading	20	3,000	1,160

*Now Aminoil.

Note: The area excludes Los Angeles and Long Beach Harbors.

Source: California Department of Conservation, 1971.

While no new producing platforms have been completed since 1972, Exxon is currently erecting one of the world's tallest platforms to exploit one of the three separate fields obligingly combined by Interior as the Santa Ynez Unit; and the Union-SOCAL combine which won the high royalty tracts in the December 1975 OCS-35 sale is drilling in the offshore San Pedro area with reportedly good production results.

Leases

The holdings of the 106 oil and gas leases in the Santa Barbara channel area, prior to the December 1975 federal lease sale, are set out in Table 4.18. The top four leasing groups held 53.8 percent of the leases and the top eight 69 percent. These groups comprised companies that alone and with others held another 27 leases (25 percent).

TABLE 4.17

Offshore Production Facilities in Southern California

Operator	Facility	Function	Oil Pipelines[a] Diameter (inches)	Length (feet)	Volume (barrels)
Phillips	Platform Harry	Crude oil	6	10,000	330
Texaco	Platform Herman;	Crude oil	6	12,000	440
	20 ocean floor wells	Flow lines	$2\frac{1}{2}$	39,000	200
Texaco	Platform Helen	Crude oil	6	11,500	420
Atlantic-Richfield	1 ocean floor well	Flow lines	3	22,500	180
Standard-Shell	2 ocean floor wells	Flow lines[b]	3	24,000	Gas Condensed
Standard-Shell	1 ocean floor well	Flow lines[b]	4	13,000	Gas Condensed
Standard-Shell	5 ocean floor wells	Flow lines[b]	4	13,000	Gas Condensed
Phillips-Pauley	4 ocean floor wells	Flow lines[b]	4	57,600	Gas Condensed
Atlantic-Richfield	Platform Holly	Crude oil	6	26,400	960
Atlantic-Richfield	3 ocean floor wells	Flow wells	2	34,700	140
Standard	Platform Hilda	Crude oil	6	26,400	960
	Platform Hazel	Crude oil	6	26,400	960
	2 ocean floor wells	Flow lines	4	9,200	130
Union[c]	Platforms A and B	Crude oil	12	62,000	8,700
Phillips	Platforms Houchin and Hogan	Crude oil	10	32,500	3,150
Standard	Platforms Hope and Heidi	Crude oil	10	21,500	2,090
Atlantic-Richfield	Rinson Island	Crude oil	6	3,000	110
Thums Long Beach Co.	Islands Grissom, White, Chaffee, and Freeman	Crude oil	6	3,700	130
			3	21,300	1,320
			12	20,400	2,860
			14	4,400	840
Humble	Monterey Island	Crude oil	3	8,700	30
Standard	Island Ester	Crude oil	12	8,000	1,120
Union	Platform Era	Crude oil	3	18,000	1,120
Signal	Platform Emmy	Crude oil	14	7,000	1,190

[a]There are other submarine pipelines connecting the offshore production facilities to shore and serving various functions.

[b]Flow lines from Standard-Shell and Phillips-Pauley facilities carry a combination of gas and crude oil (the volumes given [sic] are for the crude oil portion only).

[c]USGS notes that Operator should read "Union & Sun" and "Facility" should read "Platforms A, B, and Hillhouse."

Note: While utilities have unloading facilities, these are for fuel oil, not crude. If crude were sold for direct combustion, other products would be foregone and prices would be charged as if for crude.

Source: U.S. Department of the Interior, Environmental Impact Deterrent to Proposed Plan of Development, Santa Inez Unit, Santa Barbara Channel.

TABLE 4.18

Santa Barbara Channel Oil and Gas Leases

Lessee	Number of Leases	Percent of All Leases
Atlantic Richfield		
Alone	1	0.9
With Exxon and SOCAL	10	
With Marathon and Getty	1	
With Mobil Oil	2	
With Mobil and Aminoil	1	
With Aminoil	2	
With SOCAL	4	
Total	21	19.8
Exxon		
Alone	21	
With Phillips	1	
With SOCAL	18	
With Texaco	1	
With Union Oil	1	
Subtotal	42	39.6
With ARCO	10	
Total	52	49.0
Gulf		
With Mobil and Texaco	8	
With Mobil and Union	2	
With Union	1	
Total	11	10.4
Others*	12	11.3

Lessee	Number of Leases	Percent of All Leases
Standard Oil of Calif. (SOCAL)		
Alone	4	3.7
With Shell Oil	4	
Subtotal	8	7.5
With ARCO or in group with ARCO	14	
With Exxon	18	
Total	40	37.7
Mobil		
With Gulf Oil, Union and Texaco	8	
Union Oil Co. and Gulf Oil	2	
With Union Oil	4	
Subtotal	14	13.2
With ARCO	3	
Total	17	16.0
Shell Oil Co.		
Alone	4	3.7
With SOCAL	4	
Total	8	7.5
Union Oil		
Alone	4	3.7
With Exxon	1	
With Mobil	14	
Total	19	17.9

*These include two leases for Texaco and two for a group consisting of Cities Service, Continental Oil, and Phillips Petroleum, three for a group headed by Ashland Oil, and one for a group headed by Marathon Oil.

Note: The top four firms, alone, and the groups they are in, control 80.1 percent of the leases; the top five firms alone and in groups control 90.5 percent of the leases.

Source: Data from Department of Interior, Environmental Statement to proposed Plan of Development, Santa Ynez Unit, Santa Barbara Channel (1975).

Significant segments of offshore production are controlled by joint ventures of the major oil companies: for example, the Santa Ynez* Unit's HAS group (Humble [now Exxon], ARCO, and SOCAL) and THUMS of Long Beach--Texaco, Humble (Exxon), Union, Mobil, and Shell.[65]

On December 11, 1975, offshore leases in southern California were sold at auction. According to a newspaper article, "The U.S. Geological Survey has indicated that there may be three billion to five billion barrels of recoverable oil in the area covered by today's lease sale, but the Western Oil and Gas Association, an industry group, has used a working figure of 14 billion barrels."[66] Joint bidding groups headed by Shell had high bids totaling $122.8 million and groups led by SOCAL had high bids of $111.2 million. There was a $105.2 million bid on one tract alone by a group composed of SOCAL (39 percent), Union (26 percent), Getty (22 percent), and Skelly (22 percent). This last tract was one of three on which base royalty is 33-1/3 percent of production, double that of other offered tracts. Texaco bid $93.5 million in a two-thirds/one-third venture with Champlin Petroleum; and Exxon bid $29 million to be high bidder on 12 of the 70 tracts bid on.

	Tracts	Bids rejected[67]
SOCAL alone	6	2
SOCAL-Union Oil	2	2
SOCAL-Getty	2	1
Oxoco group	3	
ARCO	11	
Exxon	12	3
Shell Oil	1	
Shell group	6	
Texas-Champlin	5	
Gulf	6	
Mobil group	7	3
Marathon	1	
Challenger Oil & Gas	5	1
Amoco (Std of Ind.)	1	1

As this list shows, the majors will continue to dominate lease holdings offshore of California.

---

*Published reports indicate that recoverable reserves for the Santa Ynez Unit may be as high as two billion barrels of oil and one trillion cubic feet of gas; however, reasonably accurate reserve estimates can be calculated only after additional drilling and some production has taken place. The unit encompasses the separate Sacate, Pescado, and Hondo fields.

The prevalence of joint enterprises in the development of OCS properties is noted in an FTC staff report,[68] and by the Senate Judiciary Committee.[69] In reviewing the joint ownership of federal offshore leases, the committee noted that "Only Shell, Exxon, Standard Oil of California, and Gulf held more than half of their leases independently. Mobil independently held 6 of the 52 leases in which it had an interest; Amoco held 3 out of 60; Arco had 3 out of 94; Cities Service, 1 out of 100; and Getty, 1 out of 119."[70]

Offshore leasing has a longer history in the Gulf than elsewhere, and may thus indicate the ownership patterns that will be characteristic of future offshore development on the West Coast. The ownership of OCS production capacity from federally leased tracts in the Gulf of Mexico is concentrated, while installed productive capacity (MPR) is substantially less than the maximum efficient recovery (MER) rates at which production could be maintained. Table 4.19 summarizes this information.

Alternative Sources of Crude Oil: PAD District IV

PAD District IV, the Rocky Mountain states, is largely self-sufficient. The FTC Staff Report indicates that the top four producers in District IV, SOCAL, Standard Oil of Indiana, Shell, and Marathon accounted for 36.7 percent of production; and the top eight companies accounted for 54.5 percent. The report states that these data show a moderately concentrated crude oil industry,[71] and also notes that with the recent acquisition of Pasco's crude operation, Standard Oil of Indiana's share of production would rise from 11.2 percent to 13.7 percent, causing the top four companies to have 39.2 percent of production and the top eight 57 percent.

District IV is not likely to provide a competitive source of oil supply for District V.* On the contrary, given the overwhelming quantity of Alaskan production, oil flows are likely to be to the East.

Alternative Sources: Oil from the Alaskan Arctic

The onset of deliveries of local oil through the Trans-Alaska pipeline from Alaska's North Slope region could herald a great change

---

*There is a small pipeline (65,000 barrels per day) carrying crude oil from the Southwest to southern California. The Four Corners Line was owned by ARCO (10 percent), Continental (10 percent), Gulf (20 percent), Shell (25 percent), SOCAL (25 percent), and Superior Oil (10 percent). It has recently been wholly acquired by ARCO which is reversing the direction of line flow.

TABLE 4.19

Ownership of Federal OCS Production Capacity, Gulf of Mexico
January 1, 1976

MER Rank	Firm Operating Leases	Maximum Efficient Recovery Rate (MER) (barrels per day)	Percent of All OCS Oil MERs	Maximum Production Rate (MPR)	Percent of All OCS Oil MPRs
1	Shell	215,108	16.52	148,563	15.56
2	Exxon	168,053	12.91	138,827	14.54
3	Continental	144,414	11.09	105,701	11.07
4	Chevron	143,848	11.05	145,773 (sic)	15.27
5	Gulf	114,427	8.79	73,580	7.71
6	ARCO	70,535	5.42	35,914	3.76
7	Marathon	63,760	4.90	52,895	5.54
8	Pennzoil	60,080	4.61	49,839	5.22
Total			75.29		78.67
Total of all lessees*		1,302,040		954,736	

*Other lessees include Union Oil (34,147 barrels per day MER), Mobil (41,145 barrels per day MER), Amoco (36,220 barrels per day MER), Texaco (32,514 barrels per day MER), Tenneco (42,896 barrels per day MER), Citgo (6,190 barrels per day MER), Sun (3,960 barrels per day MER), and Burmah (4,440 barrels per day MER).

Source: USGS, Metairie, Louisiana, List of Approved Maximum Efficient Rates for Reservoirs and Maximum Production Rates (January 16, 1976).

in West Coast patterns of petroleum supply. The Alaskan North Slope in the Prudhoe Bay area is believed to hold as much as one-third of domestic oil reserves. Production could run to 22 million barrels per day from presently known resources. The Trans-Alaska Pipeline (TAPS), delivering this oil, is planned to be able to ship two million barrels per day in 1982.[72] Production levels ultimately could rise to over five million barrels per day.[73]

Forecasts of crude oil production vary among enterprises and are shown in Table 4.20.

TABLE 4.20

Current and Future Crude Oil Production in PAD V
(millions of barrels per day)

Respondent	Current	1978	1980	1982	1985
Interior	11.06	--	--	--	--
FEA	1.07	1.334	1.708	--	1.928-2.328
FPC[a]	--	1.679	2.565	--	3.490
Exxon	1.074	2.3	2.7	--	4.1[b]
SOHIO	1.076	1.2-1.4	1.4	1.3	--[c]

[a]Based upon data supplied by SOHIO.

[b]Include a substantial volume of crude from yet-to-be-discovered reserves and must be viewed as speculative.

[c]Have not attempted to forecast beyond 1982 since so many unpredictable variables would have to be considered.

Source: Interview with member of Senate Interior Committee.

Most of the oil in Prudhoe Bay appears to be owned, in proportions still subject to the outcome of the unitization discussions, by SOHIO (BP), ARCO, and Exxon.* These unitization discussions

*Besides offshore California and the North Slope, the other new source for West Coast refineries is Elk Hills, where production is expected to be able to reach to 160,000 barrels a day. However, it should be noted that 20 percent of the Elk Hills reserves is owned by Standard Oil of California, which also owns the only pipeline now connecting the field. Legislation is necessary to permit opening of full production from the reserve.

have been taking place privately among these major firms. They envisage operation of the North Slope fields as a single coordinated unit with two operators planning production with and for the other lease holders. The single unit is curious, as the North Slope appears to be composed of a number of pools and fields; moreover, the draft unitization agreement submitted to the state of Alaska does not specify how the shares of the companies in the unit are to be developed. Alaskan North Slope shares are shown in Table 4.21.

TABLE 4.21

Alaskan North Slope Shares
(percent)

Ownership of North Slope Oil		Ownership of TAPS Pipeline
SOHIO	53.2	33.3
BP	--	15.8
Exxon	20.3	20.0
ARCO	20.3	21.0
Mobil	2.1	5.0
Phillips	2.0	1.7
Chevron	0.08	--
Ten others	1.3	--
Union	--	1.7
Amarada Hess	--	1.5

Source: New York *Times*, June 19, 1977, p. 33.

With the opening of the TAPS line, Standard Oil of Ohio (SOHIO), a new entrant into western production, will become the leading District V crude oil producer. However, the percentages of District V crude controlled by the top four crude producers may rise sharply. Moreover, since SOHIO will be controlled by BP, for many years the world's largest crude oil producer, this new entrant may not significantly alter the crude supply rules.*

---

*The larger Prudhoe Bay producers--BP/SOHIO, ARCO, Exxon, Mobil--do not seem to be very active in geothermal. The major oil companies involved in geothermal energy, SOCAL, Phillips, and Union for instance, have limited interests in the North Slope. It

According to FTC, in 1978 SOHIO will have 29.9 percent of District V crude oil production; ARCO, 14.1 percent; Exxon, 11.7 percent; Texaco, 2.9 percent; and Aminoil (Burmah), 2.7 percent. This would result in a decline in the production shares of the current top four companies on the West Coast. Aminoil's share would decline from 5.6 percent to 2 percent, Getty's from 9.3 percent to 3.2 percent, and Shell's from 9.9 percent to 3.5 percent. SOHIO's would go from zero to 32 percent, while SOCAL's would decline from 15.1 percent to 5.3 percent. However, since the majority of West Coast production will come from joint ventures in which these companies hold closely linked interests, particularly THUMS, Prudhoe Bay, and Santa Ynez, the individual shares do not reflect the total competitive impact.

SOHIO is reported to have estimated that it owns 53 to 54 percent of North Slope leases.[74]

TAPS

The effect of North Slope oil on West Coast fuel markets could be conditioned by how the TAPS line operates. Although TAPS is required to operate as a common carrier, it may not be subject to federal pipeline rate regulation.[75] The TAPS system will be jointly owned, with each owner separately filing a tariff with respect to its time-cycle share of TAPS capacity.* This means that there is no central pipeline company to which outside shippers may tender oil for shipment on this line. Rather, they will have to go to a particular owner's pipeline operating division to arrange transportation. If the potential shipment is large, it is easily conceivable that the outsider would have to make separate tender arrangements with each of the owners. Even with federal pipeline rate jurisdiction, joint tariff divisions would not be revealed.[76] The ownership of the Trans-Alaska pipeline system is set out in Table 4.22.

There does not appear to be any governmental authority to require expansion of the TAPS line; and producers may become reluctant to increase production at a rate that would put downward pressure on crude oil prices.[77]

---

will be interesting to see how their geothermal efforts continue after the opening of offshore leasing in California for which both have recently submitted high bids.

*In the operation of an undivided-interest pipeline, each owning company is assigned a period of use of the basic facility corresponding to its basic ownership share.

TABLE 4.22

TAPS, Present Ownership

	Percentage of Ownership	Design Capacity (barrels per day)
ARCO Pipe Line Company (formerly Atlantic Pipe Line Co.)	21.00	252,000
SOHIO Pipe Line Company* (formerly BP Pipe Line Corporation)	33.34	400,080
Exxon Pipeline Company (formerly Humble Pipe Line Company)	20.00	240,000
Amerada Hess Corporation	1.50	18,000
Mobil Alaska Pipeline Company (formerly Mobil Pipe Line Co.)	5.00	60,000
Phillips Petroleum Company	1.66	19,920
Union Alaska Pipeline Company (formerly Union Oil Company of California)	1.66	19,920
BP Pipelines Inc.*	15.84	190,080
Totals	100.00	1,200,000

*With respect to the period from July 8 through July 16, 1974, SOHIO Pipeline Company's and BP Pipelines Inc.'s percentages of ownership and barrels of design capacity set forth above were held by SOHIO Pipeline Company.

Source: California State Lands Commission, California and the Disposition of Alaskan Oil and Gas (June 1976).

The ownership, operation, construction, and expansion of TAPS are governed by the basic agreement among the co-owners, the Trans-Alaska Pipeline System Agreement, dated August 27, 1970. By the agreement, the line is designed to permit expansion of capacity to approximately two million barrels per day (Article I). A separate agreement for design and construction was signed, and a construction committee provided to oversee this work. Each owner obtained one representative in the committee, and ARCO, Exxon, and SOHIO/BP each obtained a second, in apparent recognition of their majority ownership of both the line and its shipping source. Pipeline and terminal tankage capacity ownership is allocated to owners in proportion to each undivided interest in the project (Article III).

SOHIO has proposed to build a port facility in southern California to receive Alaskan oil. The port is proposed as a common carrier, able to handle petroleum greatly in excess of SOHIO's allotment of North Slope production.[78] It also would be equipped to load, as well as unload, crude oil. Mr. Garibaldi of SOHIO, briefing the California State Energy Conservation and Development Commission on November 19, 1975, indicated that an underlying reason for plans for a port larger than SOHIO needs for itself alone is to "provide a system that can move the surplus oil off the West Coast, not necessarily to move our oil off the West Coast."[79] The proposed port could handle five to six million barrels per day.

SOHIO proposes to ship 500,000 barrels of oil per day east to Texas and the Middle West, utilizing pipeline capacity that would be retired from interstate gas service for the purpose.* Use of the line will reduce the possibility of additional natural gas entering West Coast fuel markets.

The firms producing North Slope oil have not been required to disclose their marketing plans, and the firms owning and expanding West Coast refinery capacity have not been required to disclose what refining feedstock they plan to use, and what feedstock they could employ. If the West Coast refiners do not arrange to run North Slope crude they may be forced to continue to import oil, and North Slope oil might either be delivered at high prices to other U.S. destinations, or it just might be sent to Japan, while imports from the Middle East into Atlantic coastal states rise.† This fortuitous set of circumstances happens to be an excellent way of assuring the stability of OPEC oil prices in both West Coast and world oil markets, while avoiding local gluts of crude supply.[80]

Petroleum from the TAPS system will not totally displace foreign oil. In Washington State, while ARCO will use Alaskan Oil in its 100,000 barrels per day refinery, designed expressly for such feedstock, other refiners including Shell, Texaco, and Mobil are

---

*A Northern Tier Pipeline has been proposed to move Alaskan and foreign oil to Rocky Mountain and mid-continent areas. The line is envisaged as being about 1,500 miles long, and 36 inches in diameter, with a capacity of 800,000 barrels per day. This proposal, however, is made by a transportation company that is not among the TAPS owners. There is no assurance that the TAPS owners would in fact use this facility.

†With planned additions, SOCAL's share of West Coast refining has been projected to increase.

reported to have indicated that they will not be using substantial quantities of Prudhoe Bay crude.[81] However, Canadian exports of oil are declining, and FEA is allocating such exports to the upper Middle West. Accordingly, Alaskan oil may be used by Washington refiners to replace Canadian supplies (188,899 barrels per day in 1974).[82]

The situation for the North Slope appears to be that while 11 companies have concessions in the Prudhoe Bay field, nine-tenths of the oil is owned by three groups--BP/SOHIO, Exxon, and ARCO. TAPS flow is to reach 1.2 million barrels per day by the end of 1977; the capacity is likely to be raised to 1.6 million barrels per day and then to 2 million barrels per day (about 1980). This two million barrels per day figure compares with West Coast oil consumption which will reach 2.8 million barrels per day in 1980[83] if it rises by some 5 percent per year. Considering California production of about 1.2 million barrels per day, other Alaskan production of 0.2 million barrels per day, and estimated imports of 0.2 million barrels per day,[84] a surplus of 0.8 to 1.3 million barrels per day might occur in PAD V.

Of the three main North Slope producers, BP/SOHIO has no West Coast refineries and apparently will be selling or exchanging that portion of its production (which will total almost one million barrels per day) to be refined on the West Coast. Exxon's surplus will reach 300,000 to 400,000 barrels, ARCO will be in balance, and the state of Alaska, after present commitments to Tesoro are met, will have a supply of 250,000 barrels per day. ARCO crude oil (up to 120,000 barrels per day) may be available, by contract, to SCE.

Control of the TAPS line by a few companies with a large interest in maintaining high prices in the United States makes unlikely an expansion of this line at a size likely to lead to major price effects on the West Coast, or a major change in refining and marketing from the positions outlined here.

Alternative Sources: Foreign Oil

Foreign oil is supplied to the West Coast chiefly by Indonesia (the largest supplier), Iran, Canada, and Saudi Arabia, with lesser amounts of production from countries such as Ecuador. Table 4.23 provides sources.

In general, production in these countries has been under joint ventures, or concessions controlled by major oil companies. The pattern has been shifting toward country ownership of fields, with the former concession holders having off-take agreements, and sometimes field operating agreements. Off-take, or preferential access agreements, often give the oil company involved access to the bulk (over 90 percent) of the oil it formerly produced (as owner or concession owner), at a price lower than that available to others. The proceeds of price increases are, in most cases, split between the owning country and the marketer. See the earlier discussion of foreign residual fuel oil sources.

TABLE 4.23

Imports of Foreign Crude Oil to PAD V, 1975

Country of Origin	Volume (thousands of barrels)
Algeria	399
Bolivia	1,940
Canada*	59,694
Ecuador	19,219
Indonesia	107,820
Iran	38,398
Libya	2,685
Malaysia	1,951
Nigeria	5,025
Oman	492
Qatar	3,989
Saudi Arabia	34,901
United Arab Emirates	18,217
Venezuela	16,105
Total	310,385

*The Canadian government has declared its intention to curtail oil exports from Canada and has imposed heavy export duties.

Source: U.S. Bureau of Mines, mineral industry surveys.

VERTICAL INTEGRATION

The petroleum industry in California is substantially integrated, with six firms being among the top eight enterprises at all levels. SOCAL, Union Oil of California, Shell Oil, and ARCO constitute the top four firms at each level (save production, in which ARCO is sixth).*

*ARCO's and Exxon's respective crude oil shares in the West will increase dramatically when North Slope oil comes to market.

This same vertical integration is present throughout the American oil industry, with integrated enterprises in control of pipelines and, as joint venture partners with each other and independent firms, increasingly controlling crude oil production. Independent refiners are dependent upon these enterprises both for feedstock and for marketing products via sales and exchanges. The major petroleum firms' control of successive stages of production in California is coupled with their control of alternative petroleum sources.

This vertical control of essential facilities is exercised through a system of coordinated operations among larger oil companies. Production and transportation of crude oil and refined products are scheduled in an interdependent manner. This is required by and accomplished through exchanges and joint enterprises. This interdependence leads to common problems and, in some instances, common outlooks.

The larger petroleum companies have a vested interest in high crude prices. They have preferred access to OPEC oil (through participation agreements in which they receive a share of price rises, in itself an incentive for price maintenance), and they have non-OPEC crude oil assets whose value is tied to the price of oil. This provides an important incentive for maintaining price.

Dominating refining and transportation, they can assure markets for their own and OPEC's high priced crude without fear of being undersold in substantial volume. By controlling and balancing off-takes from OPEC countries they proration production to prevent supply gluts that could lead to price cuts. Other firms, lacking large oil supplies, cannot substantially expand sales by price cutting and can find markets for their output at cartel prices, if and for as long as the large firms are prorationing OPEC production.

A number of observers, including the Senate Subcommittee on Multinational Corporations, have concluded that the international companies prorate production for the cartel. In its 1975 report the Multinational Subcommittee concluded:

> The multinational oil companies . . . provide the OPEC with important advantages. As vertically-integrated corporations, the major oil companies guarantee OPEC members an assured outlet for their production in world markets. The primary concern of the established major oil companies is to maintain their world market shares and their favored position of receiving oil from OPEC nations at costs slightly lower than other companies. To maintain this favored status, the international companies help pro-ration production cutbacks among the OPEC members.[85]

In the Multinational Subcommittee's hearings, Professor M. A. Adelman testified that:

> The cartel governments use the multinational companies to maintain prices, limit production and divide markets. This connection, I submit, is the most important strategic element in the world oil market. The governments act in concert, the companies do not need to collude. The governments transfer oil to the companies at identical publicly announced prices. Thereby the governments can watch one another. The companies produce only what they can sell. So long as the governments are content to accept the market shares that result from the companies' sales efforts, the cartel holds. Time is on their side. As the cartelist governments get richer, it becomes easier for them to accept output limitations or even cutbacks.[86]

OPEC production capacity is currently worked at about 75 percent of capability. Lower production rates would put pressure on prices. The OPEC countries have never been able to agree on production levels for member states.[87] Major oil companies spread production. Here a major question of international market structure and operation is presented. The production control necessary for OPEC prices is maintained perhaps, in part, by OPEC member tacit consent, and definitely through major oil company concord in the off-take arrangements. To the extent that OPEC's control of prices depends upon the preferred areas given to major firms that are horizontally linked and vertically integrated, OPEC prices could not be maintained, should the oil company network break down.

If the oil industry were more diverse, with fewer joint enterprises, and more independent channels available for bidding on and taking OPEC crude, the price and supply of oil would be much more competitive.[88] Professor M. A. Adelman of M.I.T. has suggested that a community of interest between the OPEC governments and major oil companies is partially responsible for a failure in realizing more competitive conditions in international oil markets.[89]

Similar efforts are made by major firms to restrict new supply in domestic markets.[90] Plans for North Slope utilization, TAPS, and the subsequent distribution of North Slope oil have been made by the three large vertically integrated enterprises which own, jointly, that large increment to domestic supply. These companies will be entering into joint arrangements with other major oil companies in the West, enabling them to prevent destabilization of the market.* The extensive ties among financial institutions and larger petroleum companies make competitive capital formation less likely, as is set forth in Appendix C. The control of incremental supply increases at the production and pipeline levels will be jointly determined by these firms. That control, leveraged forward by vertical integration, creates the capability to effectively limit energy supply to maintain or raise its price.†

NATURAL GAS

In California and the Southwest, for many years large amounts of interruptible natural gas were burned as a principal utility fuel. The sharp decline in gas supply has resulted in substantially increased consumption of residual fuel oil in utility boilers and increased use of electric energy for space heating.‡

Utilities in California have been supplied by three interstate gas pipelines (El Paso Natural Gas Company, Transwestern Pipeline Company, and Pacific Gas Transmission), and to a smaller extent

---

*Joint arrangements are likely to include exchanges, refinery service transactions, and joint venture pipelines.

†Federal petroleum price regulation does not provide a substitute for price competition nor does it initiate the role of competition in pricing or in production planning. Federal regulation is only a temporary program to set ceilings on some of the prices charged for some domestic crude oil production and some upstream profits. Regulation does not extend to supply planning, pipeline, refinery, or well facility certification; nor does it specify prices. Indeed, many oil sales are made at prices below those permitted by regulation. Efforts to raise prices by government action may not be expected to hold up in the face of competitive pressure to reduce prices. Indeed, government action is as likely to mirror industry conditions as it is to determine them. This is so in part because federal petroleum regulation did not come into being as a response to perceptions of intrinsic abuses in the regulated industry, as did many other regulatory institutions. Being addressed to an emergency, federal petroleum regulation does not concern itself with changing industry structure.

‡Rising alternate costs of space and process heating may enhance secondary markets for geothermal energy.

by indigenous production.* A fourth interstate pipeline that served California was acquired by its major competitor. Although suit was filed in 1957 to bar this merger, the subsequent proceedings were prolonged. Divestiture did not occur until February 1974 when the Northwest Energy Company acquired the Northwest Pipeline Corporation from El Paso.

In the intervening years, the two lines to southern California, El Paso and Transwestern, alternately expanded their pipeline capacity in a load-growth-sharing "minuet" while the FPC refused to certify a proposed pipeline (PEMEX) intended to transport gas for electric utilities.[91]

The Pacific Gas Transmission (PGT) Company imports Canadian natural gas which it supplies to its parent, PG&E† in northern California. This gas is acquired from Alberta producers by a second PG&E subsidiary, the Alberta Southern Gas Company, which also arranges for transportation over lines owned in part by an El Paso subsidiary.

During the FPC proceedings for the certification of the El Paso and Transwestern Pipelines, concern was voiced about the supplies of gas of these systems. This concern has proved justified. In these proceedings the court stated that "Presently, neither El Paso nor any of the applicants for acquisition have gas supplies available to serve any part of the unsupplied demands of the California market. . . . competition among suppliers to serve the incremental demands of the California market no longer exists."[92]

Intrastate Gas in California

California utilities obtain dry gas produced in northern California, and wet gas produced in conjunction with oil and natural gas liquids, in southern California.‡

Except for limited competition from Dow Chemical, PG&E is the sole purchaser of gas in northern California. The price has risen

---

*The history of the natural gas pipelines supplying California is too extensive to be fully treated here.

†PGT is not entirely owned by PG&E: a member of the board of the Montana Power Company, another combination gas-electric utility with Canadian and U.S. production and extensive coal deposits, is on PGT's board.

‡In 1974 gas production was 52,454,000 cubic feet per day, considerably down from 60,983,000 in 1973. Natural gas reserves in California are only 1.6 percent of total U.S. reserves, and the California share declined 7.4 percent from 1973 to 1974.

in a few years from 30 cents to the current level of 75 cents per thousand cubic feet, and production has declined.[93] PG&E's great market power in northern California gas fields is demonstrated by its ability to schedule deliveries from those fields only for its higher load periods. Contracts with gas producers provide for deliveries at the lower of one-twentieth of field gas reserves or of the installed production capacity of the field or well.[94] Contracts require that the buyer pay the producer for a contracted-for amount, or proportion of capacity, whether he takes that amount of gas or not.* Individual contracts may vary according to whether PG&E or the producer owns the gathering lines.

Future gas supplies may be expected to come from utility gas exploration ventures, and from novel use of liquefied and synthetic natural gas.

"Natural" gas from SNG,[95] LNG, and Alaskan sources is likely to cost more than two dollars per thousand Btu, and will in some cases be priced with direct reference to the cost of foreign oil.[96] The possibility of high prices, plus the need to find gas to fill their pipeline systems has attracted California gas utilities, including PG&E and SDG&E, into the petroleum exploration business. The management of PG&E must also now contend with problems involved in the importation of LNG from Indonesia, and the construction of systems to transport gas from the North Slope.[97]

Currently there are plans to reduce the availability of gas pipeline capacity to California. The El Paso Natural Gas Company has requested the FPC to allow it to abandon gas service on 669.4 miles of large-diameter pipeline and to transfer this line to a new subsidiary, apparently outside the FPC's jurisdiction. The new subsidiary would lease the facilities to SOHIO, which would convert the line to carry oil from west (California) to east (Midland, Texas), where it would join lines to the Middle West. El Paso asserts that its gas supply is depleted and that an oil route into the Middle West would avoid a glut of crude oil on the West Coast, expected after the Trans-Alaska Pipeline begins operation. The leasing of facilities is divided into two phases, the second of which would require abandonment of gas service on additional portions of El Paso's system.

Rising gas prices and curtailments encourage utility diversification of power supply sources. They also, however, may encourage speculative withholding, hedging of fuel supplies, and employment of tariff rates designed to obtain customer financing for energy exploration and development by established utilities, and their joint venture partners.

---

*In the event gas is taken at a 100 percent load factor, prices are reduced by seven cents per thousand Btu.

COAL

Coal is the major factor in U.S. utility fuel supplies. The mines, plants, and transmission grids employed in coal-fired projects are and will be very large, and quite limited in number. A 1,500mw power plant requires about 4.6 million tons per year of good quality coal (12,000 Btu/lb.). Since 1965, only 18 domestic coal mines with a capacity of two million tons a year or more have started production in the United States. Thirty-six domestic coal mines produced two million tons a year or more in 1974.[98]

For northeastern Wyoming, the Interior Department made the following projections:[99]

	1974	1980	1985	1990
Number of mines	3	10	12	14
Tons per year (millions)	8	88	122	150
Cumulative tons of coal mined (millions)	--	297	858	1,543
New power plants	2	2	3	4
Gasification plants	--	1	2	2

Ergo, mines are forecast to produce an average of over 10 million tons per year each.

The pattern of ownership and control of coal resources will affect the rate at which coal resources are made available to customers. This pattern will also, among other things, affect the state of competition in the industry, and thus the interaction of coal prices with other fuel prices. The pattern of coal ownership could conceivably have several different effects on development of geothermal energy. If owners of both coal mines and geothermal energy sites perceive opportunity costs in those other holdings if cheaper geothermal energy sources are developed, they might wish to warehouse the geothermal sites. If the coal owners with geothermal resources have sufficient market power to preclude utility development of alternative fuels, they could retard geothermal development. Or if, but for ownership patterns, the state of competition and costs in the coal market were such as to reduce coal prices to levels producing significant development of geothermal energy, noncompetitive ownership of coal could foster geothermal development.

The Western Coal Resources

Coal deposits in the Pacific Coast states tend to be small and scattered.* Plans for coal generation by West Coast utilities

*Alaska, particularly in the far north, appears to have extensive coal deposits.

generally involve coal from the Rocky Mountain province which contains an abundance and a greater variety of coal than any other province in the United States.[100] One plant is proposed to be expanded for the Northern Great Plains Basin* (at Colstrip, Montana).

The tremendous coal reserves in the Northern Great Plains Basin will initially serve mainly eastern utilities.[102] Coal energy from this basin (in addition to that from the Colstrip plant) could potentially enter western utility fuel markets, but this seems unlikely until Rocky Mountain Basin deposits have first been developed.

To indicate the amount of coal that could be provided to western utilities at different levels of costs (including a reasonable profit), the Battelle Pacific Northwest Laboratories have prepared Figure 4.3 which shows the tonnage of coal that could be forthcoming at ascending costs (1976 dollars) assuming physical delivery of the coal at a California point.

Prices may diverge from a cost-based competitive level.[103] Costs presently reflect neither the unique start-up costs that would be encountered were coal mining to increase sharply, nor the future inflation that may be experienced as new mines are opened.

The selling price of western strip-mined coal is expected to be lower than underground coal mine prices. A recent USBM publication, noting the lower heat content and the shallow, thick seams of western coal (for example, averaging 55 feet in the Powder River Basin), quotes costs of three to five dollars per ton (equivalent to about 17 to 28 cents per million Btu for 9,000 Btu/lb. heat content coal.)[104]

While capital costs for western mines are low per ton, they are high per mine.[105] With increases in required investment levels, participation by small producers is deterred. Average investment cost for deep mines is estimated (in 1974) at $12 per ton of annual output and for strip mines at $6 per ton. Initial capital investment costs for western strip mines, per ton of output, are expected to be lower than the national average. All coal produced in the Northern

---

*These Montana units will be jointly owned by the Montana Power Company, Puget Sound Power and Light, PP&L, Portland General Electric Company (PGE), and the Washington Water Power Company. A 500kv line is planned to transport power west. The Northern Great Plains Basin includes the western Dakotas (Fort Union Region), northeastern Wyoming (Powder River Region), and central and eastern Montana. The Fort Union Region contains 440 billion tons of low specific-heat-content lignite--by far the largest coal resource in the United States.[101] Extensive coal beds also occur in eastern Colorado. Most of the coal on federal lands lies in the Northern Great Plains Basin.

FIGURE 4.3

Supply Curve of Coal Mined in the West Delivered to the Western United States (Central California)

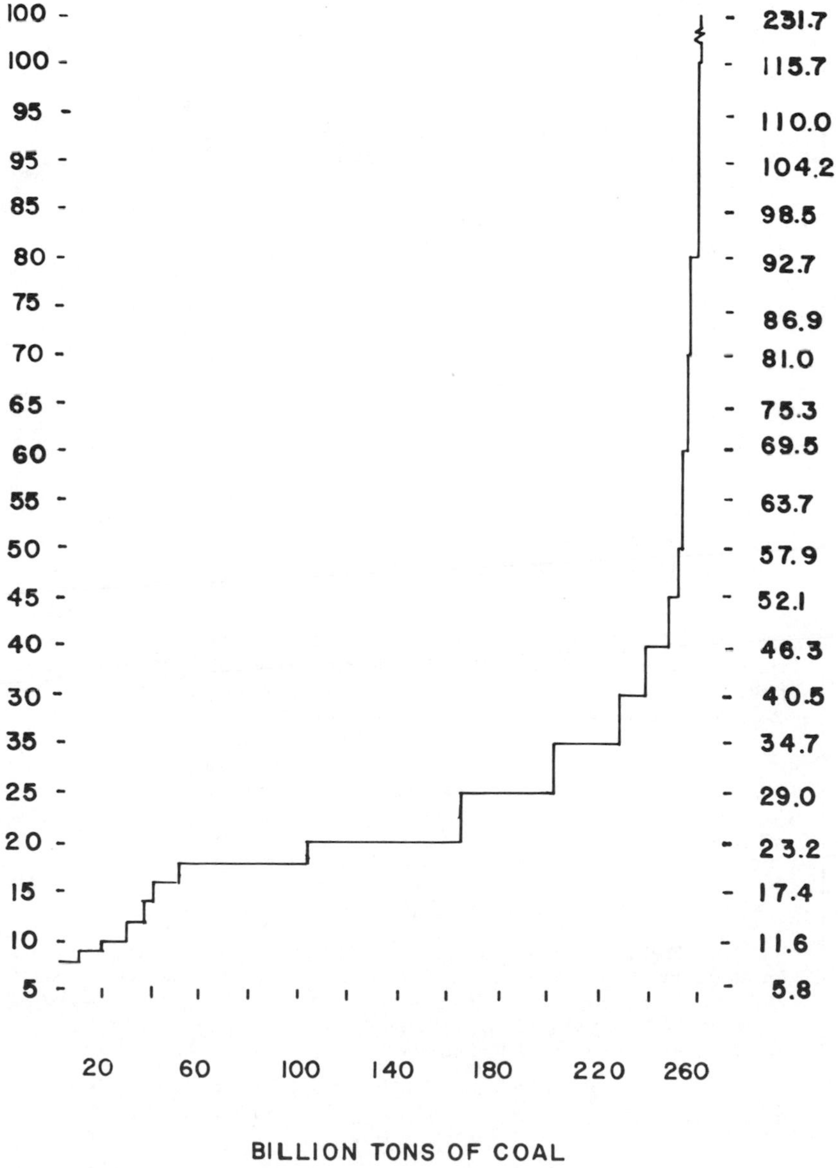

Notes to Figure 4.3

Rail transportation costs to Central California are shown below.

COAL AT MINE MOUTH = COAL DELIVERED-TRANSPORT
TRANSPORT = MILES x RATE

	Miles	Rate (mills/ton-mile)	Transportation Costs (rate x miles)
Montana	1,013.5	5.6654	$5.7419
Colorado	979.4	5.7823	$5.663
Utah	681.7	6.7823	$4.170
New Mexico	876.5	6.0782	$5.328
Wyoming	944.4	5.848	$5.523

Note: The delivered costs of coal are approximated on the supply curves for each state (using a distance, as was done for rail transport, that is from the "eyeball centroid of state coal areas to Central California".)

As an illustration, the data from the chart regarding Montana coal are arranged below.

Millions Tons Available	Cost in California of $ or Less	x Mine Cost of $ or Less per Ton
8,610	9.00	3.26
18,465	10.00	4.75
27,979	12.00	6.26

Coal costs reflect fuel quantity:
$3.26 per ton of coal at 8,000 Btu/lb costs 20.4 $/mbtu
9,000 Btu/lb costs 18.1 $/mbtu
10,000 Btu/lb costs 16.3 $/mbtu

At plant costs of 15 mills, and operating costs of 3.3 mills (ERDA, "The Economics of Nuclear Power") burning Montana coal would provide power at about 25-30 mills at the busbar in California. For nuclear power to meet this price, plant costs must not exceed $620 to $850 per kw.

The translation from cents per million Btu for coal to mills per kwh varies with the efficiency of conversion of coal to electricity (for example, the heat rate of a generating unit expressed as Btu per kwh). The heat rate tends to decline as the size of coal unit increases. Power Technologies, Inc., recently reported coal-fired unit heat rates:

Size (mw)	Full Load Unit Heat Rate Btu/kwh	Average Unit Heat Rate Btu/kwh
500	8,830	9,700
700	8,780	9,605
1,000	8,550	9,350
1,500	8,700	9,350

At a heat rate of 9,350 Btu/kwh, and rail rates of 9 mills per ton-mile for 16 million Btu per ton of coal, transportation costs are 0.5625 mills per million Btu per mile. At 11 mills/ton-mile and 21.4 million Btu/ton of coal, the costs are 0.514 mills/million Btu-mile. For each 100 miles, haul costs are about 0.5 mills per kwh by train.

Source: Power Technologies, Inc., Expansion Study of the Central & South West Corporation Electric Power System (November 14, 1975).

Great Plains, and most western coal, is strip mined. The Interior Department estimated, in 1975, that a Northern Great Plains subbituminous coal mine producing 9.2 million tons per year would require an initial capital investment of $42 million, and a total investment of over $78 million. Maximum negative cash flow is about $25 million, and with a 15 percent discounted cash flow rate of return (after taxes), a selling price of $3.99 per ton was estimated to be necessary.[106] An underground mine producing 4.99 million tons of coal per year from a six-foot thick coal bed has been estimated to require an initial capital investment of $84,027,500, including land, interest during development, and working capital. The maximum negative cash flow is $33,611,000 ($50,416,500 discounted at 15 percent).[107] Only firms with revenues in the hundreds of millions per year, such as Kennecott Copper Corporation, could be expected to be major new entrants. A court has upheld FTC findings that coal mining in the United States is becoming increasingly concentrated, partly as a result of increased capital cost requirements.[108]

Transportation Factors

Transportation charges can represent a sizable portion of the delivered cost of coal, and are an important factor in the definition of coal markets.[109] The major alternative means of moving the energy in coal to western markets are by rail, usually unit trains, and transmission of electricity from a mine-mouth plant. Coal slurry pipelines have been proposed as a third major alternative. The ICC discusses western coal shipments by rail in a recent report:[110]

> The western mines are diverse geographically and generally the coal, in order to reach its markets, must be hauled via rail greater distances than in the east or south. For example, Burlington Northern's average haul per ton for 1973 was in excess of 525 miles. Its longest trainload haul was 1,430 miles. The unit-train concept has provided a means to effectively market western coal which, by 1980, could very well reach 150,000,000 tons per year or more.
>
> Definite marketing territories for western coal are difficult to define because of varying influences in the present continuously changing market. Generally, western coal

> moves or will move into the territory east of the Mississippi because of its low sulfur content rather than a lower delivered price than for midwestern coal. This marketing boundary appears, however, to be expanding.

Unit-train or trainload coal traffic in the West for the most part originates at a single mine and terminates at a single location.[111]

The scope of possibilities for coal slurry pipeline is not yet well determined. Only one major line in the West is now in existence. Costs relative to rail are not entirely clear. Several plans are in existence, but most seem tentative.

Reflecting air quality control restraints, there is a clear tendency to use long distance transmission of electric energy generated at mine-mouth to serve the West Coast. Large sources of coal are also planned to be used for coal gasification. Because of the distance to California, inter-mountain, coal-fueled generation will require a long haul over high voltage transmission to coast markets* from southwest and Wyoming plants. High voltage lines are also projected to run from Montana coal-field plants to the Pacific Northwest.†

The costs of high voltage alternating current transmission must be determined with the aid of a load flow study (which considers how the flow is transported over parallel lines of varying impedence). It is estimated that lines of 500kv capacity will cost $280,000 to $3 million per line-mile, and that energy losses will run from 10 percent to 20 percent on extra-long lines (600-800 miles). Were a direct current circuit to be used, it might provide for lower line losses, but it would be isolated from other lines. Estimates for the cost of such a circuit were recently made by Power Technologies, Inc. for the Central and Southwest System. This study predicts that all base-load units of the Central and Southwest System's utilities will be coal-fired or nuclear. Costs for d.c. transmission for a

---

*For purposes of reliability and economics, large, expensive coal-fired units tend to be owned jointly.

†The Williston Basin and the Eastern Slope have more than enough resources for midwestern markets. To date, generation has been built or planned in the Four Corners area for wheeling to California, while unit trains have been employed to move Montana coal east. In addition to transmission losses and other transmission costs, coal in the area between the Rockies and the Sierras is unlikely to compete in eastern markets because electric grids are not coordinated between the East and the West of the mountain region.

distance of 855 miles were projected to rise from $210 million for 1,000 mw to $485 million for 3,000 mw. For like loads moving 1,315 miles, transmission costs were projected at $289 million to $641 million. It is generally understood in the trade that at least 1,000 mw of capacity must be shipped to justify long distance transportation.[112]

Recent Battelle[113] figures give unit energy transport costs (in mills per kwh per 100 miles) of up to 0.3-0.36 for 700 kv; 0.36-0.48 for 500 kv; 0.48-0.66 for 345 kv; and 0.66-0.9 for 200 kv. Battelle estimates a 765kv line to cost $300,000 per mile in 1980.* These estimates agree reasonably closely with FPC figures. Transmission costs will vary with the distance hauled and the line voltages at which transmission occurs.†

Wheeling charges for firm service are normally quoted in terms of dollars per kilowatt. The costs of high voltage transmission for mine-mouth units could be as much as 7 to 10 mills per kwh. Even so, coal-by-wire fuel costs to the West Coast (5 to 12 mills) would be about half of the cost of oil (about 24 mills). Such charges may be readily converted into charges per kwh by dividing the charge per kw of transmission per year by the hours used (24 x 365 x Usage Factor). In the Bonneville area, charges for short-term (one year) wheeling arrangements range from 0.5 to 1.25 mills per kwh (5-9 percent line losses). At high load factors, costs may be expected to be between $5 and $10 per kw per year over a few hundred miles, or 0.5 to 1 mill per kwh with rates running about 20 percent more on private grids.

These substantial capital costs for new transmission lines will set coal-by-wire prices, and they indicate that geothermal generating units must be located either near load centers or near high voltage lines. Because even large geothermal fields may be expected to be delineated and developed gradually, it is unlikely that new very high voltage lines will be built for substantial distances to serve such development. The new line capacity cost is too great.

However, given available preexisting line capacity and conditions permitting it, power could be wheeled from geothermal units. The addition of one or two mills to the delivered cost to cover transmission losses would permit base-load geothermal energy to be transmitted 400-800+ miles (80 percent load factor) at BPA rates, and 320-600+ miles at rates 20 percent higher over preexisting lines.

---

*Some components of a transmission charge are independent of wheeling distance, for example, transformers, switch gear, and metering.

†Where the service is on an interruptible basis and transmission capacity is not set aside, charges are quoted in cents per kwh.

The Pacific Northwest has a well-developed high voltage grid system that facilitates shipment of output from coal-fueled plants, such as those in Montana, to West Coast markets.*

Finally some have suggested processing coal into a synthetic crude oil, syncrude. Pipeline transportation costs (and conditions in alternate markets) would largely determine where synthetic crude oil would be sold. A 1974 study of syncrude[114] markets, which used netback prices based on the price of petroleum in various markets, indicates that syncrude from the Eastern Slope and even from the Four Corners area may command a higher price in the East. Syncrude brought south from Colorado would flow east using presently installed excess crude pipeline capacity; lesser amounts could flow to southern California, via the Four Corners Pipeline† to compete with Alaskan crude oil.

Ownership of Coal--The Federal Domain

The United States is reported to own 50.5 percent of the coal acreage and 23 percent of the mapped and explored recoverable coal reserves in seven states--Colorado, Montana, New Mexico, North Dakota, Oklahoma, Utah, and Wyoming.[115] These states in turn contain 53 percent of the nation's reserves.[116] Federally owned reserves, as a percentage of total recoverable reserves, are 59 percent in New Mexico, 82 percent in Utah, and 48 percent in Wyoming.[117]

Federal coal lands are developed by leasing to private parties. As stated in the Interior Department's Final Environmental Impact Statement,

> Federal Coal until recently has not been a major part of overall coal supply. Federal production has been approximately 1 percent of total production. Federal ownership of the coal resource in the West is approximately 60 percent but because of ownership patterns, Federal policy influences upwards of 80 percent of western coal. Federal coal is, therefore, inseparably tied to overall western coal development.[118]

---

*The Montana Power Company (TMPC), in whose service area lies much of the Montana coal fields, is synchronized with the high voltage grid in the Pacific Northwest.

†ARCO has since acquired the Four Corners Pipeline, however, with the intention of reversing the direction of flow, thus foreclosing this possibility.

Despite these facts, coal production from federal leaseholds totaled only 21 percent of western state production in 1971. Production from federal leaseholds in the West, although increasing, did not keep pace with overall growth of the western coal industry between 1960 and 1972. Table 4.24 shows the extent of federal ownership of coal lands.

TABLE 4.24

States with Major Federal Coal Acreages

	Federal Coal		Nonfederal Coal		Total
	Number of Acres (millions)	Percent	Number of Acres (millions)	Percent	Number of Acres (millions)
Alaska	23.4	97	0.8	3	24.2
Colorado	8.7	53	7.9	47	16.6
Montana	24.6	75	8.2	25	32.8
New Mexico	5.5	59	3.9	41	9.4
North Dakota	5.6	25	16.8	75	22.4
Oklahoma	0.4	4	8.9	96	9.3
Utah	4.1	82	0.9	18	5.0
Wyoming	11.8	65	10.7	35	30.5
Totals	92.1		58.1		150.2

Sources: U.S. Department of the Interior, Southwestern Energy Study, Appendix J, p. 48, 1972; Bureau of Land Management, state office estimates; Paul Averitt, "Coal Resources of the U.S., January 1, 1967," U.S. Geological Survey Bulletin (1969), p. 32; Interior Department, Final Environmental Impact Statement on Federal Coal Leasing Program (undated).

The recently lifted Interior Department moratorium and the courts have both limited leasing and development of new coal lands. However, a large number of outstanding leases have not been developed, as Table 4.25 shows, and development plans for much leased acreage is absent or leisurely.

Several factors indicate that leases were acquired for speculative or warehousing purposes. Lease bids prior to the moratorium were low, and sales frequently drew only one bidder.[119] Production has been more likely to occur on competitively let tracts than on those otherwise let.

TABLE 4.25

Federal Coal Leases, 1973

State	Number of Leases	Number Producing	
		Strip	Underground
North Dakota	24	9	
Montana	18	5	
Utah	195		17
Wyoming	94	10	3
New Mexico	28	2	1
Colorado	132	2	12
Total	491	28	33

Note: A total of 61 producing leases.

Source: Table taken from county data in Department of the Interior, Coal Lease Regulations, Final Environmental Impact Statement (undated). Leases and mines operating in more than one county are listed in each.

Recoverable reserves in the amount of 16.1 billion tons are under federal lease. Production from these leases in 1973 was 12.9 million tons. The reserves-to-production ratio was 1,248 to 1. If production grew in the future at 10 percent per year, it would take 51 years to mine 16.1 billion tons. BLM estimates the growth of production between 1973 and 1990 at 20 percent per year. At this rate, it would take 31 years to mine 16.1 billion tons.* This presumes no production from leases yct to be issued.

Of the 16.1 billion tons of leased recoverable reserves, 6.6 billion are in mines or prospective logical mining units (LMUs) which are now producing or have plans to produce. In these mines and LMUs the ratio of present reserves to planned production is 149 to 1 for 1975, and 53 to 1 for 1980.

State-by-state, the recoverable federal reserve, 1973 production, and reserves/production ratios are shown in Table 4.26 (major western states only).

---

*If a mine were to be amortized over 30 years with 50 percent of the associated coal reserve recovered, a reserve-to-production ratio of 60 to 1 would be called for--if all mines were new.

TABLE 4.26

Federal Coal Statistics for Western States
(in millions of tons)

	Recoverable Reserves	1973 Production	R/P Ratio
New Mexico	335	0.3	1,117
North Dakota	268	1.5	179
Colorado	1,650	1.8	917
Montana	1,181	1.9	622
Utah	3,604	2.4	1,502
Wyoming	9,065	5.0	1,813
Total	16,103	12.9	1,248*

*Average

Source: U.S. Congress, Senate, Interior Committee, Hearings on Federal Coal Leasing Amendments Act of 1975, 94th Cong., 1 Sess. (1975), p. 490 (Interior Department Option Paper: "Diligence Requirements for Existing Coal Leases").

1974 data presented in the Coal Leasing Regulations[120] show different, but high, leased coal reserves to production ratios. While production is slated to rise, committed acreage is huge, reflecting the tendency of huge mines to tie up tremendous reserves in an integration of mining operations with leases. Also, interim projections of future coal availability tie a committed 40-year reserve to each gasification (36 plants, 8 million tons per year, 12.8 billion tons) or power plant (3.2 billion tons).[121]

On 49 percent of the leases held by 128 separate lessees, production has occurred or is planned for the near future; 51 percent "have not produced in the past and have not indicated any plans for future production."[122]

> A major finding of the analysis is that over 50 percent of the 467 leases, which are held by 66 lessees, cover 60 percent of the total lease acreage and contain 60 percent of the total coal reserves, have never produced and have not indicated any plans for development of production before 1990. Nearly all of the leases that have no plans for production were issued within the last 20 years and over 60

percent within the past 10 years. The average age is 11 years. The leases average 200 to 300 acres larger than the average size for all leases and are two to four times larger than the average 30 year old leases. Over 60 percent of the leases that have no plans for production were obtained through the preference right method as opposed to the past and currently producing leases, which were obtained primarily through competitive bidding. Most of the coal reserves of leases with no production plans are located in Wyoming and are surface recoverable coal. However, over one half of the leases and 27 percent of the coal reserves in this category is located in Utah and is underground mineable.[123]

Federal Coal Lessees

The FTC Staff Report ranks the 20 largest holders of federally leased coal acreage in seven western states (1974). The four largest hold 28.6 percent of the acreage; the eight largest hold 44.5 percent; and the 20 largest 68.3 percent. Ranking by leased federal reserves (tons) results in the top four having 34.6 percent, the top eight, 56.6 percent, and the top 20, 81.7 percent.[124] The largest holders consist of large coal companies independent of oil and utility companies (19.5 percent),* utility systems (17.2 percent),† petroleum companies (18.0 percent),‡ and steel companies (6 percent).**

One interesting feature is that seven of these companies are not now producing from the federal lands held; and five of these seven companies are oil companies (Sun, ARCO, Carter, Bass, and Kerr-McGee). Five of the seven oil company users are not producing from their federal leases. Several of these companies have mining plans pending.

A comparable ranking for production in the seven western states is not available. A listing of the 20 largest coal producers

---

*Included are Peabody, Garland, and Utah International.

†Included are Resources Co., PP&L, Kemmerer, Western Coal, and Nevada Electric Investment. Kemmerer is a joint entrepreneur with utility systems.

‡Included are Consolidation Coal, Sun Oil, Richard Bass, Atlantic Richfield, Carter, Arch Mineral, and Kerr-McGee.

**Included are U.S. Steel, Armco, and Kaiser.

nationally shows a slightly different profile. In Table 4.27, nonoil and nonutility coal companies are seen to account for 24 percent,* utility related companies, 2.9 percent,† steel companies, 5.1 percent,‡ and oil companies, 22.9 percent.**

Table 4.28 shows from BLM reports that concentration of leaseholdings is far higher at the state level than it is nationwide.

Wyoming is slated to provide the bulk of additional western coal production.[125] Most Wyoming coal will come from the Eastern Powder River Basin from 74 leases not previously produced. This area holds 12.4 billion tons of economically strippable coal--28 percent of U.S. strippable resources.[126] See Table 4.29.

Lease ownership is highly concentrated, as is the holding of preference rights--the right to lease after prior exploration. See Table 4.30. Thus, it is interesting to note that 21 lessees (of 42 in Wyoming) have no plans to develop 62 of the 92 leases in the state. The 62 leases cover 128,070.39 acres (199,944.21 acres are leased in the state) and contain 5,126.83 million tons of strippable resources and 396.78 million tons of resources that must be mined underground.

Wyoming's 5,126.58 million tons of leased strip-minable reserves constituted 61.17 percent of the 8,381.44 million tons of federally leased strippable resources. Its 396.78 million tons of minable coal underground are 56.38 percent of the 703.8 million tons of underground mine resources.

Information on federal leases in New Mexico, a state with significant coal resources that is near the southern California markets, shows that 6 of the 13 lessees, with about 28 thousand of the approximately 41 thousand acres leased in that state, have no mining plans. These six lessees control about 192 million tons of strippable reserves, 70 percent of such leased reserves, and about 56 million of the 58 million tons of underground minable reserves there.

For Eastern Montana, where a 500kv line is planned to be extended to the Pacific Northwest coast, 17 lessees have no plans to develop 8 of 17 leases in the state containing 873.13 million of the 1,179.85 million tons leased.

---

*Peabody Coal Co., Pittston Co., North American Coal Co., Eastern Associated Coal Corp., Westmoreland Coal Corp., General Dynamics Coal Corp., Utah International, and Rochester & Pittsburgh Coal Co.

†American Electric Power Co., Peter Kiewit Sons Co. (joint entrepreneur with utilities), and Western Energy Co.

‡U.S. Steel Co., Bethlehem Mines Corp.

**Continental Oil, Occidental Petroleum, Hunt Enterprises and Ashland Oil, SOHIO, Gulf Oil Corp., and Houston Natural Gas.

TABLE 4.27

U.S. Production Concentration in Bituminous Coal, 1973

Rank	Firm or Operation Group	Bituminous Coal and Lignite Production (thousands of tons)	Percent of Total
1	Peabody Coal Co. (Kennecott Copper)[a]	70,172	11.9
2	Consolidation Coal Co. (Continental Oil)	60,477	10.2
3	Island Creek Coal (Occidental Petroleum)	22,879	3.9
4	Pittson Co.	18,796	3.2
5	Amax Coal Co. (American Metal Climax)[a]	16,404	2.8
6	U.S. Steel	16,222	2.7
7	Bethlehem Mines Corp.	14,129	2.4
8	Arch Mineral Corp. (Hunt Enterprises and Ashland Oil)	12,539	2.1
9	North American Coal Corp.	12,501	2.1
10	Old Ben Coal Corp. (SOHIO)	10,847	1.8
11	Eastern Associated Coal Corp.	10,640	1.8
12	Westmoreland Coal Corp.	8,809	1.5
13	General Dynamics Coal Co.	8,670	1.5
14	Pittsburgh & Midway Coal Mining Co. (Gulf Oil Corp.)	8,064	1.2
15	Utah International, Inc.	7,389	1.2
16	American Electric Power	6,563	1.1
17	Peter Kiewit Sons Co.	6,113	1.0
18	Rochester & Pittsburgh Coal Company	4,666	.8
19	Western Energy Co. (Montana Power Co.)	4,600	.8
20	Ziegler Coal Co. (Houston Natural Gas)	4,272	.7
Total		324,752[b]	
Total, U.S.		591,000	

[a]Assuming the production of the Gibraltar Coal Co., a joint venture between Peabody Coal Co., and the American Metal Climax, can be allocated on a 50 percent basis to each company.

[b]May not add to total due to rounding.

Note: Four largest comprise 29.2 percent of U.S. total, the eight largest 39.2 percent, and the 20 largest 54.9 percent.

Sources: "U.S. Coal Production by Company . . . 1973," Keystone Coal Industry Manual (New York: McGraw-Hill, 1974), pp. 9-11; and Coal Facts, 1974-75, National Coal Association, Washington, D.C., p. 52.

TABLE 4.28

Concentration of Federal Coal Leaseholds
Top 5 and 10 Lessees, by State

State Federal Lessee	Federal Lease Acreage in State			Percent of Total Federal
	Number of Acres	Number of Leases	Percent	Coal Lease Acreage
California				
Dixie & Reeves	80	1	100	0
Colorado				
Kemmerer Coal Co.	16,269	10	13	2
Industrial Resources, Inc.	14,929	6	12	2
Peabody Coal Co.	10,306	8	8	1
Consolidation Coal Co.[a]	10,015	7	8	1
U.S. Steel Corp.	9,471	15	8	1
Top 5 total	60,990	46	49	7.8
Utah International, Inc.	8,071	6	7	1
Atlantic Richfield	7,462	3	6	Less than 1
Mid-Cont. Coal & Coke	6,065	8	5	Less than 1
Garland Coal & Mining Co.	5,433	3	4	Less than 1
United Electric Coal Co.	4,842	2	4	1
Top 10 total	92,864	68	76	11.9
State total	121,470	113	100	15.6
Montana				
Decker Coal Co.	13,610	3	38	2
Western Energy Co.[b]	7,073	2	20	Less than 1
U.S. Steel Corp.	5,096	2	14	Less than 1
Peabody Coal Co.	4,307	1	12	Less than 1
Pacific Power & Light Co.	3,067	2	8	Less than 1
Top 5 total	33,153	10	92	4.2
State total	36,232	17	100	4.6
New Mexico				
Western Coal Co.	12,289	6	30	2
Consolidation Coal Co.	9,303	5	23	1
Gulf Oil Corp.	8,156	4	20	1
Seneca Oil Co.	6,336	1	15	Less than 1

(continued)

TABLE 4.28 (continued)

State Federal Lessee	Federal Lease Acreage in State			Percent of Total Federal Coal Lease Acreage
	Number of Acres	Number of Leases	Percent	
Peabody Coal Co.	2,044	1	5	Less than 1
Top 5 total	32,128	17	95	4.9
State total	40,958	28	100	5.2
Utah				
Peabody Coal Co.	43,160	31	16	6
Resources Co., et al.[c]	39,355	20	15	5
El Paso Natural Gas	27,019	15	10	3
Consolidation Coal Co.	25,533	11	10	3
Consol. Coal & Kemmerer Coal Co.	18,746	10	7	2
Top 5 total	153,813	87	58	20
Utah International, Inc.	16,157	20	6	2
Kaiser Steel Corp.	14,617	9	5	2
Nevada Electric	10,377	8	4	1
North American Coal	8,905	8	3	1
Jesse H. Knight	7,850	4	3	1
Top 10 total	211,719	136	79	27
State total	268,555	197	100	34
Washington				
Wash. Irrigation & Dev. Co.	521	2	100	Less than 1
North Dakota				
Knife River Coal Co.	7,792	6	48	Less than 1
North American Coal Co.	2,843	3	17	Less than 1
Kaukol-Noonan, Inc.	2,486	2	15	Less than 1
Kerr-McGee Corp.	2,034	1	12	Less than 1
Consolidation Coal Co.	601	3	4	Less than 1
Top 5 total	15,756	15	97	2.0
State total	16,235	18	100	2.1
Oklahoma				
Galand Coal & Mining Co.	37,115	21	43	5
Evans Coal Co.	12,622	8	15	2

Lone Star Steel Co.	10,172	6	12	1
Petroleum Int'l, Inc.	9,110	3	11	1
Cameron Coal Co.	4,464	3	5	Less than 1
Top 5 total	73,438	41	84	9.4
State total	87,014	53	100	11.1
Oregon				
Pacific Power & Light Co.	4,866	2	90	Less than 1
Mandrones et al.	538	1	10	Less than 1
State total	5,403	3	100	0.7
Wyoming				
Pacific Power & Light Co.	27,146	15	14	3
Peabody Coal Co.	23,761	7	12	3
Richard D. Bass	20,701	1	10	3
Carter Oil Co.[d]	15,491	3	8	2
Sun Oil Co.	14,680	1	7	2
Top 5 total	101,779	27	51	13
Atlantic Richfield Co.	11,724	3	6	2
Ark Land Co.[e]	11,656	8	6	1
Kerr-McGee	11,255	6	6	1
Reynolds Mining Corp.[f]	9,418	5	5	1
Energy Development Co.[g]	8,683	1	4	1
Top 10 total	154,515	50	77	20
State total	199,944	91	100	26

[a]Continental Oil Co.
[b]The Montana Power Company.
[c]Resources Co. owned by Arizona Public Services Co.
[d]Exxon.
[e]Partially owned by Ashland Oil.
[f]Affiliate of R. J. Reynolds (Aminoil) which recently purchased Signal Oil & Gas Properties from Burmah.
[g]Arizona Public Services Co.

Source: U.S. Department of the Interior, Bureau of Land Management, Coal: An Analysis of Existing Federal Coal Leases (March, 1976).

TABLE 4.29

Eastern Powder River Basin Coal Leasing

Action Status	Number of Leases	Acres
Issued federal coal leases	42	93,075
Preference right coal lease applications	44	96,517
Outstanding coal prospecting permits	28	64,252
Subtotal	114	253,844
Competitive coal lease applications	20	157,861
Total	134	411,705

Source: Department of the Interior, Eastern Powder River Basin Environmental Impact Statement, 1975, p. I-21.

TABLE 4.30

Eastern Powder River Basin Federal Coal Leases

Lessee	Federal Lease Acreage	Percent of Total Acres	Number of Federal Leases	Percent of Leases
Atlantic Richfield	11,684	12.55	3	7.14
The Carter Oil Company (Exxon)	15,490	16.64	3	7.14
Pacific Power & Light	14,440	15.51	8	19.05
Kerr-McGee	8,695	9.34	5	11.90
Peabody	17,281	18.57	6	14.29
Mobil	4,000	4.30	1	2.38
Wyodak Resources Development Co.[a]	1,920	2.06	4	9.52
Meadowlark Farms[b]	5,960	6.40	2	4.76
Humac Corp.	3,359	3.61	3	7.14
Sun Oil	6,560	7.05	1	2.38
Others (4 parties)	3,686	3.96	6	14.29
Total	93,075	99.99[c]	42	99.99[c]

[a]Black Hills Power & Light.

[b]A subsidiary of AMAX, Inc. (Standard Oil of California).

[c]Not equal to 100 percent due to rounding.

Note: Of these 14 lessees, the top four hold 63.27 percent of the leases, and the top ten 96.04 percent.

Source: Department of the Interior, Environmental Impact Statement on Eastern Powder River Basin Coal Mining, 1975 Annual Summary.

Private Domain--Particularly Montana

Data are not available that are comparable to much of the foregoing for private coal development throughout the West. Some noncomparable data follow: Private and state leases in the western states (Arizona, Colorado, Montana, New Mexico, North Dakota, Utah, and Wyoming) add up to almost 1.75 million acres.[127] Some of the largest lease tracts are reported in Table 4.31.

TABLE 4.31

Largest Lease Tracts in Western States

Company	Estimated Reserves (in billions of tons)
Burlington Northern Railroad	11
Union Pacific Railroad	10
Exxon	7
Texaco	5
Pacific Power & Light	1.6
Western Energy	1
Utah International Const. Co.	1.1
Kerr-McGee Corp.	1.5
American Metals Climax	4
Peter Kiewit Const. Co.	N.A.*
Peabody Coal Co.	N.A.*
Continental Oil Co.	N.A.*

*Data not available.

Source: Environmental Policy Center, Facts About Coal in the United States (Washington, D.C.: February 1975).

In Montana, coal companies and a subsidiary of TMPC hold large lease tracts for mining as well as for future coal gasification facilities. Some reported private holdings are set forth in Table 4.32. Others include the Burlington Northern which holds large tracts.

The coal holdings of western railroads such as the Burlington Northern, the Union Pacific, and the Santa Fe seem likely to be employed for utilities that take delivery by rail. They seem unlikely candidates to supply coal to power plants located near mines (mine-mouth plants), the plants that will serve utilities in the mountain states or on the West Coast.[128]

TABLE 4.32

Some Private Coal Acreage in Montana

	Coal Leases	Surface Rights	Exploration and Option	Other	Total
Consolidation Coal (Conoco)	35,050	67,195	71,280	4,960	178,485
Western Energy (TMPC)	--	--	205,178	4,400	209,578
Phillips Petroleum	12,526	--	--	--	12,526
HFC	59,400	--	--	--	50,400
Tenneco	48,116	56,588	--	--	104,704
Sun	29,416	--	--	--	29,416
Chevron	20,131	--	--	--	20,131
Wesco Resources	40,696	45,757	11,555	6,785	104,793
Norsworthy Reger	35,400	24,720	4,000	--	64,120
Sentry Royalty	87,480	--	--	--	87,480
Valley Camp Coal	--	--	5,060	--	5,060
Peabody Coal	2,080	--	--	5,280	7,360
Westmoreland	640	--	9,680	--	10,320
Amax Coal Company	3,840	--	--	--	3,840

Source: U.S. Congress, Senate, Interior Committee, Action for Eastern Montana, in Hearings on Federal Coal Leasing Amendments Act of 1975 (1975), pp. 167-76.

Coal Holdings of Western Utilities

As previously noted, western utilities such as PP&L and TMPC produce and sell coal from large reserves. Table 4.33 details the situation of one of the largest coal holding utilities, PP&L.

Montana Power Company's subsidiary, Western Energy, has coal leases for 610 million tons of recoverable reserves at Colstrip; 490 million tons are committed under contract. Pending federal lessee applications at Colstrip cover approximately 180 million tons.

Western Energy also has coal leases in eastern Montana, containing an estimated 250 million tons, while a subsidiary of Western Energy at Colstrip totaled 3,212,000 tons of which 505,000 were sold to Montana Power Company. By 1977 production is planned to reach 13 million tons per year.[129]

TABLE 4.33

PP&L Coal Holdings

Recoverable Coal Reserves	Assigned or Dedicated (millions of tons)	Unassigned or Undedicated (millions of tons)	Percentage Sulfur Content by Weight (average percent)
Washington			
Centralia field near Centralia	Centralia Plant[a,b] 70[c]		0.7
Wyoming			
Jim Bridger coal field near Rock Springs	Jim Bridger Plant[a,b] 133[c]	33[c]	0.6
Dave Johnston near Glenrock	Dave Johnston Plant[a,b] 117		0.5
Antelope northeast of Glenrock		300	0.4
North Antelope[d] northeast of Glenrock		145	0.7
Cherokee west of Rawlins		250	1.8
Montana			
Decker near Decker[f]	Decker Coal Company[f] 165[c]	210[c,e]	0.4
West Decker near Decker		400	0.5
Total	485	1,338	

Notes to Table 4.33

[a]See "Property and Power Supply."

[b]The company considers that the respective reserves assigned to the named plants are sufficient to provide fuel to these plants for their economically useful lives.

[c]Excludes reserves controlled by other participant in project.

[d]Nine noncontiguous reserve areas.

[e]Controlled by Decker Coal Company, but not subject to contract for sale.

[f]Decker Coal Company is a joint venture, one-half owned by PP&L's subsidiary Western Minerals Inc.

Note: Recoverable coal reserves represent the portion of total reserve estimates which, in the opinion of the company, is substantiated by adequate information, including that derived from exploration, mining operations (in some cases), outcrop data, quality testing, and knowledge of mining conditions. Reserve estimates are subject to adjustment as a result of continuing engineering evaluation, additional exploratory and development information and as a result of changes in economic factors affecting the marketability or utilization by the company of such reserves.

Source: Pacific Power & Light Company, prospectus of September 4, 1975, for sale of common stock.

Much of the utility involvement in coal production is in joint enterprises with other parties. Two such ventures are instructive. Western Coal Company, a joint venture between Public Service Company of New Mexico and Tucson Gas and Electric Company, owns extensive resources, and supplies coal to the San Juan Plant under a mining agreement with Utah International.

Another such joint venture is among, on the one hand, fuel subsidiaries of SCE (Mono Power Company), SDG&E (New Albion Resources Company), and APS (Resources Company);* and, on the other, Kaiser industries. This was to supply the proposed Kaiparowits Project.

---

*APS has several coal subsidiaries: Bixco, Energy Development Co., and Resources Company.

Utility acquisition of coal reserves is spreading. PG&E has acquired coal reserves in Utah from Island Creek Coal Company, a subsidiary of Occidental Petroleum Company. A group of public power and REA cooperatives have formed Western Fuels Inc. in order to acquire fuel resources for planned units. Similarly, Nevada Power Company (NPC) is planning two coal-fueled plants, with LADWP, and St. George's, Utah, respectively; both plants are to use coal produced by Utah International and an NPC subsidiary. This coal is to be transported by a slurry pipeline.

Power generation and the conversion of coal to synthetic fuel require sizable water flows. The following water requirements have been hypothesized.[130]

Facility	Acre-Feet of Water (per year)
Gasification plant (2 million cubic feet per day)	7,000
Power plants (water cooled)	11 per mw
Slurry pipeline (25 million tons of coal per year)	15,000
Per 1,000 population increase	200

In the West, rights to water are acquired by appropriation or by contract from the Bureau of Reclamation. One obtains an appropriation by filing with state officials, and then using the water claimed. For the rich coal beds in northeastern Wyoming and southeastern Montana, water rights have been secured largely by the oil industry and a few large utilities, as shown in Table 4.34.

Utility Purchasers of Coal

Some purchasers of coal consider the market as a sellers' market in which supplies are tight. Publicly owned systems and REA cooperatives report particular concern about coal supplies. Mr. Kenneth Holum, general manager of Western Fuels Association, Inc., a joint enterprise of municipal and cooperative power systems to secure coal, testified that "We did find it exceedingly difficult to secure coal that we needed from western sources."[131]

In the future large increases in coal-fueled capacity are foreseen, requiring substantial increases in western coal production. These increases may include the first coal-fired capacity (750 mw) in the area that includes PG&E and will include 48.9 percent of new capacity for southern California between 1975 and 1984. See Table 4.35.

TABLE 4.34

Industrial Water Appropriations, Requests, and Options in the Yellowstone River Basin (in acre-feet per year)

River/Company	Appropriations Filed	Bureau of Reclamation Options	Bureau of Reclamation Requests	River Total
Powder				
Utah International	80,375	--	--	
Reynolds (Lake DeSmet)	36,000	--	--	
Unknown (Moorhead Dam)	--	--	220,000	336,375
Tongue				
Montana Power	4,175	--	--	
Norsworthy & Reger	223,000	--	--	227,175
Big Horn				
Exxon	--	50,000	--	
Peabody Coal	--	80,000	--	
Gulf Oil	--	75,000	--	
Shell	--	48,000	--	
Westmoreland Reserve	--	30,000	--	
Kerr-McGee	--	50,000	--	
Reynolds	--	50,000	--	
Colorado Interstate Gas	--	30,000	--	
Ayrshire (AMAX)	--	30,000	90,000	
Panhandle Eastern Pipeline	--	30,000	--	
Norsworthy & Reger	--	50,000	10,000	
Cardinal Petroleum	--	50,000	92,000[a]	
Sun Oil	--	35,000	35,000	
Weld-Jenkins	--	50,000	50,000	
Mobil Oil	--	50,000	--	
Conoco	--	--	530,000	
Montana Power	--	--	50,000	
Atlantic Richfield	--	--	50,000	
Northern Natural Gas	--	--	20,000	
Pacific Power & Light	--	--	30,000	
(Unknown)	--	--	308,000	1,973,000
Yellowstone				
Tenneco (Intake Water Co.)	80,650	--	--	
Montana Power:				
Forsyth	181,000	--	--	
Billings	283,600	--	--	
Basin Electric	36,200	--	--	
Hunt Oil	144,800[b]	--	--	
Getty Oil	92,000	--	--	824,250
Grand total for Yellowstone drainage	1,167,800	708,000	1,485,000	3,360,800

[a]Intermountain Reservoir.

[b]Approximately 6,000 for irrigation.

Source: Compiled from the Bureau of Reclamation and County Courthouse files by Northern Great Plains Resource Program: U.S. Congress, Senate, Interior Committee, Hearings on Federal Coal Leasing Program (1974).

TABLE 4.35

Major Planned Coal-Fired Projects of Western Utilities

Coal-Fired Project	Size (mw)	Co-owners and Their Shares
Southwest		
San Juan No. 1	326	Public Service of New Mexico, 50 percent; TG&E, 50 percent
San Juan No. 3	466	Public Service of New Mexico, 50 percent; TG&E, 50 percent
San Juan No. 4	466	Same as San Juan No. 3
H. Allen 1	500	Nevada Power and LADWP
H. Allen 2	500	Nevada Power and LADWP
Warner Valley 1 and 2		Nevada Power, LADWP, and St. Georges, Utah
Navajo No. 2	750	SRP, 21.7 percent; LADWP, 21.2 percent; APS, 14 percent; NPC, 11.3 percent; TG&E, 7.5 percent; USBR, 24.3 percent
Navajo No. 3	750	Same as Navajo No. 2
San Bernadino	760	SCE
	760	SCE
Mountain Colstrip 2 and 2	360 ea.	The Montana Power Co.; Puget Sound P&L
Colstrip 3 and 4	770 ea.	TMPC; Puget Sound P&L; PGE; Washington Water Power; PP&L
Cray Station 1 and 2	380 ea.	Colo-Ute Elec. Assoc.; SRP; Tri-State Generation and Transmission Assoc.; Platte River Power Authority
Huntington Canyon 1 and 2	430 ea.	UP&L
Bridger 1 and 4	500 ea.	PP&L; Idaho Power
Wyodak 1	330	PP&L; Black Hills Power and Light Co.

Source: Western Systems Coordinating Council.

Utility Coal Fuel Contracts

Present and planned large western units are fueled with utility-owned coal or coal under long-term contracts from dedicated large mining facilities. Coal supply contracts for these large generating units run for the life of the unit (generally 30 years) and provide for price adjustments to reflect the cost of labor, material, supplies, and taxes. Table 4.36 sets out some of the major long-term coal supply contracts now extant in the West.

The use of large-quantity long-term contracts in the West results in a utility's being served by only a few coal suppliers. * The

---

*The Public Service Company of Colorado reports that while it receives coal under eight contracts it has an option to purchase 100 million tons from one firm (AMAX). This compares with estimated future coal requirements of 165 million tons for all of its plants in being or under construction.

UP&L burned about 3.9 million tons of coal in 1975 (1.0 million tons at its Carbon, Gadsby, and Hale Plants; 1.8 million at Naughton; and 1.1 million at Huntington first unit). Future needs include Huntington second unit and Emery first and second units, 1.2 million tons each; Naughton fourth and fifth units, 1.4 million each when in full operation. Coal for the Carbon, Gadsby, and Hale Plants comes from UP&L's lands and leases (1974 estimated recoverable assigned reserves, 18 million tons; sulfur content about 1/2 percent by weight). These reserves are estimated to suffice for the lives of these plants.

UP&L contracts and options initially cover about 330 million tons of 1/2 percent sulfur coal. One supplier sells UP&L coal for the three existing Naughton units and has agreed to provide all the fuel for those units up to the 80 million tons estimated as needed for the life of the units; the same supplier has agreed to a similar arrangement for the fourth Naughton unit (50 million tons), and has granted UP&L an option on a similar supply for the fifth Naughton unit (also 50 million tons). The fourth Naughton unit has been postponed three years, and UP&L is negotiating for a delay of deliveries.

Coal for the Huntington Plant is, and for the Emery Plant will be, bought from Peabody, which has agreed to provide all fuel up to the 150 million tons estimated to be needed for the life of units totaling 2,000 mw, for a 35-year period following commencement of commercial operation of the second Huntington unit.

UP&L agreed to pay a penalty if less coal is purchased than would be required to operate the total plant at 55 percent of rated capacity, unless the outages are unscheduled. The utility will also

TABLE 4.36

Utility Coal Supply Contract Arrangements in the West

Coal Company	Contract Duration (years)	Production (millions of tons)	Location Consumed
Arizona			
Peabody	35	5.0	Bullhead City, NV
Peabody	35	8.0	Page, AZ
Colorado			
Utah International		3.0	Craig, Colo.
Montana			
Western Energy[a]	30	0.7	Billings, Mt. (smaller plant)
Western Energy	30	3.0	Colstrip
New Mexico[b]			
Utah International	30	8.5	Fruitland
Western Coal[c]	30	1.0	Fruitland
Utah			
Peabody	30	1.2	Huntington
UP&L	30	1.2	Energy
UP&L	30	1.2	Carlson or Energy
UP&L	30	1.2	Carlson or Energy
Resources Co.	30	3.0	Kane
Washington			
PP&L	35	4.8	Centralia
Wyoming			
PP&L	30	3.5	Glenrock
PP&L	30	31.0	Pt. of Rocks
PP&L	30	2.5	Pt. of Rocks
PP&L	30	1.9	Pt. of Rocks
PP&L	30	0.4	Pt. of Rocks
Kemmerer Coal		2.5	Kemmerer
Wyodak	30	1.7	Gillette

[a]Owned by TMPC.

[b]The El Paso Coal subsidiary's contract to supply a gasification plant in New Mexico is for 8.8 million tons per year.

[c]Owned by Tucson Gas & Electric Corporation and Public Services of New Mexico.

Note: Additionally, AMAX has a contract to supply coal for 20 years to an Oregon plant of Portland General Electric.

Source: Security Exchange Commission.

prevalence of long-term contracts is coupled with the use of a number of clauses that shift production risks from mining companies and their investors and lenders to coal-purchasing utilities. These clauses assure that debts connected with a coal project's financing will be repaid, and assure coal firms that their costs will be recovered. Financial assistance given to a limited number of firms having long-term sales tends to place later entrants into mining at a capital cost disadvantage.

In the period between 1960 and 1974, "This rather dramatic shift from small to large mines, and the development of numerous new, large strip mines was financed with a combination of internal cash flows, the sale of both short and long-term production payments financed by banks, and some private placements of long-term debt."[132]

> Production payment financing, arranged by commercial banks, is rapidly becoming the favored way of structuring new mine development financing. . . . In general, . . . a typical arrangement provides for the advance sale of a stated amount of future income in the form of a production payment either from existing mines, new mines or a combination, to an arms length third party who pays for the production payment with a bank loan, usually taken down in installments as the mine owner needs the funds for development. Repayment, plus interest, is from stated percentages of future mine income, usually for a total term (including the take down period) of 8 to 10 years. The lender is secured by a mortgage to the production payment and an assignment of the proceeds of production, backed up by strong commitments of the mining company to develop and continuously operate the mines, usually pursuant to the terms of a coal sales contract for a significant part of the output. . . . Thus the financing is based upon the value of the mines

---

pay Peabody's unrecovered fixed charges if, in the first 15 years of the contract, purchases are less than required for generation at 25 percent of rated capacity in any consecutive 12-month period (again, except for unscheduled outages). The company has an option to purchase an additional 80 million tons of coal for the Huntington and Emery Plants. This amount is estimated to be sufficient for two 400mw units for their entire expected lives.

> out of which it has been carved, and future mining income retires the debt. Aside from the financial burden of the specific obligations to fully develop and continuously operate the mines . . ., the mine owner usually is not directly responsible for the bank debt incurred. . . .[133]

Small producer capital needs, for equipment and working capital and only rarely for mine development, can best be provided by local or regional banks, lender credit, and occasional private sale of equity. ". . . Many of the smaller producers would not have the financial resources needed to support the customary mine completion and operating covenants of typical production payment arrangements. When small producer requirements exceed the local banks' resources, the services of a larger correspondent bank usually can be arranged."[134]

Difficulty in obtaining production loan financing results in an indirect competitive disadvantage on top of direct financing impact. Production loans are viewed as comparable to direct debt in assessing a mining company's credit. Their effects on book earnings are nominal, and general credit and other assets are unimpaired. Accordingly firms with such financing may add materially to their debt capacity. These large mines financed by production were often owned by companies not traditionally in the coal industry.

Coal companies are reported to require take-or-pay or even hell-or-high-water clauses* and to demand terms consonant with investment payout, as well as for nonremote sites and some price speculation. Under long-term contracts, the utility sometimes guarantees mining company notes and purchases coal on a cost-plus basis, under agreements that may include most-favored-nation clauses.[135] These shifts of risk to utilities enable producers to use a very high proportion of bank debt in mine financing; the agreements also assure the producer of stable demand markets. The coal contracts usually, if not invariably, have price escalation provisions that are quite extensive.

Table 4.37 provides a nationwide sample of long-term contracts and reveals many techniques of automatic price adjustments. Most typical was sole reliance on national price and wage trends.[136]

---

*Under a "take-or-pay" clause, purchaser is liable to pay for tendered contract quantities; under a hell-or-high-water clause, purchaser is liable even if seller does not tender contract quantities.

TABLE 4.37

Techniques of Escalation of Electric Utility Long-Term Contracts

	Coal Burn Tonnages of Reporting Companies	
	As Percent of Burn by Reporting Companies	As Percent of Industry Burn
Predominately by national trends	39.6	27.3
Predominately by mine costs	20.3	14.0
Predominately by costs and national trends	18.6	12.8
Predominately by national trends and mine productivity	17.6	12.1
Use combination of national trends, mine costs, and mine productivity	3.8	2.6
Total reporting	100.0	68.3

Source: Nuclear Regulatory Commission, Facilities License Application Record, March 31, 1976.

The larger the coal requirements of an electric utility, the more likely this utility is to prefer contract purchases over spot purchases under normal circumstances for reasons of stability of supply and predictability of price. Nineteen of the 22 utilities interviewed prefer contract to spot purchases.*

Only two respondents mentioned lower prices as the reason for preferring contract to spot purchases. Four of these 19 companies would like their purchases to be 100% contract. Fifteen companies prefer to have a large percentage of their purchases to be contract and a small percentage to be spot purchases. The

*The 22 utilities reported on consume 48 percent of domestic utility coal consumption.

> respective percentage of each type of purchase varies among these companies anywhere between the combinations of 75 percent contract to 25 percent spot and 95 percent contract to 5 percent spot. These figures are not derived from any specific rule.[137]

The preference for long-term contracts is shared by coal producers, the overwhelming majority of which prefer to have a large percentage of their tonnage (70-90 percent) sold by long-term contract, with the small remainder sold on the spot market. Only three producers and two coal sales companies prefer the spot market.

Large producers need long-term contracts to obtain bank financing. Most coal sales companies also prefer long-term contracts for the demand security, while they prefer to keep a small percentage of their production on the spot market in order to enjoy high prices during good market conditions.

Coal ownership and production from large mines have come increasingly to be controlled by enterprises not principally in the coal business. During 1974 the 50 largest coal mines collectively produced nearly one-fourth of the industry's total output; only five of these mines are owned by independent coal producers. The other 45 large mines are owned by subsidiaries of electric utilities, conglomerates, and companies whose principal businesses are oil, steel, and other types of mining. The ownership of vast coal resources has come under the control of large oil companies.[138] It has recently been reported that major oil companies control about one-third of leased coal.[139] This growth in the holdings of major companies stems in part from federal leasing practice and in part from numerous acquisitions of coal enterprises.[140]

What the Coal Industry Can Indicate about Geothermal Energy

The experience of the coal industry is instructive regarding geothermal development possibilities in a number of ways:

Utilities show substantial concern with security in their fuel procurement. This produces a tendency to buy from well-financed, established enterprises.

A favored procurement mode, large-quantity long-term contracts, leads to favoring the large sellers who have better access to production payment financing.

The major buyers, a limited group, have a heavy planning commitment to coal procurement.
Procurement plans appear to evince an expectation of higher fuel prices in the future.
Only a limited number of utility and raw fuel enterprises have coal mining plans.
The matching of major generating stations with long-term coal contracts removes a large sector of the utility generation growth plan from access to geothermal energy.
The ownership of prime coal resources is in substantial part in the hands of major oil companies and their development of production has not been rapid.

The channeling of commerce into large fuel procurement arrangements with a limited number of suppliers is not conducive to geothermal development. A novel venture, with innovation risks and substantial capital requirements, is apt to need substantial financial backing. If external capital markets tend to be oriented toward larger, more conventional projects, the anomalous ventures may tend to become, perforce, joint arrangements with those companies which have capital and skills but which also have their major interests in other energy sources.

The present study indicates that the bulk of coal available to the West is in the hands of large oil companies, a few utilities, and railroads. This may tend to prevent lowest-cost development of coal resources. If the coal market were sufficiently large and diverse, holdings of companies from less competitive fuel sectors might be expected to have little effect on the coal market's competitiveness. However, with relatively few entities in major coal markets and submarkets, oil-coal cross-ownership effects, and noncompetitive relationships among utilities may be expected to reduce competition in supplying and in purchasing utility fuel. To the extent that coal prices are high, geothermal energy has a greater market opportunity, but this is the case only if geothermal prices are not similarly higher than necessary, and utility fuel purchasers are cost-conscious and not otherwise constrained in their fuel choices. It is obviously not in the national interest to have high coal prices for the purpose of stimulating geothermal development.

NUCLEAR ENERGY

Nuclear units are run at relatively steady rates for base-load service, where their low fuel costs and high capital charges are best employed. Perhaps, unlike coal-fired units, nuclear plants can be erected in California. These units have substantial economies of scale and are built in very large sizes* requiring substantial transmission services. Their large size and cost have led to a number of joint endeavors among utilities and to proceedings concerning smaller system access to ownership or unit output shares.

Portions of the industry providing fuel and equipment are highly concentrated, and entry is costly. Firms engaged in coal and petroleum hold major positions in essential aspects of nuclear fuel supply. High capacity costs, and, relatedly, schedule delays, as well as rising fuel cost and some fuel supply uncertainty, have dampened nuclear generation growth.

Power Plants in Operation or Planned

West Coast nuclear generation began in 1967 with a small unit in Humboldt Bay.[141] There are now five operating units with a total capacity of 2,770 mw, six construction permits in effect, and application pending for an additional nine units in the western states. See Table 4.38 below.

Most West Coast nuclear units will be jointly owned.† See Table 4.39.

Generating Unit Costs

Costs for nuclear generating units are rising rapidly and by the mid-1980s are expected to reach $1,200 to $1,400 per kw.‡ Major factors driving up costs are construction delays, fuel prices, and forced outages.

Schedule slippages of both fossil and nuclear units are summarized in Table 4.40. In addition to those listed, many utilities apparently have made other, unannounced cutbacks. In the first half of 1976, only one nuclear generating unit was ordered and no plans were announced for additional units.[142] Delays are especially costly on the more expensive nuclear units, since both interest during construction and escalation rates apply against larger costs per unit of capacity.

---

*About 1,100 mw for nuclear units and about 800 mw for large coal units.

†Public Service of Colorado alone owns the relatively small (330mw) Ft. St. Vrain unit--the only high temperature gas-cooled reactor in the West.

‡30.9 to 36.0 mills at 18 percent and 7,000 hours.

TABLE 4.38

Status of Nuclear Projects in the West

Project	Owner/Applicant	Maximum Dependable Capacity (mwe)	Type
Licenses in Effect			
Humboldt Bay	PG&E	63	BWR[a]
Trojan	Portland General Electric	570	PWR[b]
Fort St. Vrain	Public Service Company of Colorado	330	HTGR[c]
Rancho Seco	SMUD	817	PWR
San Onofre 1	SCE	430	PWR
Construction Permits in Effect			
Diablo Canyon 1	PG&E	1,084	PWR
Diablo Canyon 2	PG&E	1,106	PWR
San Onofre 2	SCE	1,100	PWR
San Onofre 3	SCE	1,100	PWR
WPPSS 2	WPPSS	1,103	BWR
WPPSS 1	WPPSS	1,218	PWR
Palo Verde 1, 2, and 3	APS	1,238 each	PWR
Pebble Springs 1 and 2	Portland General Electric	1,260 each	PWR
Skagit 1 and 2	Puget Sound P&L	1,277 each	BWR
WPPSS 3 and 5	WPPSS	1,242 each	PWR
WPPSS 4	WPPSS	1,218	PWR

[a]Boiling Water Reactor.

[b]Pressurized Water Reactor.

[c]High Temperature Gas-cooled Reactor.

Note: At Richland, Washington, ERDA has an 850mw unit selling steam to WPPSS for power generation. Nuclear powered generation is considered in the utility industry to have significant economies of scale.

Source: Nuclear Regulatory Commission, Facilities License Application Record, March 31, 1976.

TABLE 4.39

Ownership of Planned Nuclear Projects

Project	Owners	Percent
Trojan	Portland General Electric	67.5
	PP&L	2.5
	Eugene Water and Electric Board	30.4
Skagit	Puget Sound Power & Light	100.0
WPPSS 1 and 2	WPPSS	100.0
WPPSS 3	WPPSS, PP&L, Puget Sound Power & Light, PGE, Washington Water & Power	70.0
WPPSS 4	WPPSS and perhaps others	
San Onofre 2 and 3	SCE	80.0
	SDG&E	20.0
San Joaquin	SDG&E, LADWP, Others	
Palo Verde 1, 2, and 3	Salt River Project	28.1
	Arizona PS	28.1
	TG&E	15.4
	Public Services of New Mexico	10.2
	El Paso Electric	15.8
	AEPC	2.4
Rancho Seco 1	SMUD contract with PG&E	100.0
Rancho Seco 2	SMUD and others (indefinitely deferred)	
Diablo Canyon 1 and 2	PG&E	100.0

Source: Compiled from annual reports and interviews.

Nuclear units have been subject to a high rate of forced outages. These outages have reduced planned availability of nuclear units, even after shakedown periods, to 70 percent from earlier projections of 80 percent.

TABLE 4.40

Deferrals and Cancellations of Planned Fossil and Nuclear Units

	Number of Units Affected	Capacity (mw)
Deferred 1 year or less	30	25,095
Deferred 2 years	5	6,060
Deferred more than 2 years	1	1,150
Indefinitely deferred	3	3,420
In-service date uncertain	2	2,400
Cancelled	8	8,220
Planning suspended	2	1,540
Total	51	47,885

Note: In some cases, the deferrals reported are not the first in the history of the unit (for example, some are listed as having been deferred for "another year"). Likewise, some cancellations are of units that had already been deferred in previous years. These cutbacks are those announced in the second half of 1975 (with the exception of one 1,100mw nuclear unit whose deferral was announced early in 1976).

Source: Derived from data in the January 19, 1976 issue of Electrical Week.

Nuclear Fuel Prices

Nuclear fuel prices have climbed quickly since 1974. The average price per pound of uranium oxide was $7.90 in 1974. An ERDA survey reports a mid-1975 average contract delivery price of $8.45 which rose to $10.50 by January 1, 1976.[143] Estimates for 1982 price are $19.20 (in 1975 dollars).[144] Prices of around $40 per pound have been indicated for 1978 delivery.[145] The apparent effort at cartelization[146] and the Westinghouse announcement that it would not meet its delivery contracts push fuel prices higher.*

*The Intercontinental Energy Corporation (IEC) recently announced an agreement to supply uranium to PG&E, beginning in 1978. Under the announced agreement PG&E will advance IEC $12 million as a prepayment on future deliveries for use in acquiring leases and commencing production. The first 843,750 pounds of uranium

Price and supply after 1985-90 are uncertain, reflecting in part pending decisions as to the future role of government in the area of fuels enrichment, and uncertainty as to which means of preparing and reprocessing nuclear fuels will be used, the quantities available, and the associated costs. In response, some power companies are entering the uranium ore milling business and a few are acquiring mining properties.

While U.S. uranium fuel supply, through the year 1975, came from domestic sources,[147] commencing in 1978, uranium from other countries is expected to supply a part of domestic requirements. Current high uranium prices, which run considerably ahead of the supply cost of $15 per ton estimated as adequate by ERDA, may rise further when importation begins. These ore prices are not inconsistent with the structure of the supply industry.

In 1973, 33 open pit mines accounted for about 63 percent of domestic uranium ore production. The remaining 37 percent of the ore was produced in 122 underground mines. Domestic mines with a capacity to produce about 9 million tons of ore a year are mining less than 7 million tons annually. About 75 percent of the ore is mined in New Mexico and Wyoming; an additional 15 percent is produced in Colorado and Utah.[148]

Uranium ore is milled to produce a semirefined product. Haulage and other costs often cause mills to be located near the mines, in relatively remote areas. Recently, PG&E and SCE (Mono Power) have acquired positions in uranium milling.[149]

Ownership of uranium reserves appears to be somewhat concentrated. The significance of information about reserve ownership is unclear. Rapidly rising prices may encourage new entry, but it is not yet known how diverse and numerous the entrants will be. Table 4.41 summarizes ownership data.

Recent data from ERDA as to the capacity of uranium milling facilities is provided in Table 4.42.

Petroleum companies have been reported to be doing the bulk of exploratory drilling for uranium.[150]

---

production are to be sold for $40 a pound and further production from the subject leases is to be sold to PG&E for $40 to $50 per pound, the price being related to the market value of uranium at the time of delivery. Production is expected, for the subject lease, to be at a rate of 300,000 pounds of uranium per year.

TABLE 4.41

The Uranium Industry: Control of Uranium Reserves, 1971

Companies	Percent of Low-Cost Uranium Reserves*
Anaconda Co.	
Getty Oil Co.	
Gulf Oil Co.	
Exxon Corp.	
Kerr-McGee Corp.	
United Nuclear Co.	
Utah International, Inc.	
Subtotal for 7 companies	70.0
Atlas Corp.	
Continental Oil Corp.	
Cotter Corp.	
Dawn Mining Co.	
Federal-American Partners	
Homestake Mining Co.	
Rio Algom Corp.	
Susquehanna-Western, Inc.	
Union Carbide Corp.	
Western Nuclear, Inc. (subsidiary of Phelps Dodge Corp.)	
Subtotal for 10 companies	20.0
Total for 17 companies	90.0

*Low-cost reserves are those from which U_3O_8 could be obtained at a price of $8.00 per pound or less.

Note: Companies listed in alphabetical order.

Source: Appendix A of the Testimony of Commissioner Clarence E. Larson, Atomic Energy Commission, in U.S. Congress, House, Select Committee on Small Business, Subcommittee on Special Business Problems, Concentration by Competing Raw Fuel Industries in the Energy Market and Its Impact on Small Business, Hearings, 92d Cong., 1 Sess. (1971), p. 214.

TABLE 4.42

U.S. Uranium Production Plants Operating as of January 1, 1976

Company	Location	Nominal Capacity (tons of ore per day)
Anaconda Company (ARCO)	Grants, New Mexico	3,000
Atlantic Richfield	George West, Texas	--[a]
Atlas Corporation	Moab, Utah	1,000
Conoco-Pioneer	Falls City, Texas	1,750
Cotter Corporation	Canon City, Colorado	450
Dawn Mining Company	Ford, Washington	400
Exxon Company, USA	Powder River Basin, Wyoming	
Federal-American Partners[b]	Gas Hills, Wyoming	950
Kerr-McGee Nuclear Corp.	Grants, New Mexico	7,000
Rio Algom Corporation	LaSal, Utah	700
Union Carbide Corporation	Uravan, Colorado	1,300
Union Carbide Corporation	Gas Hills, Wyoming	1,200
United Nuclear-Homestake Partners[c]	Grants, New Mexico	3,500
Uranium Recovery Corp.	Mulberry, Florida	--[d]
Utah International, Inc.	Gas Hills, Wyoming	1,200
Utah International, Inc.	Shirley Basin, Wyoming	1,800
Western Nuclear, Inc. (Phelps Dodge)	Jeffrey City, Wyoming	1,200
Total		28,000

[a]Uranium obtained by solution mining.

[b]A joint enterprise in which Federal Resources Corp. holds 60 percent and American Nuclear Corp. holds 40 percent.

[c]A joint enterprise in which United Nuclear holds 70 percent and Homestake Mining Co. holds 30 percent.

[d]Uranium recovered from phosphoric acid.

Note: The corresponding concentration ratios for milling capacity are, in percent of total U.S. nominal capacity: 4-Firm--58; 8-Firm--87.7.

Source: ERDA.

Fuel Fabrication and Reactors

Fabrication into fuel rods is done by Exxon Nuclear Company and by four manufacturers of light water reactors. Two firms, McGee Nuclear Corporation and Nuclear Field Services Inc. (now shut-down plant owned by Getty Oil Company), provide chemical conversion services; Kerr-McGee also has a limited capability to make fuel assemblies.

Potential reprocessors are General Electric, which has recently acquired Utah International; Allied Chemical Nuclear Products/Gulf General Atomics; and a joint venture of Atlantic Richfield and Gulf General Atomics. The reactor fabrication industry itself is limited to five firms. As of 1971, only nine firms were selling fuel for light water reactors.[151]

While utilities may be able to secure captive uranium mines and mills, they are still faced with concentrated industries in the reactor manufacturing and reprocessing segments into which entry is very expensive.

THE SECTORS CONSIDERED TOGETHER

> Uranium, coal, oil, gas and geothermal energy all serve the identical function in an electric utility powerplant--namely to produce heat for steam which turns a turbine-generator to produce electricity.
>
> Considered as a whole, it appears to us that there is a high and increasing level of interfuel competition such that it would seem entirely appropriate, for the purpose of assessing the competitive impact of energy mergers, to consider the energy sector as a single relevant market.[152]

Choices are made among fuels principally when generating units are being designed. A unit will use either nuclear energy, fossil fuels, or geothermal energy. Geothermal energy could be coupled with nuclear or fossil fuels, for example, for preheating service, if a suitable site were located for both cycles. This would entail finding geothermal energy at a location where boilerfeed and condenser water could be otherwise provided for. Fossil fuel units may be designed to use a variety of fuels, or they may be located or designed so as to effectively preclude fuel changes.

The decision on the choice of fuels, and the specific suppliers of fuel for a generating unit, is often an infrequent, or even a one-time occurrence; but with many plants under construction at any given time, the number of initial purchase choices presents competitive opportunities for energy sales. This competition occurs among firms selling a fuel, and among fuels.

Fuel costs are a major component of utility production costs. A choice among fuels should be heavily influenced by the fuels' relative costs. Western coal and geothermal energy could be forthcoming to western utilities at costs (including return on investments) below those now paid for oil. However, supply cost curves do not reflect opportunity costs. Without fuel price regulation producing the approximate effect of a competitive market, coal and geothermal energy will come to market at prices lower than oil, only if their suppliers compete with other suppliers of oil and with each other. The probabilities for this turn on whether suppliers of coal or geothermal energy are situated in relation to one another and to suppliers of other fuels in such a way as to be inclined to compete rather than to accommodate one another.*

At present, firms seeking to be competitive entrants in the geothermal energy sector are confronted with the economic power of the integrated oil and utility companies which respectively control many of the better resource sites and the high voltage power grid.

Petroleum

The large California utilities procure most of the resid consumed for electric generation in California. They purchase the bulk of this oil under very large quantity, long-term contracts, principally with five very large oil companies: SOCAL, Union Oil Company of California, Exxon, Phillips, and ARCO.† The ownership of West Coast refineries that produce most of this resid is concentrated in the hands of firms which also own the Caribbean and Singapore refineries that could provide alternative resid sources.[153] The crude oil flowing to West Coast refineries, and, to a limited degree, directly to utility boilers, is quite likely to flow through facilities owned by the large petroleum refiners.

---

*Fuel choices are limited by a number of factors, including air pollution control requirements and the availability and cost characteristics of various fuels. The latter characteristics are affected by their distribution in nature, transport costs, capital requirements for development, and so forth.

†The oil supply arrangements of PG&E, SCE, and SDG&E are more fully described in the Appendix.

West Coast production is largely shipped to market through an interconnected network of non-common-carrier lines owned by these refiners. Oil imported into California is likely to come through mooring facilities owned by these same enterprises, or in the future through the TAP system from northern Alaska.* TAPS will be owned, in part, by large petroleum companies established on the West Coast, and, in part, by SOHIO (controlled by BP).†

Refineries obtain crude oil and distribute products through exchange agreements as well as by direct transport. Under these agreements Company A will deliver to Company B an amount of crude oil at one point in return for delivery of crude oil by Company B (or even C in more complex deals) to it at another point. The exchange system is necessarily dominated in California by the private carrier pipelines which have the facilities through which most such arrangements must be made.

Crude oil supply to the West Coast comes increasingly from the Pacific Outer Continental Shelf, Alaska, and imports from Indonesia and from other locations where most production is by the major petroleum companies. Foreign oil is usually obtained under concession or operating agreements with foreign nations which effectively cause the additional revenues from oil price increases to be shared between the host country and the concessionaire company. The countries producing the low sulfur crude oil demanded on the U.S. West Coast are OPEC members.

National oil companies (for example, Pertamina), a novel feature in international oil operations, may seek to increase their markets. Supply and price ramifications of their actions are unclear. However, they do not seem likely to supply large amounts of oil to the U.S. West Coast, where future import levels are very likely to decline with the advent of production from the Alaskan North Slope and increases in offshore California production.

Its holdings at various stages of production augment the market power a large refiner has at any one stage of production. Regional independent refiners are sometimes dependent upon these large companies for crude oil supply and transportation. Independent producers

---

*Some southern Alaskan production now reaches California through the Kenai pipeline jointly owned by ARCO and Union, and the Cook Inlet pipeline.

†SOHIO lacks West Coast refineries and requires crude oil for its Ohio refineries. Accordingly it is expected to seek this feedstock through exchange agreements and a proposed pipeline from the West Coast east to established lines in Texas; that is, it will not enter California refining.

in California are quite likely to have to sell their output into one of these refiners' private carrier lines. The dominance of these refiner-pipeliners is shown by the fact that only three companies post the field prices for crude oil in California.

The role of the independent refiner on the West Coast appears to be on a decline as new refining capacity is planned to be installed largely by the major refiners.[154] The role of the independent producer has similarly declined as a portion of total production. *

The new field production and pipeline facilities are frequently developed by joint enterprises among the large petroleum refiners. Some joint enterprises, such as Aramco, Stanvac, and Caltex go back many years. Newer California offshore production ventures such as THUMS and HAS and the unitization of Prudhoe Bay production in Northern Alaska continue the history of joint action among major oil companies.

The pervasiveness of joint enterprises in petroleum production is extensively set out in a Senate Interior Committee report.[155] According to this report, 69 percent of wells in which a major company held an interest in 1973 were jointly owned. On average, major companies obtained 45 percent of the product of the wells in which they owned interests.[156]

The combination of ubiquitous joint ventures in which major companies coordinate exploration field work, production, and pipelining (which lead to extensive sharing of marketing information, including plans)--together with extensive use of exchanges--all make it necessary for larger petroleum companies to extensively coordinate shipping and supply plans. The extensive joinder of company supply and marketing arrangements affects the performance of the petroleum industry. It also has implications for that industry's relationship to and effects upon the geothermal energy industry. The major interstate crude pipelines and proposed deepwater ports are invariably joint enterprises among major firms.

Onshore petroleum or geothermal exploration is frequently undertaken by joint enterprises among large and small petroleum firms. Small firms traditionally look to the large ones for subleased tracts to explore and for financing. In return, the small firms give the large firms a substantial share of any resulting production, normally coupled with an expectation of a purchasing option on the balance of production.[157]

---

*As older California fields have become depleted, an increasing amount of production is coming from offshore, Alaskan, and foreign fields held by the major West Coast refiners.

A financial practice of the petroleum industry, being replicated in geothermal work, is for small firms to locate a resource and then, if they have not previously done so, enter into a joint venture or royalty arrangement with a large firm that has the capital needed for field development. Development can go forward only at the rate agreed upon by the firm providing financing; small firms are consigned to the acreage passed over or cast off by larger ones. Joint venturing and related practices such as drilling contributions tend to create a convoy system in which all firms steam together at one rate.

Large petroleum companies are acquiring extensive and choice land positions through higher bonus bids and by financing drilling by smaller firms. Data from the Census Bureau[158] indicates that in 1974 the top 24 provided 78.2 percent of total expenditures for undeveloped acreages, 76 percent of the predrilling prospect evaluations, but only 41.3 percent of the exploratory drilling outlays. The eight largest firms* made 52 percent of the expenditures for acquiring nonproducing oil and gas leases.

Geothermal field development costs are high, and cash flow lags are substantial, so that obtaining a carrying partner is especially important to small firms; utility carrying partners seek lower risks of development and so require access to prime sites.

If a small firm can successfully acquire such a site,† it must still compete for any production loan financing that may be available for a newer technology. Bankers are more likely to permit such debt financing of a newer technology if they are familiar with the borrower--and then in all likelihood only for development, not exploration.‡ Large banks, capable of larger loans, are more likely to be familiar with the large oil companies. The larger oil companies dominate the production sector while smaller firms have sought quicker pay-outs and tax shelters. Larger firms shelter income with tax credits from foreign trade which they dominate. As and to the extent that this system is replicated in geothermal energy, development will be a sideline for larger oil companies.

---

*And firms they have recently acquired a controlling interest in--Anaconda and, especially, AMAX.

†The Bountiful Power & Light System (Utah) bid unsuccessfully against larger oil companies for geothermal leases--desiring to build a plant. It feels that noncompetitive lease sites entail too great a risk; it is interested in geothermal energy to avoid high oil prices but has sought terms of purchase from Phillips Petroleum.

‡Loan guarantees are not likely to affect the borrower's situation since loans are made on the likelihood of pay-back, not the prospect of realizations on foreclosure.

Coal

Large western coal mines with huge associated reserve fields are expected to be a principal source of energy for western power generation in the future. Much western coal is owned by the federal government. Large amounts of the federally owned lands have been leased to private entities. Coal holdings are dominated by large petroleum companies, western utilities, and railroads, together with holdings of several large independent coal companies. Coal holdings, according to federal lease data, appear to be moderately concentrated, when the western United States as a whole is taken as the relevant market. Concentration ratios are higher for statewide or regional areas.

Major oil companies have leased some of the very finest tracts, particularly those of the Eastern Powder River Basin in Wyoming, and have acquired extensive water rights from USBR. Recently, several nonoil companies with large coal holdings have been acquired by large West Coast oil refiners.* The FTC recently approved the sale of the largest independent coal company to a group of oil and gas firms.† As a result of the extensive leasing and acquisition activities of oil companies and the coal companies they control, coal firms owned by oil and gas companies now constitute seven of the fifteen largest coal companies in the United States. Petroleum companies, as a group, accounted for about 23 percent of 1973 U.S. coal production from 7.1 percent of identified reserves. However, major oil companies such as Exxon, Texaco, Shell, Mobil, and ARCO had little or no production from very large reserves.

In the western states, coal development to date has been largely undertaken either by utility-affiliated fuel companies or by independent coal companies. Low coal-production-to-reserve ratios of large oil companies are a prominent feature of the situation, and are a matter of concern to those seeking competitive fuel markets. See Table 4.43.

Because of the substantial involvement of utility companies and railroads in western coal leasing, and the large unleased federal tracts remaining, petroleum companies do not completely dominate coal markets at this time. However, their long-range position is an important one. Utility coal holdings may be available for sale to

---

*ARCO acquired Anaconda and SOCAL acquired a controlling interest in AMAX. Additionally, a major manufacturer, General Electric, has acquired another large coal company, Utah International.

†Kennecott has sold Peabody to a group headed by Newmont Mining.

other western utilities only in conjunction with joint development projects involving the coal-holder. Railroad holdings tend to be situated where they are most likely to be shipped to middle-western and south-central plants. Entry into the large-scale mining operations required to serve utility fuel contracts requires large capital outlays and three to five years of development time. With these aspects of the market in mind, it can be said that petroleum company holdings are a significant market factor.

TABLE 4.43

Large Oil Company Production and Reserves, 1973

	Reserves (millions of tons)	Production (millions of tons)	Reserves/ Production Ratio
Continental	12,058	60.6	198.98
El Paso	4,954	0	∞
Exxon[a]	7,000	2.7	2592.59
Gulf	979	8.1	120.86
Shell	5,000	0	∞
Texaco	2,340	0	∞
SOCAL (and AMAX)[b]	5,400	16.7	323.35
Kerr-McGee	1,000	--	NA[c]
Sun Oil	2,000	--	NA[c]

[a]Exxon has recently indicated that it owns 12 billion tons of reserve while Mobil has indicated it owns 3 billion tons. Mobil has no mines; Exxon plans to mine 40 million tons per year in 1985. Wall Street Journal, May 20, 1977.

[b]Before joinder of these firms.

[c]Data not available.

Source: U.S. Congress, Senate, Interior Committee, The Structure of the U.S. Petroleum Industry, Serial No. 94-37 (92-127), 94th Cong., 2 Sess. (1976).

Nuclear

The supply of nuclear fuel, like that of western coal, comes from an industry in which oil companies have substantial holdings. Unlike coal, uranium resource control is heavily concentrated and future supplies may be affected by cartel efforts. The holdings of large petroleum companies are quite substantial and large electric utility companies have begun to acquire positions.

The conversion of uranium ore into fuel, and the reprocessing of used fuel, will require very high capital cost facilities likely to be very few in number. Major petroleum companies are undertaking all but one of the major initiatives looking toward future uranium ore enrichment and reprocessing facilities.

Geothermal Energy

While the geothermal industry is still taking shape, it is clear that major petroleum refiners will control, as owners and as operators of joint ventures, many of the limited number of better sites. Other participants, principally small utilities and independent petroleum companies, are not likely to outbid the major oil companies for choice new prospects on any regular basis.

Independent geothermal companies tend to perform service functions, relying on large petroleum companies or utilities for capital. The small utilities, which are most likely to find geothermal-sized units attractive, have problems in obtaining access to bulk power coordination services requisite to development; and they must obtain a share of leases at good sites in order to have an opportunity to seek lower-cost energy at a level of risk they can bear.

While the capital costs of geothermal development are high, they appear to be less than those for many other new supplies of energy such as offshore drilling, or large coal projects. Risk, however, is present, and to alleviate it, control of prime sites becomes a key consideration, as does access to joint venture capital and intelligence. By acquiring geothermal resources, large petroleum companies assure their participation in the drilling programs of others. This is the case because drilling programs are more difficult to finance if adjacent acreages are proved up and produced--in a way, draining the tract sought to be drilled. Large firms with extensive acreage must often be included in drilling efforts to avoid this situation.

Once in a joint exploratory program, a firm obtains knowledge as to the plans and perhaps the capabilities of its partners. It may obtain the right to participate in drilling on tracts obtained by its partners in a broad area of interest, set out in the joint venture agreement. New firms do not have this access to jointly held information.

Thus, the marketing plans and modes of operation in the petroleum industry may be expected to have a substantial effect on the methods of operation, rate of development, and patterns of development of the geothermal energy sector.

Integration among Energy Companies and Fuel Sectors

Integration--common ownership of otherwise competitive supply facilities and successive stages of supply--has been a major feature of the petroleum industry and a major center of public concern for nearly a century. The last decade has seen a new extension of this integration by major petroleum companies--their expansion into energy sources competitive with petroleum.

This additional form of integration has, for obvious reasons, created additional public concern. Petroleum markets appear to have been, and are, far from competitive. Alternate energy sources offer major competitive benefits, increasing energy supply and lowering costs. If dominant petroleum firms own or control such alternative energy sources in a way that can substantially diminish the competitive thrust from the alternate concerns, there will be less energy at higher prices.

The acquisition of energy resources for future extraction is a logical course of conduct to an enterprise thinking of itself as an energy company. Large petroleum companies have money to invest and know the fuels business.

By vertically integrating, petroleum companies act to assure themselves of supply* for facilities at subsequent stages, or outlets for supply. By integrating horizontally, or into new and/or alternative energy sources, the company tends to extend its life in the fuel sector, and to hedge against competitive risks. Possibly it can help maintain prices.

Large integrated enterprises are able internally to mobilize substantial amounts of capital. They are able to achieve any economies of scale available in the securities markets. With supply security they may be better able to plan larger, more economic facilities for transportation and refining, and to support their own technical staffs, providing competition for outside engineering firms.

Integration through acquisition of different fuel resources does not necessarily provide expertise to the acquired sector.[159] The coal operations of integrated petroleum companies are not the recipients of any special mining expertise. Exploratory services are

---

*Even if a firm does not employ the raw fuel it produces, as is frequently the case for refiners, raw fuel production makes it possible to trade with other producers on more even terms.

available from independent engineering firms. The mining technology involved is not similar to that in the petroleum or electric utility industries.

The technology involved in converting coal to an oil or gas may be applied to only a portion of the reserves acquired; is largely obtained from engineering equipment firms; and is sought to be financed through government research grants and loan guarantees.* The research being done to improve processes for making liquid or gaseous fuels from coal is supported by government, not petroleum industry money.[160]

For geothermal energy, technology is available from independent service organizations, is developed with government money, and in the future is to be subsidized with federal loan guarantees. No refining operation expertise is involved; the amount of private research appears small.

Of 18 patents in the Patent Office's subclassification for geothermal power generation,† five were held by oil companies: Texaco two, Chevron Research one, Pure (now Union Oil) one, and Gulf Oil one.‡ One patent is held by Mitsubishi, one by Magma Energy Inc., and two others by small geothermal enterprises.

Enterprises such as Toshiba and General Electric Company have worked to adapt the technology of marine turbines to geothermal service. Several small enterprises have sought to promote the use of small units and binary systems.

"Recent efforts in geothermal energy recovery have been mainly directed to systems for increasing the efficiency of recovery. In the recovery of heat energy from hot brine within the earth, systems have been developed which reduce the corrosive action of minerals in the brine on energy conversion equipment. . . ."[161] There is some question whether more research should be devoted to exploratory methods.[162]

Financial Support for Energy Sector Development

Integration may provide only limited benefits in the way of lower capital costs due to access to the credit of large enterprises and may lead to economically wasteful allocations of capital.

---

*The question of whether large petroleum companies have an incentive to rapidly develop new forms of energy is dealt with elsewhere.

†Class 60, Sub 641, Power plants - natural heat.

‡Shell Oil Company holds a patent for use of geopressured energy in a total flow system.

Integrated firms raise capital for new projects by the sale of securities based on the credit and earnings of the firm as a whole, by the sale of securities based upon the earnings from a specific project, and by internally raising cash (for example, through depreciation reserves, deferred taxes, or retained earnings). If project financing techniques are used, integrated firms should have no particular advantage in raising capital. Rather, the credit-worthiness of the project in question should be determinative.

If capital is raised through internal cash flows or through general corporate financings, reliance is being placed on earnings in other areas. Such reliance--for example, on crude oil profits--precludes the discipline that competitive capital markets might otherwise impose, thus leading to wasteful investment allocations. When capital is allocated by a corporate group and not by capital markets, the choice of projects to be financed may be limited to one set of investment goals. Such internal allocations of capital can contribute to a deterioration in the liquidity of capital markets for energy projects.

Limitation of capital flows to a few integrated channels may restrict the flow of information to capital markets. This may detract from the performance of capital markets which otherwise might be as efficient as or more efficient than the company hierarchies. Large, diversified capital markets can match up multiple capital sources and multiple investment opportunities and perform a substantial risk evaluation function.

Integrated energy enterprises can channel large sums into alternative energy sources. It cannot be said that they will do so with unique efficiency; or indeed that primary reliance on them would yield as much capital, as well allocated, as reliance on market devices. The extensive use of highly leveraged project financing indicates that petroleum companies are not introducing substantial amounts of new equity money into coal.[163]

Large integrated oil companies are reported to have higher overhead and drilling costs than independents. The large enterprises also have a practice of warehousing prospects, whereas smaller enterprises must economically produce or abandon a prospect. Higher development costs and warehousing of prospects appear to be socially inefficient uses of capital.

Adverse Effects of Integration

The adverse effects of major West Coast refiners' participation in other fuel sectors stem from the probability that this participation causes development of these resources to occur in a manner that results in less competition among fuels, and among the major refiners and other fuel suppliers. Less competitive development can result in a slower pace of new energy source development than might otherwise be the case.

Diminution of competition can arise in at least two distinguishable ways. First, a company or set of companies with a major investment in one sector can be expected to phase in new operations so as to maximize overall profit, and thus protect the profits of the sector in which it already has investment. Secondly, vigorous extension of such companies into alternative supply sources can create barriers to development by rival entities, by diminishing their opportunities to acquire factors of production.

Oil companies will not be eager to promote geothermal energy sales which return less profit than sales of other forms of energy foregone by reason of geothermal sale. Recent testimony by a petroleum company witness regarding coal holdings illustrates the point:

Senator Abourezk (continuing). And ask you a question that goes right to the heart of these hearings and the heart of this legislation that we are considering today.
Would Continental Oil, owning 2 percent of the petroleum market, controlling 2 percent of that market and 9 percent of the coal market, would Continental Oil direct its coal subsidiary, Consolidation Coal Co., to undersell to utility companies that are buying oil from Conoco--would you direct your coal subsidiary to undersell them?
Mr. Hardesty. To undersell?
Senator Abourezk. On a BTU equivalent basis. Say you had a public utility, or a private utility, that was thinking of converting to coal, if it could find coal at a somewhat cheaper BTU equivalent than it buys oil for. Would you tell your coal subsidiary--or would you permit your coal subsidiary to undersell your oil subsidiary?
Mr. Hardesty. No, sir, under no circumstances.
Senator Abourezk. Thank you. Please continue.
Mr. Hardesty. I think I want to enlarge on that, and I hope that our discussions will explain what we do with our profits. Prices will reflect our costs; we recapture both our investment and our operating costs. These are returned to the corporation to put us into a position to replace that capacity or that resource as it might be worked out. That would be different for each of those energy sources, Senator, and they are not one related to the other.
Senator Abourezk. If on a cost basis, coal could sell for somewhat cheaper than a BTU equivalent of oil, would you still not direct your coal subsidiary to undersell your oil subsidiary?

Mr. Hardesty. You are dealing with actual circumstances. We would not direct a coal subsidiary, a nuclear subsidiary, to have its price changed, modified in any way, so as to either compete readily against, or not compete against, another form of energy. I think that broadly answers your question. We are not going to play one source of energy against the other.[164]

This control over intradivisional rivalry can be extended to other enterprises over which the company has substantial influence. Such firms are those needing access to joint ventures for exploration and development, unitization agreements (including logical mining units), pipelines, moorings, refineries, and exchange agreements with their competitors. In the geothermal area, such firms could include those using petroleum company capital for exploration and development, and firms involved in joint field development by reason of unitization, information exchanges, or other reasons.

In the case of the petroleum industry, extensive interrelationships among major firms can create a situation of commonality or relative uniformity in the entire sector's approach to energy resource development. A recognition of widely shared joint interests can lead to a tendency not only not to gore one's own ox, but also to avoid goring its yoked neighbor. *

The petroleum industry has a history of being concerned with overproduction. It has underproduced OCD tracts and supported state prorationing. Larger companies appear to have acquired large undeveloped blocks of coal which is being held for future use.

Development financing by large oil refiners has been preferentially channeled to overseas ventures, where antitrust investigators have reported a history of supply restraint activity aimed at maintaining or raising prices.[165]

The large integrated oil companies' pursuit of their profit goals is coupled with the knowledge that unutilized domestic resources are money in the bank. These resources are secure and their value appreciates. This is not the case for overseas resources (or for domestic oil or gas fields subject to drainage). Other incentives for

---

*If one Aramco partner were to increase his share of the West Coast utility fuel market by geothermal sales, there could likely be a reduction in oil imports to the detriment of both Aramco and Indonesian ventures of Aramco partners (for example, Stanvac and Caltex). The disruptive partner is likely to be crude-short elsewhere; or perhaps dependent upon exchanges with other partners to market in a region; or perhaps dependent upon a partner's concurrence to drill a prospect in the Gulf of Mexico (or Siam).

favoring foreign production also exist. Exploitation of foreign production often permits acquisition of oil under off-take agreements giving special price concessions not available to others, while preventing that oil (and concession or agreement) from being transferred elsewhere. Also, under the Internal Revenue Code, the portion of the purchase price paid as taxes overseas is a dollar-for-dollar credit against federal taxes. Taxes on income accruing to partially owned overseas enterprises and tanker subsidiaries may be deferred for long periods of time. These incentives for preferring foreign production would tend to delay development schedules for domestic geothermal energy development.

All this suggests that major petroleum companies may follow a pattern of phasing production from their extensive geothermal holdings in a manner geared to maximizing overall producer yields from their various operations, rather than--as an independent developer might--in a manner yielding the best possible return from each of their resources considered in isolation. (If there were no horizontally integrated companies, this last model would be equivalent to competitive pricing.) To the extent large petroleum companies influence the pace of geothermal energy development, that pace can be expected to be retarded by these considerations.

Dr. Paul Davidson has stated to the Senate Antitrust Subcommittee that the current oil and gas supply situation and cartelized fossil fuel supply worldwide were due to the energy enterprises "becoming engulfed in speculative as well as monopoly practices." As a cure, Dr. Davidson advocated convincing petroleum producers that their current monopoly cannot last, thus encouraging current production. The energy companies' present ability to reap the benefits of monopoly depends on the price inelasticity of demand for their product. This inelasticity, he stated, depends on two specific elasticities: income elasticity and substitution elasticity. Dr. Davidson stated that income elasticity for fossil fuels is approximately one--that is, a 1 percent increase in income will result in about a 1 percent increase in fossil fuel consumption. He found that at present there are no substantial amounts of substitutable goods competing with fossil fuels, and hence substitution of elasticity is small. As a result, overall price elasticity is also small.

> A high substitution elasticity requires independent producers who have no major vested interest in maintaining or improving the capitalized value of oil crude reserves in the ground. This requires breaking up the conglomerate energy companies in order to permit alternative energy supplies to be produced by independent firms

> that can have expectations and objectives which differ from the major oil and gas producers. . . . Now the cartel and the domestic producers of oil and gas value the oil and gas in the ground at a certain price, say $11 a barrel. That is the capitalized value, that is their assets.
>
> If the price of oil falls, they will take a capital loss on all that underground oil. If coal were to come in at a lower price, then that would imply a capital loss on the value of assets in the oil divisions of the same company.
>
> A rational producer observing that one of his actions causes the value of some of his assets to be lowered would not engage in selling that asset, and not take that capital loss. On the other hand, if these were independent producers, the fellow producing coal would have no compunction about worrying about imposing a capital loss on another industry.[166]

Dr. Davidson summarized the matter by stating that the object of the conglomerate energy company "is to get a single price per BTU for all these things and maximize the profits of producing them all."

Barriers to Entry

Cross-fuels integration by a set of very large firms can retard development in a new sector, lessening the variety of development and the pace of activity, by limiting the access to factors of production available to nonintegrated and nondominant firms. Only by obtaining access to such factors can new entrants substantially alter market conditions.

The factors of primary concern are land (leases) and capital. It has been previously noted that development of geothermal fields requires that willing developers have access to fields (for example, leases) that appear to present high quality resources and lower risk. It may be assumed that these sites are more conspicuous than others and will be among the sites first discovered and leased.

The geothermal industry is dominated by lease holdings of petroleum companies or of companies controlled by firms or persons actively engaged in the petroleum industry.* The importance

---

*The better sites in Utah (Roosevelt Hot Springs), New Mexico (Valles Caldera), Nevada, Heber and Niland in the Imperial Valley,

of such a strategic position can be very considerable. The firms that get the choice prospects may get a permanent advantage of major consequence.

The pattern of bidding on geothermal leases indicates that petroleum companies can and do outbid other entities for choice sites. This presumably reflects a differential in available resources and market opportunities between the large integrated oil companies and others. Many prospective investors may not seek high prices based on costs of alternative fuels, or have equivalent profit streams to shelter or invest. This situation is accentuated by federal leasing practices. The bonus-bid system requires more capital from bidders than a system geared to return from production. This favors capital-rich firms, while the lack of due diligence requirements promotes warehousing.

A complementary exclusionary effect arises from the fact that independent firms are largely performing lease-broker and service functions for larger enterprises or are otherwise largely dependent upon them for financing--for example, joint enterprises or bottom-hole contributions.* This service relationship is also a barrier to expansion by small firms. It curtails their independence in exploiting or otherwise dealing with the resources in which they have a proprietary interest.

Smaller firms in the geothermal sector have considerable concern about access to capital, for several reasons. First, they note, and the data show, that the higher prices bid for properties by petroleum companies drive up the costs for entering the business. Second, they note that larger firms established in energy markets have lower costs of capital than smaller firms. This advantage may come in part from the wider and more extensive contacts with financial mar-

---

and the Geysers are entirely or largely in the hands of major oil companies' ventures. Only at Mono Lake, at some Imperial Valley locations, and at some Geysers locations do independents have holdings on very promising sites.

*Recent federal lease data from Nevada, where a number of noncompetitive geothermal leases have been rapidly issued, indicate that a number of persons are acting as lease brokers picking up leases to trade to others, or obtaining assignments from larger oil companies concerned about exceeding acreage limitations. Oil firms such as Chevron, Mobil, Al-Aquitaine, Anadarko Production, Phillips, Getty, Southern Union Production Co., Sun Oil, and Union Oil have substantial holdings, and could exceed state maximum acreage limitations if they were not to depend upon such brokers to hold tracts the oil firms were not yet ready to explore.

kets possessed by major established companies, particularly petroleum companies. The advantage may come in part from lower risks associated with "deep pockets"--the availability of large capital streams to remedy shortfalls or misadventures on which smaller enterprises break up. It derives in part from tax treatment enjoyed by large international petroleum companies and existing firms in general as opposed to new entrants.*

The prospect presented is a possible two-tier market in geothermal energy--a higher-cost, relatively thin market for smaller and less established firms, and a larger, lower-cost market for larger, established petroleum-based firms. Such a situation would complement difficulties new entrants experience in obtaining desirable leases.

A related concern is the possible influence of the petroleum sector on financial institutions' willingness to invest in independent operations in markets for sales of limited partnerships and other new equity offerings. A new entrant is required to take his plans to financial houses that have ownership, directorate, and business-deal relationships with his competitors. According to testimony of an officer of a very large bank that is heavily engaged in such financing, petroleum-company-affiliated directors of banks are on bank boards because of their expertise; in the witness's bank, smaller petroleum loans are summarized for these directors and larger loan applications are presented to them in detail.[167]

Municipal systems do not receive tax reductions and accordingly must absorb all of their dry-hole costs. In the event of a joint enterprise with a private firm, the municipal system may or may not be able effectively to share tax reduction derived savings. Beyond this, there may be fear that the views of large financial houses (interlocked with the major oil companies) as to when and to what extent a new energy source will be developed may tend to proceed in

---

*The international petroleum companies have effective tax rates far below the general average. At present our national tax laws provide substantial benefits to large international companies; DISC, Tanker Subsidiary, and foreign tax credits are but a part of this picture. For domestic operations, investment tax credit, and accelerated depreciation and intangible drilling cost deductions are available only to firms established in drilling, while depletion deductions were (and as now limited still are) available only to producers. Exploration enterprises are encouraged to sell resources they locate to producing companies by IRC Section 632 which places a 33 percent limit on the tax rate for their gains--which occur after some intangibles have been expensed.

step with each other.* The resulting tendency for lockstepped progress may be especially strong as financing requirements rise and syndicates become more necessary to float energy project financings.

High profit expectations, growing out of the profit levels earned in noncompetitive sectors where production may be withheld, can make difficult the task of a new firm seeking to raise money to fund a project to produce lower cost energy. Much of the financing available comes directly from firms with a major stake in other fuels. To the extent a drilling or development firm is dependent for capital on such a firm, its own pace in the geothermal market may be constrained.

CONCLUSION

Geothermal energy's attraction is as a domestic source of energy and as a competitive stimulant in fuel markets. Its development can best go forward in the context of a diversified and unconcentrated industry. Channels for investment also need to be diverse.

Small firms have played a major innovative role in developing techniques for locating and engineering geothermal prospects. The continuation of such work, and their participation in high cost development of fields, will require that small firms have access to land, capital, and purchasers. Without small firms, the number of different approaches--and perhaps even the enthusiasm with which these approaches are tried--is likely to decline. Reliance for development on a relatively small number of large organizations is likely to diminish the scope of innovation and decrease the pace of enterprise. Firms with a diversity of profit goals and perceptions of future returns, including firms seeking energy for use in production of their goods, may be expected to find a wider variety of prospects appealing. The ability and extent to which small firms participate in geothermal development--in more than a service company role--will accordingly, it is believed, have a substantial effect on the rate of development of this resource.

---

*The financial institutions on whose boards petroleum company executives sit have substantial direct investments in energy projects and, if banks, they control very substantial trust holdings; many of these investments are project financings, such as joint venture participations, production loans, or debt primarily secured and to be repaid by revenues from specific facilities such as pipelines or tankers. These institutions are not likely to invest in projects that might jeopardize their return on prior investments.

The recent series of mergers between major oil refiners and uranium and coal holding companies tends to extend their common viewpoint further, as do their extensive financial linkages. At the same time, capital costs for entry into the fuel supply market on a scale large enough to serve utility contracts have dramatically risen so that established participants in a sector are less likely to be challenged by others. This synthesis of attitudes and capital can be seen as tending to rationalize the investment and major operations of the large concerns involved in a noncompetitive manner.

Federal leasing and research programs do not appear to be organized in a fashion militating against unnecessary concentration of geothermal energy supply.

Horizontal integration across energy markets is an issue of substance in the geothermal energy sector; government policy may appropriately be aimed at preventing the harmful effects of such integration.

NOTES

1. According to a Bureau of Mines News Release, "Annual U.S. Energy Use Drops Again" (April 5, 1976), in 1975 electric utilities consumed 20.1 quadrillion Btu, some 28.3 percent of the 71.1 quadrillion Btu consumed in the United States that year.

2. USBM reports residual fuel oil demand, not consumption. Total domestic consumption in 1975 of bituminous coal and lignite was 558 million tons (USBM, Minerals and Materials, January 1976). Domestic demand for residual fuel oil in 1975 was 887,963,000 barrels. (Deliveries to steam-electric plants totaled 433,568,000 barrels [Nos. 5 and 6 fuel oil].) USBM, Crude Petroleum Products, and Natural Gas Liquids (December 1975) and FPC, Annual Summary of Cost and Quality of Steam - Electric Plant Fuels, 1975.

3. Federal Power Commission, Staff Report on Proposed Generating Capacity Additions, December 9, 1975.

4. "Staff Summary of Electric Utility Expansion Plans for 1976-85," FPC News Release No. 22493 (July 16, 1976).

5. Figures for 1975 taken from Federal Power Commission, "Summary of Cost and Quality of Steam-Electric Plant Fuels, 1975" (May 1976).

6. Fuel adjustment clauses are based on variations from base fuel cost in either the imputed cost of fuel burned per kwh generated or the cost per Btu of fuel acquired. See discussion, statement of Bierman and Stover, U.S. Congress, Senate, Committee on Government Operations, Hearings on the Utilities Act of 1975, 94th Cong. 1 Sess. (1975).

7. Fuel adjustment clauses have been thought to tilt the choice between higher fuel costs and capital equipment toward use of more fuel. This is reported to have been the case prior to 1973. FPC, 1973 Annual Report, p. 17. Also see U.S. Congress, Senate, Committee on Government Operations, Hearings on General Utility Rate and Fuel Adjustment Clause Increases, 1974, 94th Cong., 1 Sess. (1975).

8. The Office of the Auditor General of the State of California in a report entitled "Adjustment of Electric Rates for Fuel Cost Charges" (August 1975) reported that charges by PG&E, SCE, and SDG&E under fuel adjustment clauses exceeded actual fuel cost increases by $270.6 million.

9. Staff Audit Report. This audit did not attempt to review the propriety of SCE fuel procurement practices.

10. Hearings on the Utilities Act of 1975.

11. See Department of the Interior, Environmental Impact Statement on Eastern Powder River Basin Coal Mining, 1975 Annual Summary. (EIS)

12. Electrical World, December 1, 1975, p. 31. It was also reported that "Next year's oil requirement is about 55 million bbl, or 62% of the utility's overall fuel needs."

13. Ibid. Appendix B contains a more detailed description of fuel procurement agreements of PG&E and SCE.

14. SCE's Form 10-Q filed with SEC for the first quarter of 1976.

15. On June 30, 1975, PG&E owned and operated 30 thermo-electric generating plants. PG&E's prospectus of October 1, 1975, for the sale of $175 million first and refunding mortgage bonds, p. 10.

16. The prospectus of October 1975, pp. 14-15, relates how shortages of natural gas together with delay in completion of nuclear units, have increased PG&E's reliance upon higher priced, low-sulfur fuel oil.

17. P.U.C. Application 55541, direct testimony of R. P. Benton.

18. R. R. Cowan, Union Oil Co., to R. P. Benton, Pacific Gas & Electric, December 16, 1974 (Exhibit 2, California PWC Application 55541).

19. Letter dated February 3, 1975, to FEA from Malcom H. Furbush, PG&E associate general counsel.

20. Tesoro prospectus of December 17, 1975, p. 24.

21. FPC Form 1-M for year ending June 1975, filed 1976.

22. PGE prospectus of December 11, 1975, for sale of bonds.

23. TG&E prospectus of September 17, 1975, for sale of preferred stock.

24. APS prospectus of November 20, 1975, for sale of bonds.

25. PSCC prospectus of October 22, 1975, for sale of bonds.

26. In January 1976 and December 1975, West Coast refinery yields of resid were 21.7 percent and 21.4 percent, respectively. U.S. total of yields was 11.2 percent and 10.4 percent. USBM, Mineral Industry Survey.

27. USBM, Mineral Industry Surveys.

28. FTC Staff Report, Report to the Federal Trade Commission on the Structure, Conduct and Performance of the Western States Petroleum Industry (September 1975), p. 12. Referred to hereafter as FTC Staff Report. This study, however, was concerned only with record ownership of refinery facilities, and did not consider functional control of independent facilities by through-put, crude supply, or product purchase contracts, nor did it examine the regional variations, as, for example, the San Joaquin Refining market which is highly isolated.

29. FTC Staff Report, p. 15.

30. Tidewater Oil Co. v. United States, 418 US 906 (1974).

31. United States v. Phillips Petroleum Co., 1973--2 Trade Cases 74789 (C.D. CA., 1973), aff'd without opinion, 1974--2 Trade Cases 75143 (Sup. Ct., July 8, 1974)--discussing competition by threat of entry in a market (motor gasoline), which few could afford to enter. Toscopetro has recently entered into an agreement to purchase crude oil from Natomas, after having acquired the Signal refinery and pipeline use rights from that company.

32. Petroleum refineries in the United States and Puerto Rico, January 1, 1976. USBM, Mineral Industry Survey Series, Capacity of Petroleum Refineries in the United States and Puerto Rico, Table 4.

33. Reported in Business Week, January 12, 1976, p. 94.

34. The Bahamas refinery source is owned 50/50 by SOCAL and New England Petroleum Company. See International Petroleum Encyclopedia (Tulsa, Okla.: Petroleum Publishing, 1974). Mobil and Exxon historically have been associated for onshore Indonesian production in the P.T. Stanvac joint enterprise.

35. Congressional Research Service, A Study of the Relationships Between the Government and the Petroleum Industry in Selected Foreign Countries: Indonesia, Senate Interior Committee Serial No. 94-26 (92-116), 94th Cong., 1 Sess. (1975).

36. Petroleum Economist, April 1976, p. 130.

37. International Petroleum Encyclopedia, 1974.

38. Subsidiary relationships from Skinner, Oil and Gas International Year Book (London: Financial Times, 1973).

39. See FTC Staff Report.

40. U.S. Congress, Senate, Judiciary Committee, Hearings on The Industrial Reorganization Act, Part 9, The Energy Industry,

94th Cong., 1 Sess. (1975) presents a series of reports on aspects of the petroleum industry in California.

41. Ibid., pp. 97-98. The Joint Committee's Report, "Crude Oil Pipelines in California," is hereafter referred to as the California Report.

42. Department of the Interior, Office of Oil, Vulnerability of Total Petroleum Systems, prepared for the Defense Civil Preparedness Agency, Washington, D.C., 20301, May 1973, p. 17.

43. California Report. The ton-mile scale understates the pipeline's competitive position versus barges, since the latter are confined to waterways whose winding configuration increases mileage.

44. USBM, Crude Oil, Petroleum Products, and Natural Gas Liquids: 1975 (Final Summary), February 24, 1977, Table 14.

45. FTC Staff Report; California Report, esp. "California Crude Oil Market Control," p. 24 et seq.

46. Ibid.

47. FEA recently issued special price rules for District V (California and Alaska) permitting adjustment of the gravity differentials. 41 F.R. 48324 (November 3, 1976). This rule authorized only partial relief for the problem.

48. FTC Staff Report, p. 42.

49. California Report, p. 22.

50. FTC Staff Report.

51. 30 USC 185.

52. U.S. Congress, Senate, Interior Committee, Hearings on Market Performance and Competition in the Petroleum Industry, 93rd Cong., 1 Sess. (1973), Vol. 3, p. 904.

53. The 1967 report of the attorney general, regarding the Interstate Oil Compact Commission.

54. Preliminary FTC Staff Report on Investigation of the Petroleum Industry (1976), p. 6.

55. U.S. Congress, Senate, Judiciary Committee, report on the Petroleum Competition Act of 1976, Part I, Rept. No. 94-1005, 94th Cong., 2 Sess. (1976), pp. 25-26. Hereafter cited as Petroleum Industry Report.

56. 87 percent of domestic inputs moved by pipeline. USBM, Annual Petroleum Statement: 1974 (Final Summary), Table 14.

57. In one case, it is alleged that a firm was denied access because of oil quality to a line to ship crude from a field whose output had previously been purchased by the pipeline owner. Testimony of C. Siess, U.S. Congress, Senate, Judiciary Committee, Part 8, Hearing on the Industrial Reorganization Act, pp. 6241-42.

58. Testimony of George Stafford, three hearings on Market Performance and Competition in the Petroleum Industry, 93rd Cong., 1 Sess. (1973). Hereafter cited as the Stafford Testimony. As the

Petroleum Industry Report notes, individual interest systems may be similarly exempt. ICC Chairman Stafford referred to The Pipeline Cases, 234 US 548 (1914). See also Valvoline Oil Co. v. U.S., 308 US 141 (1939). The Hepburn Act is part of the ICC Act, 49 USC 1-6.

59. Report of the attorney general pursuant to Section 7 of the Deepwater Port Act of 1974 (November 5, 1976).

60. Section 5(c) of the Outer Continental Shelf Act, 43 USC 1334 (c).

61. Levy, The Regulation of Offshore Crude Oil Pipelines and the Consequences for Competition (Washington, D.C.: National Science Foundation, 1975); Petroleum Industry Report, p. 27. These document the powerful control of production held by OCS pipelines.

62. Provided by the Stafford Testimony, p. 910.

63. Petroleum Industry Report, Table 4.

64. The National Petroleum Council, U.S. Energy Outlook, "Domestic Oil and Gas Availability (1972)," Ch. 4, Table 38, forecast that of the 384.2 billion barrels of oil in place, 112.3 billion barrels were located offshore (or in south Alaska), 96 billion barrels on the North Slope, and 176.9 billion barrels onshore. For the onshore Pacific Coast region, 21.9 billion barrels were forecast as discoverable, while offshore 47.7 billion barrels were so forecast.

On a worldwide basis, in 1973 some 14.8 percent (8.48 million barrels per day) of total oil production came from offshore. Petroleum Economist, "Offshore: Over 20 million b/d in the 1980's?" (February 1976), p. 49. This article reports a forecast by the Scottish Council that in 1980, some 24 percent of production will be from offshore areas (21.92 million barrels per day out of 89.76 million).

65. See Testimony of William John Lamont, U.S. Congress, House, Small Business Committee, Hearings on Energy Data Requirements of the Federal Government, Part III (1974), pp. 272, 303.

66. High bids totaled $438.2 million. Wall Street Journal, December 12, 1975, p. 8.

67. Wall Street Journal, December 22, 1975.

68. FTC Staff Report, Concentration Levels and Trends in the Energy Section of the U.S. Economy (1974), p. 42.

69. Petroleum Industry Report, pp. 28-32.

70. For large petroleum companies, and even more so for smaller ones, it appears that a substantial portion (over half the 20 firms with the largest sales, and 80 percent for smaller firms) of petroleum production occurs in situations of joint ownership. See U.S. Congress, Senate, Interior Committee, The Structure of the U.S. Petroleum Industry: A Summary of Survey Data, Ser. No. 94-37 (92-127), 94th Cong., 2 Sess. (1976), pp. 48-49.

71. FTC Staff Report.

72. See U.S. Congress, Senate, Interior Committee Staff Report, "The Trans-Alaskan Pipeline and West Coast Petroleum Supply, 1977-1982," Series no. 93-51 (92-86), 93rd Cong., 2 Sess. (1974).

73. According to Chuck Champion, the state of Alaska's pipeline coordinator (November 21, 1975), as reported in California State Lands Division, "A Preliminary Analysis of the SOHIO Project for the State Lands Division" (January 1976).

74. Ibid.

75. Section 28 (r) of the Mineral Leasing Act, as amended. See Report of FTC Western Task Force, Appendix A.

76. The attorney general's November 5, 1976 Superport Report characterizes these aspects as involving "inherent anticompetitive problems" (p. 103), and, particularly, that "Undivided interest Pipelines present severe anticompetitive problems" (p. 112, emphasis supplied).

77. See Senate Interior Committee Staff Report. Again, in the November 5, 1976 Superport Report, the attorney general reiterates that the "ICC's powers are severely circumscribed . . . [with] no power over the form of corporate organization . . . no power to order elimination of restrictive provisions in underlying agreements . . . no power to order frequent share distribution . . . [and] no power to order expansion of facilities."

78. California State Lands Commission, op. cit., p. 10 and cf. p. 11.

79. Ibid., p. 15.

80. Although reporting that SOCAL says that its new 300,000 barrels-per-day refinery expansion is designed only for imported, high-sulfur, light crude oil, the State Lands Report, citing an article in the Oil and Gas Journal of April 7, 1975, observes that expansion built into the design would indicate that these refineries are capable of refining Alaskan crude oil. Ibid.

81. California State Lands Commission, California and the Disposition of Alaskan Oil and Gas--A Working Paper (June 1976), Ch. 3.

82. Ibid.

83. SOHIO/BP projects a crude oil surplus on the West Coast of 300,000 to 600,000 barrels per day by 1978, and E. Stanley Tucker, writing in the November 1976 issue of the Petroleum Economist, "Markets for Alaskan Crude," projects an 800,000 barrels-per-day surplus in 1980.

84. Ibid.

85. U.S. Congress, Senate, Committee on Foreign Relations, Subcommittee on Multinational Corporations, Hearings on Multinational

Oil Corporations and U.S. Foreign Policy, 93rd Cong., 2 Sess. (1975), p. 95. And see U.S. Congress, Petroleum Industry Report, pp. 43-44.

86. Multinational Subcommittee, Hearings, Part II, p. 3.

87. The meeting of OPEC in Bali, Indonesia, is generally believed to have foundered on the inability of OPEC countries to devise price differential and prorationing schemes. This subject was a topic of frequent discussion in the June 2, 3, and 8 hearings of the Joint Economic Committee on Multinational Oil Companies and OPEC: Implications for U.S. Policy, 94th Cong. 2 Sess. (1976).

88. See statement of Senator Frank Church to House Antitrust Subcommittee, U.S. Congress, Judiciary Committee, Hearings on Energy Industry Investigation, Part 1, 94th Cong., 1 Sess., p. 420 (Serial no. 48, 1976).

89. See article by M. A. Adelman in the Washington Post (May 1, 1976), quoted in U.S. Senate, Judiciary Committee, "Report on Petroleum Industry Competition Act of 1976," Report no. 94-1005, 94th Cong., 2 Sess. (1974), p. 44; testimony by Adelman, "Multinational Hearings," op. cit.

90. As evidenced in the attorney general's Deepwater Port report.

91. Transwestern Pipeline Company, 36 FPC 1010.

92. United States v. El Paso Natural Gas Company, 358 F. Supp. 820, 827 (D. Colo., 1972).

93. The 75 cent price, which reflects the price of alternative fuels, prompted the issuance of a show cause order by the PUC as to why California producers should not be regulated. Decision 84616, July 1, 1975.

94. Interview with the late Colin Garrity, California PUC, Gas Section. The California producers have sought to escape from their status as peaking resources for PG&E, in part by opposing PGT's applications to the FPC for authority to expand its throughput of Canadian gas. In one case, the producers attempted to persuade the FPC that PG&E was abusing its monopsony power. See California Gas Producers Association v. FPC, 421 F. 2d 422 (CA 9, 1970).

95. Contracts for synthetic gas from this project set prices at the cost of service plus a 15 percent return on vendors' equity, irrespective of actual deliveries. The hell-or-high-water provisions were asserted to be necessary to project financing. Cf. Transwestern Coal, FPC Docket No. CP 73-211.

96. Pending before the FPC (Docket No. CP 74-207) is a proposal by Pacific Lighting to import Indonesian LNG, to be purchased from Pertamina (joint producer with Mobil) under a 20-year contract, whose price is tied to the non-spot-market export price for Indonesian crude oil and the Index of Fuels and Related Products and Power.

97. PG&E's vice-president for gas supply, John Sproul, testified before the California PUC that he believed the price of Alaskan natural gas delivered in the PG&E market area will be about equal to the cost at that time of an equivalent amount of fuel oil in the same market area. Information taken from 1975 proceedings on advance payment from PG&E to Edison.

98. Presented by Gerald Gambs of Ford, Bacon & Davis, Inc., to Energy R&D Conference of the Atomic Industrial Forum, Washington, D.C., February 1976, as reported. Weekly Energy Report (February 16, 1976), p. 4.

99. Eastern Powder River Basin EIS, p. I-56, 57.

100. These areas are described in the Interior Department's Draft Environmental Statement on Proposed Surface Management of Federally Owned Coal Resources, DES 73-53 (October 1, 1975).

101. Ibid., pp. II-108-09.

102. Eastern Powder River Basin EIS reports that Campbell and Converse Counties, Wyoming, have estimated economically strippable coal reserves of 12.4 billion tons, that 13.3 billion tons are in the Northern Great Plains of Wyoming, and that 36.5 billion tons (coal and lignite) are in the Northern Great Plains of Montana, North Dakota, South Dakota, and Wyoming. The national strippable coal reserve was estimated by the U.S. Bureau of Mines in 1971 to be about 45 billion tons. The national reserve has been increased by new coal discoveries (sic) since 1971, but the Eastern Powder River Basin contains a significant portion of the nation's economically recoverable strippable coal reserves. Pp. I-22-23.

103. Coal fuel costs were predicted to be 16.1 mills at western mine-mouth plants for plants going in between 1978 and 1985. Edison Electric Institute, "Cost Comparison of Nuclear and Coal-Fired Plants," March 1976.

104. T. C. Campbell and S. Katell, "Long-Distance Coal Transport: Unit Trains or Slurry Pipelines," U.S. Bureau of Mines, Information Circular 8690.

105. Joseph P. Mulholland and Douglas W. Webbink, "Concentration Levels and Trends in the Energy Sector of the U.S. Economy," Staff Report, Federal Trade Commission, Washington, D.C., March 1974, pp. 83, 84.

106. Sidney Katell et al., "Basic Estimated Capital Investment and Operating Costs for Coal Strip Mines," U.S. Bureau of Mines, Information Circular 8703 (1976). (If debt costs 11 percent and constitutes 75 percent of project financing, equity would earn 27 percent.) Mulholland, op. cit., p. 84, referring to National Petroleum Council, op. cit., Vol. II, p. 136.

107. Katell, "Basic Estimated Capital Investment and Operating Costs for Underground Bituminous Coal Mines," Revision of Information Circular 8682A, USBM (1975). Katell uses United Mine Workers wage scale which may not apply in nonunion western mines. He derives a selling price, using a 15 percent after taxes return on investment, of $11.64 per ton. Where coal seams are thick--seven feet or more--costs tend to be reduced. Assuming a 95 percent recovery, recovery of about 1,644 tons of coal per acre foot is feasible. Seams of 52 feet are reported by Decker and Westmoreland, of 33 feet by Peabody, 28 feet by Western Energy, and 16 feet by Knife River. With a 24-foot seam, 48 acres yield 19,100 tons.

108. In Kennecott Copper Corp. v. Federal Trade Commission, 1972 Trade Cases 74157 (CA10, 1972) the court upheld FTC findings of a nationwide line of commerce consisting of bituminous, sub-bituminous, and lignite coal, which was becoming increasingly concentrated. "During the period 1954 to 1967, the top four companies increased their share of production from 15.8 percent to 29.2 percent. While during the same period the market expanded by 40.9 percent, the share of the top four companies grew by 160.5 percent. . . ." The court noted that "a number of factors militated against new entry. . . . The fact that the big demand was for consumption by utility companies involving long-term contracts, extensive reserves and ready ability to deliver all contributed. Thus, experience, know-how and equipment are essentials." It later held that only a company the size of Kennecott could enter the industry, noting that experiences of two other firms demonstrated a basic time requirement of from 10 to 15 years plus large utility contracts for large-scale operations. Kennecott was required to divest itself of Peabody Coal Company which it had acquired. This divestiture has recently been completed. In United States v. General Dynamics, 341 F. Supp. 534 (ND. IL, 1972), affirmed on other grounds, 415 US 486 (1976), the district court found that the relevant market is for energy, not for coal alone. Coal submarkets were viewed as being regional.

109. "Because of its bulk, the cost of transporting coal has historically been an important part of the total cost to the consumer. This relationship, however, has not remained constant and today rail transportation costs are of declining importance in the marketing of coal. . . ." Interstate Commerce Commission, Investigation of Railroad Freight Rate Structure--Coal, Ex Parte No. 270 (Sub. No. 4) (decided December 3, 1974; Service Date March 14, 1975), p. 94. (ICC Report).

110. Ibid., pp. 210-24.

111. Ibid., pp. 212-13.

112. Costs presented by Power Technologies, Inc., in November 1975, for a scheme to transport energy from 1000mw units in western low-sulfur coal fields over 500kv d.c. lines include line costs of $170,000 per mile and terminal costs of $65 per kw. Line losses range from 6.5 mw per 100 miles at 1000 amperes to 14.6 mw per 100 miles at 1500 amperes. Terminal losses are 2.5 percent of kw loading. Costs of rail transportation of coal to Texas from Wyoming, 9 mills per ton-mile in 1977 rising 6 percent per year, and from Colorado, at 11 mills per ton-mile rising 15 percent per year, may make high voltage transmission the most economical way to move coal.

113. Battelle Pacific Northwest Laboratories, "Regional Analysis of the U.S. Electric Power Industry" (BNWL-B-415), done for ERDA, July 1975.

114. By Foster Associates, Inc. for the then Office of Coal Research, "Prospective Regional Markets for Coal Conversion Plant Products Projected to 1980 and 1985," 1974.

115. U.S. Department of the Interior, Draft Environmental Impact Statement on Proposed Federal Coal Leasing Program, 1975, p. 208.

116. Federal Trade Commission, "Staff Report on Federal Energy Land Policy: Efficiency, Revenue, and Competition" (Washington, D.C.: October 1975), Chapter 9.

117. FTC Staff Report, p. 543 A.

118. See Department of the Interior, Coal Leasing Regulations, Final Environmental Impact Statement: Proposed Federal Coal Leasing Program, pp. 1-2, 7a (undated).

119. Ibid., pp. 1-80.

120. Ibid.

121. Ibid., pp. 1-81.

122. BLM, Coal: An Analysis, p. 23.

123. Ibid., p. 2.

124. FTC Staff Report, pp. 631-32.

125. BLM, Coal: An Analysis, p. 28.

126. USBM, Strippable Reserves of Bituminous Coal and Lignite (1971). Wyoming holds 31 percent of such reserves.

127. Environmental Policy Center, Facts About Coal in the United States, revised February 1975, p. 16.

128. The Hepburn Act's prohibition against a railroad's manufacturing, producing, or mining a product it then hauls does not apply to railroad holding companies. 49 USC 8, construed in United States v. Elgin, J&E R. Co., 298 US 492, 80 L. Ed 1300 (1936), even though this may not seem entirely cogent. Should the sales be limited to delivery by rail, it could be held a violation of law.

129. The Montana Power Company, Prospectus of December 10, 1975, for sale of First Mortgage Bonds.

130. Interior Department, Eastern Powder River Basin EIS.

131. U.S. Senate, Interior Committee, Hearings on Federal Coal Leasing Amendments Act of 1975, 94th Cong., 1 Sess. (1975), p. 253.

132. Wallace W. Wilson, Mine Development Financing for the Coal Industry During the New Decade (March 1, 1976); referring in part to an unpublished study by Bankers Trust Company, "Capital Resources for Energy through the Year 1990" (1976).

133. Ibid.

134. Ibid.

135. Office of Coal, FEA, "Analysis of Steam Coal Sale and Purchases, 1975" (April 1975). Numerous filings under the Public Utility Holding Company Act detail utility financial support to firms opening mines.

136. Ibid.

137. FEA Report, "Analysis of Steam Coal Sales and Purchases" (April 1975).

138. Two of the top ten owners of coal reserves are Mobil (3 billion tons) and Exxon (12 billion tons). Mobil is just starting to develop its first mine while Exxon's mining plans call for the production of 40 million tons by 1985--a reserve to production ratio of about 225:1. See Wall Street Journal, May 20, 1977, pp. 3, 16.

139. Washington Post, May 22, 1977, p. 1.

140. U.S. Congress, Senate, Judiciary Committee, Testimony of Peter Max, Hearings on Interfuel Competition, 94th Cong. 1 Sess. (1975), pp. 31, 40-41.

141. Owned by PG&E. As of June 1976, 60 nuclear reactors were operable, representing 8.1 percent of U.S. installed generating capacity. An additional 178 were planned or being built. ERDA News Release, No. 76-246 (July 28, 1976).

142. Ibid.

143. ERDA News Release 75-178, September 9, 1975.

144. ERDA, Survey of U.S. Uranium Marketing Activity (April 1976); News Release 75-178.

145. Recent allegations of a cartel fixing overseas uranium prices add to price insecurity. The United States is opening its doors to foreign uranium and will be importing substantial quantities. As in petroleum, the low cost resource is predominantly located overseas. See Edison Electric Institute, 1 Nuclear Fuel Supply 42 (1976) and 2 Nuclear Fuel Supply, Table 4.23, p. 72; Figures III, IV.

146. See, for example, Washington Post, June 17, 1977, p. 1. The House Commerce Committee is currently investigating the alleged cartel and the participation therein of Gulf Oil Company.

147. USBM, Minerals in the U.S. Economy (1965-74).

148. Atomic Industrial Forum, "The Nuclear Fuel Cycle: U.S. Capital and Capacity Requirements 1975-1985 (1975)." (AIF Report)

149. PG&E obtained an option to acquire a 35 percent interest in a uranium joint venture with Minerals Exploration Co., a subsidiary of Union Oil Company. The venture involves Wyoming properties. Moody's Public Utility News Reports, June 25, 1976.

SCE's Mono Power Co. subsidiary and Rocky Mountain Energy Co., a subsidiary of the Union Pacific Corp., are in a joint venture involving a 1,000-ton-per-day uranium mill in Wyoming whose output will go to SDG&E. Moody's Public Utility News Reports, December 30, 1975).

150. U.S. Congress, Senate, Interior Committee, Testimony of Hon. Dixy Lee Ray in Hearings on Market Performance and Competition in the Petroleum Industry, Part 2, 93rd Cong., 1 Sess. (1973).

151. Ibid., p. 612.

152. U.S. Congress, Senate, Judiciary Committee, Testimony of Peter Max, Hearings on Interfuel Competition, 94th Cong., 1 Sess. (1975).

153. This concentration is also present at the national level. Petroleum Industry Report, Table 9, p. 49.

154. See FEA, "Trends in Refinery Capacity and Utilization" (December 1975).

155. U.S. Congress, Senate, Interior Committee, The Structure of the U.S. Petroleum Industry, Serial No. 94-37 (92-127), 94th Cong., 2 Sess. (1976).

156. Ibid.

157. The Petroleum Industry Research Foundation, Inc., in its recent report, "The Role of Major Oil Companies and Independent Producers in Domestic Exploration Activities" (June 1976), refers to "the practices of majors funding predrilling preliminary work, with others taking over the drilling phases--with the majors drilling the prime prospects themselves and farming out other acreage on which independents drill."

158. Census Bureau, Annual Survey of Oil and Gas 1974, MA-13K (74-1) (January 1976).

159. See Testimony of John F. O'Leary, Hearings on Interfuel Competition.

160. Washington Post, May 20, 1977, p. 1.

161. U.S. Department of Commerce, Patent and Trademark Office, Technology Assessment and Forecast, Sixth Report (June 1976).

162. The oil industry is a leader in geological exploration work and drilling. A recent article suggests that drilling success ratios are lower than would result from random sampling. "Scientific Uses of Random Drilling Models," Science 190, October 24, 1975.

163. Cf. Testimony of John F. O'Leary, Hearings on Interfuel Competition, p. 88.

164. Ibid., p. 194.

165. See, for example, J. Blair, The Control of Oil (New York: Pantheon, 1976); and U.S. Senate, Report on Multi-National Oil Corporations and U.S. Foreign Policy.

166. See Testimony of Paul Davidson, U.S. Senate, Hearings on Interfuel Competition; Statement of Paul Davidson, appearing at pp. S1263-65, Congressional Record, January 24, 1977.

167. Testimony of Wallace Wilson, U.S. Congress, Senate, Judiciary Committee, Hearings on the Industrial Reorganization Act.

CHAPTER

5

GOVERNMENT AND ENERGY

While this study has dealt with technical as well as market factors, market factors have been emphasized. This is because market factors have major effects on the rate and scope of geothermal energy development.

The public means for controlling markets are antitrust law, regulation, and public ownership. While antitrust action helps create or maintain competitive markets as resource allocative devices, administrative action can substitute public for private central planning as another allocative device.

ANTITRUST LAW AND POLICY: PERSPECTIVES AND ACTIVITY

> The Sherman Act was designed to be a comprehensive charter of economic liberty aimed at preserving free and unfettered competition as the rule of trade. It rests on the premise that the unrestrained interaction of competitive forces will yield the best allocation of our economic resources, the lowest prices, the highest quality and the greatest material progress, while at the same time providing an environment conducive to the preservation of our democratic political and social institutions.[1]

This passage from a leading Supreme Court antitrust decision highlights the two essential functions of antitrust legislation and enforcement: economic efficiency and the maintenance of pluralistic and democratic institutions in an industrial society. The first of these goals is familiar to all who have even a passing acquaintance with antitrust doctrine; the second receives less emphasis in most discussions, but is at least equally important.

Americans have generally, and correctly, regarded it as axiomatic that economic power is translated into political power and, ultimately, government action. Concern about concentrating economic and political power in a few hands has been repeatedly voiced throughout our history.

Firms with economic power gain political power in part from the control of information that follows control of resources. Government cannot act responsibly without reliable information; accordingly, inaction may follow if information is kept secret, or if those possessing information publicize only such data as they deem likely to elicit government actions favorable to their interests. Control of resources also makes the public dependent on the plans and investments of the controlling firms. These plans and investment actions determine future supply; when sufficiently encompassing, they can make it difficult or impossible for the public markets or government to change industry conditions without disrupting vital supply lines. The threat of such disruption, when voiced by the firms that control both resources and the information needed to evaluate their extent and that of alternative resources, is a powerful influence on public opinion. In many areas--not least in that of energy--Americans have become used to abundance and are unwilling to sanction political experiments that they are told may curtail their usage of resources currently enjoyed. Under these conditions dominant firms may use their economic power and control of resources to obtain and exercise broad discretion in planning their future conduct. There is no guarantee that this conduct will not diverge from the public interest. And the political power they can acquire as a result of their economic strength can be used to thwart or subvert government antitrust or regulatory action. Competitive organization of industry, therefore, can and should be regarded as an essential element in maintaining an industrial democracy.

Antitrust policies are instructional in considering how conditions in the petroleum and bulk power industries affect geothermal development. A brief recapitulation of some major antitrust approaches to industry structures and practices will readily point to applications in areas with which this study is concerned.

Antitrust law has long been concerned with the acquisition by one firm of sources of substitutable products, or the joining together of erstwhile competitors.[2] Such horizontal joinders are recognized as being capable of creating future as well as present antitrust problems.[3] Competition can be curtailed by the acquisition or misuse of control of supply or sales outlets. This is particularly possible where there has been a merger movement. Problems arising from vertical integration into control supply or outlets have been recognized.[4]

Preserving potential competition has been of particular concern in energy cases, where one is hard put to find actual competition,[5] and has been of concern in horizontal joint venture cases.[6] Entry through acquisition by a large firm which potentially could enter on its own has been thought likely to discourage other entrants, and raises antitrust problems.[7] So also does acquisition by marketers of a well-established fuel of the resources essential to a potentially competitive energy source that is obviously developable but not yet in general commercial use.*

Joint action, whether vertical or horizontal, is of concern even where done through a commonly owned enterprise or a joint venture.[8] Similarly, "The intimate association of the principal American producers in day-to-day manufacturing operations, their exchange of patent licenses and industrial know-how, and their common experience in marketing and fixing prices may inevitably reduce their zeal for competition inter sese in the American market."[9] Use of monopoly power, however lawfully acquired, to foreclose competition, to gain a competitive advantage, or to destroy a competitor is unlawful.[10]

A body of precedents dealing with bottleneck situations, in which control of a key economic function is used to extend market control to other sectors, is closely related to the law regarding vertical integration. These cases teach that a party controlling a facility whose use is essential to reach markets cannot lawfully use its monopoly power to exclude others.[11] Supply restraints will be allowed if they are found to be "fairly necessary" to achieve a legitimate end.[12] If vertical power is employed to require purchases of other products from suppliers, and this exercise of power is considered to be very detrimental to competition in the tied product, the "restraint of trade can be justified only in the absence of less restrictive alternatives."[13]

---

*Described as obviously developable in order to eliminate the suggestion that proceedings could be maintained on the basis of highly speculative impacts on trade. Although acquisition of novel forms of energy may affect their rate of development, until these forms of energy reach a state where they are clearly developable commercially, antitrust action against their acquisition would be quite difficult to maintain.

Exclusive dealing arrangements, another way of extending market power into other markets, normally provide that a purchaser agree to purchase exclusively or for a significant period of time from one supplier.[14] Such arrangements, not always anticompetitive, are not considered illegal per se. In determining their legality, courts look at the degree to which competition has been foreclosed and the substantiality of the share of the line of commerce affected,[15] seeking to determine the agreement's net effects upon competition.[16] Control of supply or outlets may also be used to encourage reciprocal trading at the uncontrolled stage.

Finally, the tying device is employed to use market power or leverage in one market to appropriate a position in another market. (This latter market frequently has a parallel or horizontal relationship to the tying firm's originating market.) A tie-in is an arrangement in which the sale of a product or service over which the seller has substantial market control is conditioned upon the sale of a separate product or service from the seller or a designated party.[17] These arrangements are viewed with great suspicion under antitrust laws, as they are thought rarely to serve purposes beyond the suppression of competition.[18]

Tying arrangements "deny competitors free access to the maret for the tied product, not because the party imposing the tying arrangements has a better product or lower price but because of his power or leverage in another market. At the same time buyers are forced to forego their free choice between competing products."[19] The economic vice of tying arrangements is that they permit a party to extend its power in one market into another, by noncompetitive means. Profits from combined sales may be higher than those for separate sales (even by two monopolists). Tying also provides an opportunity for price discrimination. Other reasons for seeking tying arrangements are that a vendor may seek to maintain quality control over a product it is associated with,[20] or to obtain economies of jointly producing or distributing the two products. Tying may be sought to evade government controls over one product.[21]

The foregoing has dealt with the body of doctrine or approach to market situations--built up in litigation under the general antitrust laws. A number of statutes have been interpreted as requiring an accommodation of antitrust law and policy;[22] and some have required a judgment as to whether particular situations or practices are inconsistent with antitrust law and policy.[23]

Responsibility for enforcing competitive policies is distributed in numerous places in the government. The Justice Department and the FTC are directly responsible for enforcement of antitrust laws. As noted, the SEC has substantial antitrust responsibilities under the Public Utility Holding Company Act, while other agencies--the

ICC (oil pipelines, trucks, and rail hauling), the FPC (gas sale and transport, and sales of electricity for resale), the FEA (price, access to and marketing of crude oil and oil products), the NRC (nuclear construction and operation permits, nuclear fuel transportation), the Department of Transportation (deepwater ports), and the Interior Department (mineral leasing, OCS development, public land rights-of-way and administration of western federal power facilities) have statutory obligations to consider antitrust implications of their actions.[24]

Antitrust issues arising in the petroleum and bulk power supply industries can involve matters under the jurisdiction of federal and state regulatory agencies. In petroleum cases, issues may arise involving matters within the jurisdiction of foreign governments. Determination of these issues first requires choice of decisional forums.

The legal doctrines for determining decisional hierarchies as among courts and agencies are discussed in Appendix D, Primary and Exclusive Jurisdiction, State Action, and Foreign State Action Defenses. The doctrines constitute a serious impediment to those seeking redress under the antitrust laws, as plaintiff's case may be deferred while administrative matters drag on for years; litigation is extended when efforts are made to defend practices complained of, on the grounds that the actions are permitted or mandated by state law, or are due to the sovereign actions of a foreign government.

Currently a defendant can assert that a matter should be tried before an agency, and not the court, or that the court should defer its case until agency proceedings are over. It is recommended that an antitrust defendant wishing to have the primary jurisdiction* doctrine invoked be required either to request that an agency assert and support the application of the doctrine, or, alternately, to request that a court ask the agency to set out and support its position on primary jurisdiction. In this way, the court would not have to wait to deal with cases in which the agency doubts its jurisdiction so that a referral to it would be futile, or where the agency does not feel it can contribute something substantial to the resolution of the problems at hand.[25] Such an allocation mechanism will also help assure that timely consideration of antitrust issues actually occurs. This is in line with the case law which reflects an intention to

---

*By this is meant a claim of need for preliminary reference of matters to an agency for determination before court action proceeds, rather than the question of whether the agency jurisdiction includes any court action at all--exclusive (agency) jurisdiction.

require antitrust aspects to be considered, either by refusing to allow courts to be ousted of jurisdiction,[26] or by requiring agencies to consider antitrust aspects.[27]

Use of factual evidentiary proceedings would lessen the dangers of primary jurisdiction being employed in a stereotyped manner in response to abstract expertise.[28] This recommendation could be implemented by uniform court action or by legislative direction. That primary jurisdiction is a judicial doctrine of long standing does not make it inappropriate for Congress to modify it. For example, after the Supreme Court's decision in Banco Nacional de Cuba v. Sabbatino,[29] Congress expressly forbade the courts to invoke the well-established "act of state" doctrine as a bar to inquiry into the official acts of foreign governments unless the secretary of state specifically so requests.[30]

When antitrust courts mandate actions for which tariffs or contracts must be filed with the FPC, courts may be able to retain jurisdiction to make sure that the defendant and the agency actions comply with the court orders.[31]

APPLICATION OF PROCOMPETITIVE POLICY TO MAJOR MARKETS INVOLVED IN GEOTHERMAL ENERGY DEVELOPMENT

Procompetitive approaches are relevant to geothermal energy development in a number of ways, including practices among firms developing geothermal energy; practices of firms now engaged in supplying other fuels and becoming involved in the geothermal sector; practices of firms purchasing geothermal energy and electricity generated with such energy; and government activity which affects practices in all these sectors.

In the field of power generation, geothermal generating units could be developed by entities entering into the business of power generation, or their development could be restricted to the rate of advance prescribed by existing generating firms. A procompetitive policy would seek to foster entry so that development would not be restricted to the path prescribed by the needs and objectives of the large firms now dominating western bulk power supply.

In geothermal field exploration and development, a procompetitive policy would seek the existence of a number of producing and purchasing entities operating independently of each other, which are not constrained by considerations of how sales of one fuel might compete with sales of other fuels. The antitrust approach is concerned with both contractual and sales practices and with ownership patterns. Substantial lease holdings by a few large firms, coupled with the existence of some other firms as service companies for the larger firms, does not constitute a situation conducive to the diversity a procompetitive policy would seek. Foreclosure of entry by exclusive dealing contracts, warehousing of prime sites, unduly restrictive joint enterprise provisions, proliferation of joint enterprise relationships, or restrictions on the possibilities for smaller companies in allied fields of endeavor (for example, petroleum) are subjects for antitrust concern.

ANTITRUST ENFORCEMENT

The Petroleum Industry

Historically, the petroleum industry has been considered--in principle, at least--subject to antitrust policies. Some substantial efforts have been made to achieve competitive supply conditions[32] but some quite substantial deviations from this approach[33] have also occurred.

While antitrust enforcement did lead to reorganization of the oil industry early in this century, it was blocked for several decades from interfering with state industry rationing of domestic petroleum supply. It has been unable effectively to challenge joint pipeline ownership arrangements, even though there were several attempts at the staff level to move in this direction in the 1950s and 1960s. While some ameliorative provisions were included in the terms of the right of way permit for the Trans-Alaska Pipeline, and in deepwater port permits, no structural changes were developed.

The Antitrust Division appears to be effectively precluded from dealing with basic international petroleum procurement arrangements among producer countries and major oil companies which appear to channel a major portion of crude oil supplies into the hands of major companies.

With fundamental government policy favoring an international "safety net" supporting high petroleum prices,[34] seeking loan guarantees for synthetic fuels, and encouraging large petroleum companies to meet together, here and overseas, to plan for contingencies,[35] antitrust enforcement would have seemed anomalous, had it in fact been undertaken. Even where legislation has been enacted to restrict joint bidding for OCS leases and mandating production from federal leases, the restrictions pertain only to the very largest firms acting together and do not restrict unitization of production or joint pipelines.[36]

The Department of Justice has made an investigation of joint activity relating to the Elk Hills Naval Petroleum Reserve, and the

Pacific Outer Continental Shelf, but to date no action has been taken on these matters. The Justice Department has apparently declined to inquire seriously into possible antitrust problems arising from a joint bid by Union Oil Company and Standard Oil of California.[37]

As regards cross-fuels ownership, while not opposing coal acquisitions by larger oil companies, the Justice Department sought, unsuccessfully, to bar the acquisition of a coal company by General Dynamics. The district court in this proceeding was affirmed in its holding that the acquiring company was bringing its needed coal resources to a coal company that, having large mines, lacked the substantial uncommitted resources needed to make new large coal sales prerequisite to new large mines.[38] The court did not affirm or deny the district court's holding that an energy market, not a separate coal sector, was the relevant market.

The FTC has instituted a major action to achieve divestiture or disaggregation in the domestic petroleum industry. This suit has proceeded extremely slowly. The FTC and local U.S. attorneys have, over the years, brought numerous proceedings regarding anticompetitive local retail practices such as price fixing and wholesale tie-ins of accessories and tires to sales of gasoline to service stations.

The FTC successfully brought a proceeding to prevent a large copper company from acquiring Peabody Coal Company;[39] Kennecott is now preparing to sell Peabody, pursuant to court order, to a consortium including several large independent petroleum firms.

Antitrust activity in the petroleum industry has generally not penetrated to major structural arrangements in the last several decades.[40] This reflects the difficulty of assembling an adequately knowledgeable, far-sighted staff in a federal agency--a major problem for decades--and national and international concern over security of vital energy supplies.

The petroleum industry is notable for extensive vertical integration among stages of the industry and for a pervasive web of interfirm linkages. These facts, together with the industry's large size and its wealth, have created major concern for decades. This concern currently is reflected in, among other things, divestiture bills designed to break up major petroleum firms, introduced in Congress.

While there are pronounced differences between major oil companies--particularly between those that have crude oil supplies in excess of refining capability and those that do not--and the relative market positions of majors do change, the competition that does occur among them is quite constrained. The present study recommends either cross-fuel divestiture by major energy companies or, alternatively, that cross-fuel ownership be restricted, if such restriction can reasonably be expected to limit cross-fuel holdings so as to avoid substantial foreclosure of other firms, or diminution of interfuel competition. Furthermore, the restrictive approach should be chosen only if it can be expected to be effectively implemented in view of the prior lack of government action in petroleum matters. This implementation would, in all likelihood, require a nonpolitical administration of the antitrust agency in charge.

So long as the petroleum industry is not competitive, and is commanding high returns on investment, geothermal development by petroleum firms is likely to be at a very deliberate pace and at prices set with reference to other fuels, not to costs of production. To the extent that prime geothermal resources are in the hands of large petroleum companies, this pace will largely govern geothermal development.

Attention should be called, in any divestiture program, to the need to deal with petroleum company-financial market relationships, and with directorate relationships among major oil companies and large financial institutions. Were petroleum industry divestiture legislation enacted, such legislation should provide for relatively impersonal, arms-length, and competitive underwritings of securities of the spun-off entities. For example, public offerings should be sold to underwriters at competitive bidding (not negotiated) sales, and should be sold in batches small enough to permit a number of firms to be the lead underwriter.

In addition, it is suggested that antitrust enforcement bodies and the Congress take a longer-range view of the potential for competition among energy sources, and the need to preserve interfuel competition. Initial oil company acquisitions of coal companies were passed up by antitrust enforcement agencies at least in part on the hypothesis that coal and oil did not compete. Today that thinking can be seen to be obviously in error. In the future, all major energy sources must be potentially substitutable to provide electric power.

The Electric Utility Industry

The electric utility industry may be unique in the number and variety of trade restraints that have gone unchallenged for long periods of time.

In the bulk power industry, vertical integration of generation, transmission, and distribution, with attendant control of coordination services, provides great market power to large enterprises. This pattern of vertical integration is in large part the result of historical evolution in power technology. Early power systems, using small-scale generation equipment and lacking any means of long distance transmission, were local in character and consisted of generating capacity connected more or less directly to low-voltage distribution lines. As long distance, high voltage transmission technology developed, carrying with it the possibility of capturing scale economies by coordinated use of larger and more efficient generating units to serve a number of local distribution areas, the pattern of combining bulk and distribution facilities under a single ownership was maintained.

Such a pattern, however, is no longer required by technological factors. At least two other patterns are possible under modern conditions: separation of the bulk power supply industry from the distribution industry, with competition between bulk suppliers for the distributors' business; and joint ownership of bulk facilities in a region by local distribution entities, or increased use of smaller economic units (for example, geothermal energy).

One result of adherence to the industry structure that arose under more primitive technological conditions has been, as noted above, to allow large vertically integrated firms to retain dominant market power.[41] This power and control of facilities has been used to the detriment of small systems in a long series of market-foreclosing actions. These actions reduce the ability of small systems to enter into self-generation.

The concentrated structure and restrictive intercompany arrangements in the electric utility industry impede diversity and competitiveness in geothermal energy development. The number of buyers of geothermal energy is thereby limited by factors other than technical and organizational necessity, and the market possibilities of smaller purchasers are likewise limited.

Public Policy on Electric Utility Competition

Public policy toward electric utilities has sought to achieve economies of scale, avoid redundant investment, and maintain competitive conditions. Achievement of the first two goals has been sought through state franchising of retail services, and state and federal power facility licensing procedures. A major purpose of retail service franchises and licensing is to avoid local duplication of facilities. State franchising laws have been held to exclude generation or service by new entrants, and to prevent existing enterprises initiating service in new areas.[42]

State and federal agencies have acquired or reserved lands in order to secure suitable sites for large-scale operation. The FPC is to license only facilities that use the full economic capability of a hydro site in a manner consistent with the optimum development and operation of facilities at other sites on the waterway.[43] Site conservation is also a reason that permits for transmission line right-of-way across federal lands contain provisions requiring the permittee to wheel power for the United States on the permitted line.[44]

Some aspects of federal law are double-edged, seeking to secure both scale-economies and competitive results. These include requirements that some (but not all) agreements pertaining to joint utility enterprises be subject to regulatory review. Antitrust related provisions in hydro and nuclear licensing statutes encourage coordinated planning.

Federal statutory policy for disposing of federally produced power, and for federal research, evidences a procompetitive intent. Federal statutes governing the disposal of output from government power projects have sought to maintain yardstick competition, giving a preferential right of purchase to public-power bodies.[45]

Legislation pertaining to energy research conducted by or for ERDA contains clauses related to patents, inventions, and know-how, reflecting concern about competition. Continued congressional concern is reflected in the amendments adopted by the Senate in June 1976 requiring antitrust review in conjunction with any loan guarantee for biomass conversion projects.[46]

> Despite a continuing debate, it appears that the basic goal of direct governmental regulation through administrative bodies and the goal of indirect governmental regulation in the form of antitrust law is the same--to achieve the most efficient allocation of resources possible. For instance, whether a regulatory body is dictating the selling price or that price is determined by a market free from unreasonable restraints of trade, the desired result is to establish a selling price which covers costs plus a reasonable rate of return on capital, thereby avoiding monopoly profits. Another example of their common purpose is that both types of regulation seek to establish an atmosphere which will serve to stimulate innovations for better service at a lower cost. This analysis suggests that the two forms of economic regulation complement each other.[47]

Accordingly, the problem of exclusion of smaller systems from coordinating services, and anticompetitive provisions in sales to them, appear to be topics for redress by both antitrust and public utility regulation. In like manner, regulation of the disposition and use of the public domain should seek to foster competitive markets for the acquisition of leases and their exploitation.

Judicial Enforcement of Antitrust Policy

The judicial application of the antitrust laws to the electric utility industry goes back a number of years.[48] In related cases,[49] it was decided that contracts among utilities allocating territory and customers and reducing competitive growth were, per se, violations of the Sherman Act. The court held that: "In short, the grant of monopolistic privileges, subject to regulation by governmental body, does not carry an exemption unless one be expressly already granted, from the antitrust laws or deprive the courts of jurisdiction to enforce them."[50] The same court was faced with adoption by the FPC of certain provisions of the contracts the court had previously found unlawful. A judgment holding the contracts void was nevertheless affirmed. The court recognized the public interest in the continuance of the pooling arrangement which the illegal contracts had created. It determined that further action by the FPC would be required.

> The problem of the Commission would then be to decree a plan which will affect the purposes of the Federal Power Act and at the same time conform to the antitrust statutes. This would not seem to be an impossible or unreasonable duty to perform. As was said in Southern SS. Co. v. NLRB, 316 US 31, "Frequently the entire scope of congressional purpose calls for careful accommodation of one statutory scheme to another, and it is not too much to demand of an administrative body that it undertake this accommodation without excessive emphasis upon its immediate task."[51]

The foregoing and other cases established that electric utility systems are subject to the general limits on use of monopoly power.[52]

The leading judicial decision on antitrust issues in the bulk power supply area is the Otter Tail case.[53] Otter Tail Power Company attempted to prevent municipalities it formerly served, and in particular the Village of Elbow Lake, Minnesota, from forming viable municipal power systems. To this end, Otter Tail used, inter alia, its control over subtransmission in the area as a bottleneck to keep the village isolated, refusing to sell wholesale power or wheeling service to the village. Otter Tail also employed litigation and agreements with the Bureau of Reclamation in an effort to foreclose entrant power firms. The court in Otter Tail denied, in a four to three decision, the assertion by defendants that anticompetitive action could be justified to avoid losing customers, and that the bulk power industry had a blanket exemption from the antitrust laws by virtue of FPC regulation. FPC regulation is not pervasive[54] and can be accommodated with antitrust remedies. Otter Tail, the Pennsylvania Water cases, and other decisions thus show that such principles--discussed above--as the bottleneck doctrine, the policy against use of monopoly power in one market to monopolize another, and agreements allocating customers, apply to the electric utility industry.

In discussing judicial performance under the antitrust laws, mention should be made of the constraints arising from state or local franchising.[55] The role of antitrust law in the regulated industries is circumscribed by the existence of franchises and certificates.[56] However, state franchise limitations on antitrust law enforcement relate to retail trade, and not to interstate commerce.[57]

Utility Regulation

The federal regulatory agencies with the most substantial competitive policy responsibilities in the electric utility industry are the SEC, FPC, and the NRC. The SEC administers the Public Utility Holding Company Act of 1935. Agreements among utility systems for joint enterprises to own electric utility plant are subject to approval under Sections 9 and 10 of this act, since the joint enterprise --a separate entity--constitutes a subsidiary creating a utility holding company relationship. Utility mergers involving the creation or continuance of a holding company fall under this act. The tests for permitting mergers and acquisitions, found in Section 10, are more stringent than those normally applied under antitrust law: there must be a showing of positive benefits, and not just a lack of substantial harm, or the threat of harm, to competition.

The FPC, which is now part of the Department of Energy and has been retitled the Federal Energy Regulatory Commission, is entrusted with the administration of the Natural Gas Act and the Federal Power Act.[58] Under the Federal Power Act, the FPC has the authority to license and regulate some aspects of nonfederal hydroelectric power projects. It has authority concerning transportation and rates for resale of electric energy in interstate commerce. It has some authority over the accounts of companies engaged in intrastate transmission and generation and companies with hydro licenses.

The commission can pass upon the reasonableness of contracts, rates, and terms of service subject to its jurisdiction. In limited circumstances, it can require interconnection of facilities and sales of power. It cannot require planning coordination among utilities, or wheeling, except as a condition to hydro licenses. The FPC is required to approve the rates of some but not all federal power marketing agencies.[59] It also collects and provides information and reports to Congress regarding industry conditions and problems.[60] The terms of electric utility coordination agreements (including agreements to which the Bureau of Reclamation is a party) and wholesale tariffs are subject to FPC approval. The FPC is charged with the duty of reviewing these filings under a statutory standard that places maintenance of competition in an important role.[61]

Although the FPC has no power to adjudicate antitrust violations,[62] it can refuse to permit anticompetitive exclusionary pooling agreements, refuse to approve rates tying services, eliminate no-resale or load growth restriction provisions, eliminate onerous notification terms before new sources of power may be used, and attempt to reduce the incidence of price squeezes.[63] It can also deny mergers.[64]

While it lacks the ability to mandate coordinated planning or wheeling, the FPC could require that power services not be bundled into a firm service tariff for full or partial requirements, but rather that the elements of firm service--generation, transmission, and coordinating services--be separately available. It can require that hydro licensees coordinate their units with others so as to obtain the optimum economic use of these facilities--for example, their widest use as a source of ready reserves, their placement in the peak so as to maximize their dependable capacity rating, and their optimum use in substitution for fossil fuel.[65]

Though the FPC has authority, and has occasionally acted, to restrict anticompetitive tariff provisions, it has rarely made a serious inquiry into specific power pool agreements and has permitted numerous uncompetitive tariff provisions to stand for years without inquiry. For example, the FPC declined to take up the antitrust questions raised by public power systems when the 1967 Bureau-PG&E integration contract was filed,[66] and made no independent inquiry. The commission has framed no objective standards for coordination agreements to guide companies away from anticompetitive conduct and toward conduct that would achieve efficiencies and open up diversity, with attendant flexibility and competitiveness, in bulk electric power markets. Where a matter comes before an agency, such as the FPC or a state utility commission, antitrust considerations often are given short shrift. The FPC has engaged in little factual inquiry in power pooling cases.

Access to the facts in a case is basic to achieving reasoned decisions grounded on a knowledge of conditions. A party's control of information as to the facts in a matter is often dispositive. In order for agencies to obtain the facts, they require the services of skilled personnel able and willing to understand the subject matter involved, to assemble and report data, and to analyze and consider the facts in light of competitive considerations. The current system of administration is largely passive. Information pertaining to the planning of new bulk power facilities, or even the operation of power pools is largely unfiled. For instance, neither the operating guidelines of pool dispatches, nor pool planning documents are filed with the FPC. No inquiry is conducted into the structure of the industry or its planning processes.*

Inquiry of course is pointless unless relevant information is being collected. Cost accounting under the present Uniform System of Accounts for Public Utilities does not encompass any details for fuel subsidiaries, and does not provide information on incremental costs. Accordingly, exchanges of coordinating services or economy transactions are burdened by dissimilar accounts between firms attempting to trade on the basis of incremental or decremental outlays. Even worse, there is no uniform system for determining available capacity from a generating unit, nor have uniform standards been adopted for setting load-shedding relays. Settings and capacity ratings are usually not audited.

The FPC has only a very limited number of personnel who understand power pooling, conducts few field investigations, and has developed a substantial backlog of rate cases. Antitrust inquiries, even with motivated staffs, cannot be expected to go forward at a reasonable pace unless the party having the basic data--the utility--is motivated to cooperate. The FPC practice of separating antitrust inquiries from rate, financing, or even hydro-licensing proceedings, and undertaking to grant the application sought while the separate antitrust inquiry runs on, destroys such motivation.[67]

Improved agency antitrust performance will not be achieved easily. It would be helpful to develop separate antitrust staffs within agencies, or regular interventions by antitrust agencies,[68] together with specific statutory requirements that agencies make

---

*The state commissions lack jurisdiction over extrastate members of power coordinating groups, and they do not have jurisdiction over pool agreements, as these are filed with the FPC. Accordingly, their role in planning is attenuated. FPC electric power surveys are conducted by industry advisory committees. Regional surveying is only done by regional industry advisory groups.

findings regarding competitive impacts of proposed actions, which should include considering alternatives to proposals.[69] However, agencies have a history of protecting the status quo.[70] They are not likely to change their basic attitudes, while their pro forma findings might retard access to judicial remedies (by doctrines of primary and exclusive jurisdiction, unless this development was precluded by specific statutory language).

Increased judicial supervision of regulatory agencies with the judiciary imposing more stringent requirements for reasoned agency decisions that are based on substantial evidence could improve agency decisions. However, courts are reluctant to become overly involved in agency matters[71] and agency responses to remands are often merely productive of further post hoc rationalizations.

So long as agencies tend to reflect the views of the industries they regulate, regulation of pooling seems likely to go forward only if industry views (that is, structure) are nonuniform. In that circumstance, the agency would not be a mere interloper into the affairs of its friends when it inquired about power pooling transactions or their absence.

The structure of utility rates affects how cost-conscious utilities are about energy procurement practices. Utilities making money under fuel adjustment clauses[72] and through integration backwards into raw fuels and fuel financing may lack some incentives for experimenting with potentially lower cost, alternative types of energy.

Fuel procurement and ownership of fuel companies by electric utilities has received little surveillance by the FPC. The propriety of intrasystem fuel transactions goes unquestioned, fuel contracts need not be filed, and no system of accounts is mandated for fuel subsidiaries. Only very broad statements as to procurement policy need be filed; essentially this area has been left to managerial discretion. While noting utility assertions that noncompetitive fuel supply markets are a reason for not requiring fuel contracts to be submitted, the commission has taken no action to encourage more competitive fuel procurement.[73]

While arguments could be made regarding the very participation of utilities in nonutility enterprises, with the current lack of competitive conditions it appears more sensible to insist that utility-affiliated fuel companies be true competitive entrants. This can best be done by assuring that such fuel subsidiaries and their parent companies have an incentive to seek fuel supplies at prices lower than otherwise obtainable. This would be encouraged if fuel subsidiaries were required to conform their books to a uniform system of accounts. The sales of energy to an affiliate should be at prices set by cost (including a reasonable return), but not in excess of market. Otherwise, utility energy procurement should be required to be on a competitive basis, and no more than a limited percentage (say, 10 percent) of large procurements should be purchased from one concern.

Competitive procurement is more likely to occur if it can be supervised by regulators. Such oversight requires considerably greater disclosure than is now obtained by utility commissions regarding utility procurement practice, and direct and indirect financial relationships among utilities and fuel suppliers.

The Nuclear Regulatory Commission

The Atomic Energy Act provides for a prelicensing, antitrust review on applications for nuclear power plants.[74] Nuclear license applications are sent to the attorney general for review and advice as to whether issuance of the license would create or maintain a situation inconsistent with the antitrust laws.[75]

A number of antitrust related challenges have been made in nuclear licensing proceedings since 1970. Generally, the electric utilities involved have agreed to license conditions before a hearing has been conducted.* These conditions have been negotiated by the Department of Justice, the NRC staff, applicants, and, in some cases, members of the public (including smaller electric systems that intervene, or are potential intervenors).

Table 5.1 lists 23 primary applicants for licenses to construct and operate 54 nuclear generating units and arrays license conditions. All of the applicants are large investor-owned utilities. The average size of generating units involved has been 1,041 mw. Many of the units are joint ventures between the primary applicant and other utilities.

The license conditions agreed to come under four headings: unit access, transmission service, coordination, and contractual provisions.

Unit access refers to the applicant's granting another electric system participation in a nuclear facility to be licensed. It also includes provision of such transmission service as is required to obtain, and reserve protection for, that nuclear unit's output. Eighteen of the primary applicants have agreed to unit-access conditions. Some agreed to permit ownership in a particular unit, others to permit purchase of power and energy from a particular

---

*As of March 10, 1976, there had been only three full-scale evidentiary hearings.

TABLE 5.1

Catalogue of Antitrust License Conditions Agreed Upon

		License Conditions Effected			
Applicant	Agreement Date	Unit Access	Transmission Service	Coordination	Contractual Provisions
Arizona Public Service Co.	4/3/75		x	x	x
Carolina Power and Light Co.	8/18/72		x	x	x
Commonwealth Edison Co.	2/22/74	x			x
Detroit Edison Co.	8/31/71				x
Detroit Edison Co.	3/21/74	x	x	x	x
Duke Power Co.	4/16/74		x	x	x
Florida Power Corp.	12/6/71		x	x	x
Florida Power and Light Co.	2/25/74	x	x	x	x
Georgia Power Co.	4/30/74	x	x	x	x
Gulf States Utilities Co.	1/29/75	x	x	x	x
Gulf States Utilities Co.	3/20/74	x	x	x	x
Illinois Power Co.	4/5/74	x	x	x	x
Louisiana Power and Light Co.	2/3/75	x	x	x	x
Mississippi Power & Light Co.	5/22/73	x	x	x	x
No. Indiana Public Service Co.	11/5/71	x			
Northern States Power Co.	8/9/74	x	x	x	x
Philadelphia Electric Co.	5/20/74	x	x		
Public Service Co. of Indiana	3/18/75	x	x	x	x
Southern California Edison Co.	6/6/74	x	x	x	x
Texas Utilities Co.	1/15/74	x	x	x	x
Virginia Electric and Power	11/14/73	x	x	x	x
Virginia Electric and Power	7/26/74	x	x	x	x
Wisconsin Electric Power Co.	2/7/75	x	x	x	x

Note: See text for interpretation.

Source: David W. Penn, James B. Delaney, and T. Crawford Honeycutt, "The NRC's Antitrust Review of Nuclear Power Plants: The Conditioning of Licenses," NRC Staff Report (Preliminary).

unit. In a few cases, potential owners are offered the choice of ownership or purchases of unit power either as to the unit to be licensed or as to future units.

The transmission service category involves a wide variety of conditions. Some applicants are required to wheel power for all electric systems that request this service. Others are required only to wheel for certain designated systems. All but three of the applicants agreed to transmission service conditions.

The coordination category involves any of the following general commitments to other utilities: to interconnect and coordinate reserves (including, among other things, emergency and scheduled maintenance support); to engage in other bulk power transactions such as economy and diversity exchanges; and to participate in joint planning and development or generation and transmission facilities. All but four of the applicants agreed to such coordination conditions.

The contractual-provisions category involves requirements that applicants delete restrictive or discriminatory language from existing contracts, avoid using such contract language in future dealings, and make affirmative statements regarding intentions to enter into agreements to deal. Provisions disallowed include restrictions against interconnection and coordination agreements, restrictions on power pool membership, restrictions on the use or resale of transmitted power and energy, and refusals to deal with systems engaged only in distribution.

The direct beneficiaries of the conditions agreed to by nuclear license applicants are the small utilities located in the service areas of the applicants--mainly municipals and cooperatives. The pervasiveness of the antitrust challenges in licensing proceedings and the scope of the conditions agreed to suggest a widespread general pattern in which large investor-owned utilities have systematically acted to isolate, control, and/or acquire municipals and cooperatives by a variety of forms of conduct.

The license conditions obtained in the NRC proceedings may serve to thwart certain forms of conduct by certain large utilities directed against certain municipals and cooperatives. However, their efficacy is largely untested.

Even where coordination conditions have been agreed to, there are usually qualifying statements that prescribe reliability or duration characteristics necessary for transactions involving distribution systems. These qualifying statements could require unrealistic equipment conversion expenditures by small utilities, which have the effect of negating the purpose of the license condition.[76] The net result is that many small utilities are still eagerly searching for new sources of economical power and energy.

While no NRC antitrust cases have gone completely through litigation, and three are pending (awaiting commission action after decision by a nisi prius licensing board),[77] a number have been settled. Settled cases include a proceeding involving SCE in which that firm agreed to afford smaller power systems an opportunity to participate in ownership of nuclear generation, and to provide wheeling and back-up services. In a pending NRC proceeding, where the Justice Department has entered into a limited, somewhat vague, settlement agreement with PG&E, public power intervenors continue to litigate.*

State Utility Regulation

State utility commissions influence bulk power supply in a number of ways. They, and not the federal government, certify construction of most power plants and transmission. In this capacity they can attempt to block entry or they can encourage load growth coordination.

State utility regulation encompasses directly, or through municipal corporations, the franchising of utility retail service areas. Franchise restrictions can severely retard the ability of nonutility customers to enter into self-generation, especially if they want to string lines across a public road. In the past, franchise disputes have led to territorial allocations among utilities.

In the western states, engaging in the sale or exchange of electric energy, with certain exemptions for industrial plants, makes one a public utility.[78] PUC certificates of convenience and necessity required before new plants or transmission lines can be built, may not be forthcoming because of territorial allocations of utility service areas. Such allocations have defeated efforts by new entrants, or even existing firms, to serve areas or build plants.[79]

State agencies are often small, overwhelmed by utilities, and lacking in planning evaluation abilities. In some mountain states they have been very ready to permit rapid rate increases. New Mexico has gone so far as to permit utilities to raise rates to achieve allowed rates of return without further proceedings.

State utility commissions have not required that gas or electric utilities submit tariffs to provide supplemental power to firms

---

*SDG&E has recently announced plans to sponsor a nuclear power plant, and so will be subject to antitrust review. No action was recommended in conjunction with a review of a nuclear plant LADWP sponsored.

seeking to use geothermal or solar total energy systems, for all or part of their own energy requirements. For instance, Southern California Gas Company will not back up another primary source of energy beyond 1,000 cubic feet per hour.[80] A survey of western power companies indicates that some do not offer a tariff for back-up for on-site energy sources. Among those that do, rates can be from two to five dollars per kw per month (2.7 to 6.8 mills per kwh at 100 percent load factor).

Users of total energy systems generally either restrict themselves to a partial application (that is, heating and cooling but not electric power), or they install their own back-up units.[81]

In the field development sector of the geothermal industry, state governments could employ their unitization and prorationing either to protect parties from being excluded from production units, or to restrict production in an anticompetitive manner, as has occurred with petroleum production.

The Securities and Exchange Commission (and State "Blue Sky" Commissions)

Firms seeking to publicly offer their securities* are required to register those securities with state "blue sky" commissions, and generally, with the SEC. Moreover, they must file annual financial reports with these commissions. While some state commissions pass upon the suitability of proposed public offerings, the SEC is restricted to mandating appropriate disclosure in the prospectus (or offering circular) and registration statement. Information disclosed pertains to the description of the offerer, the nature of its business, its financial statements, and proposed use of the proceeds of the offer.[82] Material documents, such as contracts of substantial importance to the offerer, are also to be filed. Information disclosure burdens, of necessity, fall heaviest upon smaller offerers. They have fewer security sales over which to allocate attendant costs, more of the details of their business are material to investors, and their shenanigans are less likely to be hidden.

While some utilities file coal supply agreements, major petroleum companies do not file their joint-venture, concession, or buy-back agreements in conjunction with either registration statements or annual reports to the SEC. Small firms are required to disclose the

---

*The term "securities" under the federal laws is broadly defined so as to include transferable instruments of debt, and various types of joint venture offerings.

profits of their one or two lines of business, and even to disclose the particular geographic area where they wish to conduct exploration. Large firms merely report aggregate financial results. The competitive advantage is obvious.

Growth of project financing may some day lead to disclosure of financial results in narrower areas; however, its initial impact is to reduce disclosure by substituting off-the-balance-sheet financings and contingent liabilities for long-term debt. Disclosure of financial results is essential if investors are intelligently to compare investment alternatives and efficiently allocate capital. They may be hindered in doing so by aggregate reports of widely integrated firms.

The quality of disclosure by petroleum joint ventures has not been universally such as to bolster investor confidence. Only recently, and then only after criticism, did the SEC require that firms disclose or discuss variations regarding reserves quantities reported to different government agencies.[83] In the area of exploration and development of energy resources it should be noted that neither SEC, FPC, nor FEA has developed a uniform system of accounts.

Lack of agreed-upon accounting systems impairs the ability to compare firms, and precludes effective cost-based regulation of transactions among affiliates.* The SEC has been directed to assure the development and use of a system of accounts for the sectors of the petroleum industry locating, developing fields, and producing crude oil.[84] This work appears to be awaiting action of the private Financial Accounting Standards Board, as is contemplated in the SEC's authority. In the electric utility area, the SEC largely relies upon the FPC to supervise utility accounting.[85]

The Executive Agencies

While other executive agencies have jurisdiction that affects geothermal development (for example, the Treasury Department through tax and loan guarantee policy, and the FEA, primarily through its informational and intergovernmental coordination responsibilities) the Interior Department and ERDA play major roles.

---

*The historic SEC policy, under the 1935 Public Utility Act, of restricting utilities to utility operations and requiring that nonutility subsidiaries be operated for the benefit of utility or pipeline companies, is being eroded. For instance, intrasystem prices may now be at market, and tax savings of nonutility subsidiaries need not be shared to benefit affiliated utilities.

The Department of the Interior

The Interior Department has a determinative role in the rate at which geothermal energy will be produced. It is responsible for carrying out the Geothermal Steam Act, for leasing alternative domestic sources of energy,* for granting and supervising use of rights of way for transmission lines and pipelines, and for managing western federal power projects. It is also responsible for obtaining and making available information concerning energy production and use, and the resources of the federal domain. Interior has far-reaching discretionary authority to affect the ability of small geothermal and electric power systems to hold their own in energy markets. In each of its areas of responsibility, the department is by law directed to maintain a procompetitive stance, or to give preference to public entities.

Several factors indicate that the Interior Department has acted in a manner inconsistent with fostering diversity in industrial organization. For many years, the Interior Department has been assigned, by a history of political action, the role of governmental patron of the petroleum industry.† The Department has historically viewed its fuel information responsibilities more as a service performed for industry than as a means of providing necessary background for public policy decisions. For example, the Interior Department for years has maintained an information system well adapted to displaying demand, stocks, and supply data in a way useful to state prorationing schemes, but did not see fit to develop the energy data reporting needed for planning energy policy.[86] The Bureau of Mines has been content with a voluntary system of sometimes stale reports in which information on reserves or exploration effort has been largely lacking.

Interior's management of federal geothermal, petroleum, and coal resources has features inconsistent with realizing competitive conditions in the lease market, or expeditious development of the

---

*The Forest Service, Department of Agriculture, must approve leases on lands within its jurisdiction.

†The oil industry has often spoken through the National Petroleum Council (NPC). The announced purpose of the NPC is to act as an advisory council to Interior. NPC working groups of company representatives meet without public notice or a transcript of their proceedings Also the Interior Department has provided advice for the maintenance of oil import restrictions through its very small Office of Oil and Gas, where policy was formulated by and with personnel on one-year loan from major oil companies. This continued for over a decade.

leased resources. In no federal leasing activity has there been sustained effort to secure production at maximum efficient levels. Leasing in both coal and oil has persistently lacked adequate compensation and due-diligence features that might secure the value of locational rents for the public and prevent concentration of leaseholding, and inventorying of speculation.* Leasing is also commonly carried out on the basis of bonus, rather than royalty, bidding. This favors large bidders with ample supplies of cash.

The GAO has reported that no effort is made to enforce maximum acreage limitations on holdings of federal onshore oil and gas leases, nor are adequate lease records maintained.[87] Similarly, and contrary to OCS leasing legislation, no effort is made to require access for nonowners to offshore pipelines.[88] Investigators' reports about Interior Department lease supervision activity on the OCS are disquieting in their implications for the geothermal area. Disarray, lack of public information, and lack of regulation of a highly concentrated group of lessees are asserted.[89] In addition, the GAO has reported that the provisions of the Organic Act of 1879,[90] prohibiting ownership by USGS employees of stock in companies with a principal interest in the mining or production of materials generally classified as mineral resources, have not been enforced. Scrutiny of the federal geothermal leasing program shows that many of these more general deficiencies are visible in the geothermal area, and may be expected to hamper speedy and sound resource development.

As with coal and petroleum, the federal government has no comprehensive program for locating and evaluating its geothermal resources.[91] Lack of such a program encourages speculation and creates information entry barriers for small entities. Speculative withholding cannot be restrained by due diligence requirements without both the basic information needed to determine when and if a resource is capable of development, and effective regulations.

Leasing of better geothermal acreage has been by bonus bidding,[92] where large firms have a capital accumulation advantage. Though leases have been widely issued, there has been no government effort made or proposed to direct them to lessees most likely to develop resources. Deadening delays encountered by firms seeking to obtain administrative approvals necessary for exploration work are particularly difficult for small firms to bear.

---

*In 1976, the Interior Department issued due diligence requirements in the coal area. Coal leasing legislative proposals were then pending in Congress.

The extensive energy resources remaining on unleased portions of the federal domain could be developed in ways that could stimulate competition and diminish the present concentration of control of energy resources. A procompetitive leasing policy in which diversity of lease ownership and diligent lease development are sought is essential if small systems and firms are to be able to attempt geothermal energy development. Such a program (akin to those sought in recent coal legislation, and in pending legislation regarding the OCS) could be implemented by Interior Department action, or, if necessary, amending the Geothermal Steam Act to require royalty lease bidding, and enhanced federal exploration that could provide information required to effectively compel every lessee either to develop his leasehold at a reasonable rate or relinquish it.

In order to reduce the barriers to entry resulting from risks of exploration, the Interior Department (and state governments) should increase the amount of government-sponsored testing of public lands, and disclose the results of such tests. Increased testing of public lands, called for in both recent coal-leasing legislation and proposed OCS legislation, would increase the ability of the public land managers to design alternative competitive bidding systems, and lower entry barriers

The Interior Department (as well as ERDA) should act to create an enlarged market for specialized geothermal services. To this end it is recommended that the United States enter into a program of purchasing geophysical studies, well logs, and related analyses. The information so purchased should be publicly filed and should relate principally to unleased public lands. The opportunities to sell information to the government should be competitively available.

Exploration leasing might advantageously be separated from development leasing, and should be tried at several prime sites. This would permit first a determination of the existence and general extent of a site, and then determination of how that site can be developed in a manner most consistent with comprehensive development of the resource. Exploration could be done by government and by holders of exploration permits.

Separation of exploration from development could have several benefits. It could permit informed federal leasing, and avoid the substantial commitment of bonus money to unproductive areas. This money could be spent on productive properties to the benefit of consumers, shareholders, and the public interest. Publication of government exploration results would permit a more informed market for development. Separation of exploration from development is practiced overseas with great success. It has been suggested for the

OCS on a trial basis.[93] If employed for geothermal leasing, it might enhance the competitive prospects of smaller enterprises. If done in an informed manner, separating development leasing from exploration permits could assure lease development. Presently due diligence is sought indirectly as a product of lease bonuses and rents. Theoretically, the Department of the Interior could commence an administrative proceeding on a lease after it had not been developed for a period of some years. However, geothermal rents and lease bonuses are generally low; lease bonus bids may represent the value appraised less imputed carrying costs for speculative holding of the tract; and due diligence proceedings must be attempted on the basis of very vague guidelines and against a burden of proof resting on the government.

If development and exploration were split, there could be competition for development. Comparative proceedings regarding development could be conducted (on the basis of written submissions to conserve time).[94] Because of their scope, such proceedings could be approximately coincident with proceedings regarding development under NEPA. For such a proceeding to work, decision makers must be available who are in a position to make difficult choices promptly. This will require that the decision maker be placed high enough in bureaucratic order to be publicly accountable, that he be not principally engaged otherwise, and that he have sufficient tenure of office to withstand political pressure. These same requirements are needed if due diligence is to be seriously pursued. Because similar decisional problems will be arising in leasing other energy sources, a tribunal independent of the Interior Department is called for. Splitting exploration and development licensing will probably require legislative amendment of the Geothermal Steam Act, as would creation of an independent licensing tribunal.

Lastly, as to leasing, in view of the highly concentrated electric utility market in the western United States, and the history of anticompetitive actions in that market, public power entities should be afforded preferential access to geothermal energy leases.

Electric Power Activities

The federal power facilities of the Bureau of Reclamation have not been so employed as substantially to diminish or ameliorate the concentrated conditions in bulk power supply in the western states. Rather, federal facilities have been employed in a manner limiting access of smaller, potentially competitive entities to alternative sources of energy, and limiting the development of bulk power supply facilities and practices that could afford competitive opportunities for new energy sources. A substantial portion of the benefits of the federal facilities--it appears an unnecessarily large portion--have gone to the dominant utilities in the area.

Federal power marketing agencies should be taken out of the Interior Department (or the Department of Energy), reorganized as independent federal corporations, and authorized to construct transmission so as not to be dependent upon their competitors for integration of federal systems. Reviews of Interior Department documents indicate that the professional staffs of these agencies have usually resisted, in vain, decisions by political appointees which have repeatedly accommodated the desire of large private utilities to be free of competitive alternatives.

Stipulations in federal right-of-way permits will pertain to almost all, if not all, new western high voltage lines. These stipulations should require that permittees offer transmission services and attendant planning and operating coordination services to others. Public transmission-line-corridor planning proceedings should be promptly undertaken to seek the best layout and use of the limited number of high voltage transmission corridors in the West. These planning proceedings, which could be undertaken in a NEPA framework, should seek to assure the development of corridor plans that are sufficient for regional needs, and that are so laid out as to permit looping of lines to meet growing loads, and access by line owners and nonowners via spurs of reasonable length.

Proposed right-of-way permit terms should be publicly noticed with opportunity to comment. Formalized planning, if undertaken in advance of needs, need not result in administrative delay, while it can discourage the tendency to waive wheeling provisions on key lines.

Power marketing authorities should review all present federal power contracts and rights of way to determine whether they are consistent with antitrust law and policy, and whether the representations given to the United States, and relied upon when the contracts or rights of way were let, have been adhered to. Where the contracts and rights of way are found to be inconsistent with antitrust law, or to contain provisions induced through false or erroneous representations, action to abrogate these agreements, as contrary to public policy and law, should be carefully considered. Particular attention should be paid right-of-way provisions waiving wheeling requirements. Major federal power marketing contracts and right-of-way permits should be granted only after public notice and opportunity for comment on economic as well as environmental aspects. The power marketing agencies should be authorized to develop geothermally driven capacity to augment and firm up their hydro capacity.

The lack of bureau efforts to study coordination among bureau systems or to provide California public entities with alternative routes to bulk power coordinating services, independent of CALPP, significantly retards the ability of industrial, municipal, or smaller private utility firms economically to develop firm geothermal bulk power supplies.

The Energy Research and Development Administration (ERDA)

ERDA is charged with the formulation and execution of most of the federal research program relating to geothermal energy and to bulk electric power supply. This authority is given under a rather broad writ regarding procedures, and very general policy guidelines.[95] ERDA-funded projects may be expected to comprise a very great portion of total geothermally related research.

ERDA states that the goal of the National Geothermal Energy R&D Program is "to work with industry to provide the Nation with an acceptable option which, if exercised, would permit the timely exploitation of our substantial geothermal resources."[96] Whether the industrial structure which is being fostered by ERDA research will permit future options is not clear. To work with industry, one first should determine the type of industry one wants to work with.

Federal energy research programs have an important subsidizing role in the development of the geothermal industry. In some of these programs, such as loan guarantees or project assistance contracts, the subsidy to some firms can be quite direct. It is obviously important that federal research be carried out in a manner that provides for equal treatment of all potential beneficiaries,* and in a manner that encourages a competitive industrial organization conducive to entry by innovators.

Care is required both in allocating subsidies and in choosing what types of subsidies should be employed. An example of problems in the allocation of subsidies is seen in the program for guaranteeing loans for geothermal projects. ERDA regulations for that program appear to preclude loan guarantees to publicly owned enterprises. An example of problems in the choice of subsidies is the orientation of ERDA programs toward supporting work by private firms, while not pursuing development of authorized government demonstration projects.[97] No effort is made to combine the carrot of federal subsidies with the yardstick of federal demonstration plants.

---

*A case can be made for a federal policy giving preferential treatment of small enterprises and publicly owned power systems.

ERDA's predecessor, AEC, instituted programs at two national laboratories to locate and develop two geothermal resources to the actual point of having two demonstration generating stations in operation. A program was begun in 1973 by the Lawrence Berkeley Laboratory (LBL) to find a site for a 10mw geothermal demonstration plant in northcentral Nevada. At the reported behest of the Office of Management and Budget, the plans to carry the project through to the demonstration plant plans were dropped when ERDA came into being.

The field stages of the project did continue. To assure the availability of adequate land for the eventual development, in 1973 some 88,000 acres of public land were withdrawn from the federal leasing program until December 1975, at which time only 5,000 acres would be retained for development work. The program proceeded to develop and test geochemical, geophysical, and geological techniques for searching for geothermal resources. With the use of these experimental techniques, interest came to center on the Grass Valley area where extensive tests were conducted. The logical culmination of these experiments would be the drilling of several wells to corroborate or refute the experimental results. This was recognized by both ERDA and LBL personnel.

The oil industry had not been sitting by, however; it had lobbied for cessation of the LBL program and against any withdrawals of federal land from leasing or any deep drilling. When the LBL program came up for assessment, a review panel was convened, consisting of several representatives of universities funded by industrial money, a representative from USGS, and two representatives from industry--one from SOCAL and one from its affiliate, AMAX.

This meeting was quickly converted by the head of the ERDA geothermal reservoir assessment program, Dr. John Salisbury, from a review of LBL's work to a general, albeit unannounced, advisory meeting on the propriety of federal drilling and lease withdrawals. Panel members generally agreed that a coordinated drilling program was in order, but that further work should first be done to pick the location and type of holes to be drilled; it was suggested that a series of moderate-depth holes precede deeper drilling.

The panel unanimously agreed that LBL had done an excellent job and had acquired an impressive data base with a very broad range of techniques. They recommended that LBL determine the cost and configuration of the holes which should be drilled and that further consideration then be given to the drilling program. The ERDA manager in charge of the LBL program recommended that the program be wound up. While urging LBL to analyze all existing data and plan final field programs so as to provide the best possible reservoir models, particularly in Grass Valley, and to recommend the location and configuration of confirmation drill holes, he stated, "It is further recommended that the laboratory plan and coordinate the drilling program with industry."[98] He recommended that all withdrawn parcels be released for public leasing, since a government demonstration plant is no longer being considered for the area.

The ERDA program manager's recommendations, and the subsequent termination of the LBL program reflect the strongly stated views of the head of ERDA's program of assessing geothermal reservoirs, Dr. John Salisbury. At the review panel meeting, Dr. Salisbury outlined petroleum company opposition to the LBL program's withdrawal and assessment of public lands, on the grounds that it would raise the price of leases for tracts explored. He then indicated his opposition to the program as it was not in step with his ideas of the ERDA program. Dr. Salisbury stressed that the orientation of ERDA was to foster industry, and that the northern Nevada program was politically a negative factor, and "clearly since it is our purpose to foster the geothermal industry we are not going to continue to pursue this program." He later stated "quite clearly [that] within the economical, political and philosophical context of the ERDA program as it should be, this Northern Nevada project is kind of an embarrassment frankly,"[99] and that the northern Nevada program did not fit into the framework of the ERDA program.

Subsequently, the LBL program budget was cut to $300,000 per year from $700,000 and wound up without drilling. The land in Grass Valley ceased to be withdrawn and is proposed to be leased by BLM without any special conditions regarding drilling or the availability of drilling results.

The lack of leasing conditions is interesting. The review panel member from Chevron (SOCAL), while urging against withdrawing large areas of public land, and favoring of reliance on industry for drilling, suggested that "where so much basic data has been obtained at public expense, a well within the block should be required to a specific depth within the first year after issuing the leases."[100] No provision has been made for obtaining drilling results from a lessee to corroborate or refute LBL's research.

The ERDA geothermal resource assessment program is proposing to provide part of the financing for drilling by private entities in southern Utah in return for data from the drilling. The proposed program would most directly aid the limited number of firms with which subsidy contracts are to be negotiated, and the time delay before acquired data is published may be expected to be a point of negotiation.

The orientation of ERDA's reservoir assessment program to the subsidization of a limited number of private wells and away from federally sponsored evaluation and demonstration development of federal lands has gone on with little comment. ERDA has embarked upon a course of working with industry, not of leading it. This approach is fraught with opportunities to turn the agency into a welfare department for chosen firms, with its employees being tempted to be zealous welfare workers.

The geothermal program's decision to subsidize, rather than to provide a yardstick or a prod, raises important questions as to the balance of approaches undertaken to foster development of new domestic energy sources. Procompetitive research programs whose benefits are accessible to entrant enterprises are more likely to occur if research programs are formulated in a framework of publicly revealed data and rationales regarding a program's competitive impact, its costs and benefits, and, relatedly, its alternatives. Such procedures might direct federal programs toward subjects having a range of capital requirements comprehending the potentials of small as well as large firms, and toward more competitive procurement practices.

Procedures for government research grants and contracts are more informal than those for other types of government procurement. This informality reflects a legal view concerning the nature of grant making and research contracting--that such grant making and contracting are matters committed to the widest discretion of the government agencies involved. Grants, unlike contracts, in this view are not considered to confer legally protectable rights upon recipients. Research contracts have been exempted from normal government procurement requirements for competitive bidding, and can be let after nonpublicly disclosed negotiations.

As the role and extent of grants and research have developed, there appears to be a tendency toward greater formalization of processes and increased use of public notice before grants or contracts are let, so as to make their procurement more competitive. The law is also starting to inquire into the evenhandedness among putative beneficiaries with which grant programs are administered.[101] The discretion vested in grant makers has been subjected to legislative criticism.

Informal processes and broad discretion may expedite decision making. Unfortunately, they appear to preclude external discipline and give the appearance of being arbitrary and even capricious. Decisions not supported by reasoned discussion are likely to be questioned where large sums of money are involved and where important policy factors should be taken into consideration. Formalization, and on-the-record proceedings, are likely to come into greater vogue for such decisions. Their use might reduce the possibilities for lobbying by one segment of an industry to the detriment of another,[102] or possibilities for conflicts of interest.

Federal law governing the acquisition of services and property has long favored competitive acquisition.[103] Similarly, a procompetitive policy has been established for the sale of federal plants and other surplus property,[104] and power output. In the research area, Congress has mandated a procompetitive policy regarding resulting patents.[105]

CONCLUSION

For geothermal energy to provide the portion of western states power generation that it economically might, development must be carried out by small utilities or industrial power users matched with small geothermal firms at sites sufficiently promising to embolden these firms to risk development capital. These firms can engage in geothermal development only if they have access to bulk power coordination and transmission services proffered on reasonable terms. Larger utilities take a more deliberate pace in geothermal efforts; large oil companies warehouse resources. Small utilities and industrial firms confront a number of problems in seeking geothermal self-generation. However, small specialized technology firms and small utility systems have shown considerable interest in developing geothermal power.

Small geothermal firms are not generally able to outbid large oil companies for leases or to match them in warehousing tracts. The holdings of the smaller geothermal exploration and development companies are dwarfed in size by the hundreds of thousands of acres held by large petroleum companies such as SOCAL, Phillips, or Gulf Oil. The independents present are either substantial petroleum companies or partners of a large entity.

The smaller geothermal enterprises are generally directed toward functioning as service enterprises for either larger petroleum companies or utilities. They also serve industrial enterprises, particularly Dow Chemical and AMAX. AMAX, however, now is subject to control by SOCAL, so that Dow, a substantial producer of oil, is the sole large industrial firm not principally involved in oil that is engaged in geothermal activity in a large manner--particularly through its holdings in the Magma companies.

Entry barriers relating to capital and land, coupled to some extent with the extension of habits carried over from other industries, have resulted in the geothermal field exploration and development sectors being organized on the pattern of the petroleum industry. Larger petroleum and utility enterprises have acted in ways that appear to increase entry barriers for smaller firms. They have preempted many of the better sites, and in joint enterprise agreements have partially limited the availability of services of technological enterprises. Entry barriers encountered in the early days of geothermal development may permanently disadvantage smaller firms. The acquirers of prime geothermal sites may be expected to have a leg up in obtaining, internally or by purchase, the highly essential know-how that will be a prerequisite to economic entry into present or future geothermal exploration and development.

The dominance of oil companies in geothermal resource holdings is cause for concern, both because it may tend to produce a concentrated structure in geothermal energy development, and because larger petroleum firms are not likely unilaterally to reduce the price of a marginal utility fuel, and thereby directly affect their returns from their primary product. Independent developers are not likely to be able to postpone production while waiting for smaller utility systems to overcome entry barriers.

The attractiveness of geothermal resources is largely related to their potential as a lower cost energy source. Both large oil companies and small independent geothermal developers seek oil-based prices for geothermal energy. Larger oil companies have preferred to prove up their geothermal reserves, on their own drilling schedules. They have not sought to enter into joint ventures with small power systems. Smaller geothermal developers must and do enter into joint ventures to obtain capital and land tracts. In such ventures they can only charge for their--and not the power system partner's--share of production. Small developers' needs for cash flows may reduce their ability successfully to bargain for prices far above costs.

Large petroleum companies exercise control over the West Coast petroleum industry, limiting the competition from others in that market, and influencing supply routes and volumes, without effective government check. They are extending their relatively concentrated influence over energy supply to the OCS and to Alaska (by leases and the Trans-Alaska pipeline agreement).

The practice of controlling supply is logically extended when oil earnings are used to acquire alternative energy sources, buttressing market power as barriers are raised to access by others--first to resources and then, as a product, to technology. Joint acquisition and maintenance of energy market power by major oil companies and large utilities, in these respective areas, has been little resisted, and on occasion actively assisted by government action. The government has conducted its resource management and its regulatory programs in ways that buttress entry barriers and retard participation by small enterprises.

Application of antitrust law and policy has been far less than that required to produce more competitive conditions. Parallel, if not collusive, conduct has been encouraged by such means as IEA, National Petroleum Council,* and FEA marketing regulations. More competitive energy supply markets would be conducive to geothermal development. Such markets may be anticipated only if the energy companies dominating utility fuel markets are reorganized by government.

In the electric utility area, power pooling (sharing of resources) and coordination practices have been such as to discourage both the development of power generation by smaller systems, and the development of federal or state transmission through which the smaller power systems could coordinate and pool geothermal generation.† Moreover, federal power marketing and right-of-way practices have accommodated efforts of large utilities to dominate bulk power supply.

Diversity and competition in bulk power supply can exist only if access to coordination services is available. Such access is necessary if more than one buyer is to be available in the Geysers area or in other geothermal regions. Without more than one buyer, contracts for the sale of geothermal energy are likely to continue--as is the case in the Geysers--to place all the risks of carrying charges for field investment associated with delays in the construction and operation of generating units on the field developer.

Access to power pooling can be opened only with the help of government. Effective government action for more competitive and efficient industry structure and practices will come about only if the government has data and articulated policies that are consistently implemented. Efforts to obtain uniform cost data are just beginning to get under way in both the fuel and utility industries.[106] Creating a government approach to energy industries that seeks diversity and competition is a massive undertaking. Such an approach must overcome agencies' habit of "getting along" with the industries with which they are in contact.

---

*This group, established as an industry advisory committee under the Interior Department, has a largely private, substantial existence. While there was for a time some dispute as to its existence as an Interior or as an FEA entity (the requirements imposed on FEA committees would have substantially affected its character and operations), it has since lapsed safely back into its role as adviser to Interior. NPC activities have involved substantial exchanges of information among its members regarding subjects such as storage capacity, transport facilities, geology, refinery capabilities, and emergency preparation.

†For example, California Department of Water Resources.

This study does not paint an encouraging picture for geothermal development. Rapid development requires that independent developers and small industrial or utility power systems develop projects not yet begun. They must overcome a number of obstacles in government and in the environment for entry and innovation in the fuel and bulk power industries. Development of new sources of energy, especially those that could provide a decentralized alternative to large projects and rising prices, requires action going counter not only to the general flow of much of present government action, but also to the perceived interests and economic power of large and dominant firms.

Development of geothermal energy for bulk power supply could provide, first, a useful new energy source; second, competition in the supply of fuel for power generation; third, diversification of ownership in the fuels sectors; and fourth, perhaps in the long run, a type of government planning that favors individual choice in decisions to consume and produce, avoiding government ownership and control of resources, and excessive political influence on the part of a relatively few private entities.

NOTES

1. Northern Pacific Railway Corp. v. United States, 356 US 1, 4 (1958).

2. In a leading case on the acquisition of a substitutable product, a large can company was prohibited from acquiring a glass container manufacturer. United States. v. Continental Can Co., 378 US 441 (1964). Joint ventures have been analyzed as akin to horizontal mergers. United States v. Penn-Olin Chemical Co., 378 US 158 (1963).

3. Brown Shoe Co. v. United States, 370 US 294 (1972) (look to industry trend); and United States v. Von's Grocery Co., 384 US 270 (1965).

4. Ford Motor Co. v. United States, 405 US 562 (1972); United States v. E. I. DuPont de Nemours & Co., 353 US 586 (1957); United States v. Bethlehem Steel Corp., 168 F. Supp. 576, 1958 Trade Cases §69189 (SDNY, 1958) (foreclosure of supply to independent manufacturers); United Nuclear Corp. v. Combustion Engineering, Inc., 1969 Trade Cases §72969 (E.D. Pa.); and United States. v. Kimberly-Clark Corp., 1967 Trade Cases §72081 (N.D. Ca., 1967) (acquisition of outlets in merger trend). Bottleneck cases are discussed hereinafter.

5. Neither the letter of the law nor its purpose distinguishes between strangling a commerce which has been born and preventing

the birth of a commerce which does not exist. United States v. General Dyestuff Corp., 57 F. Supp. 642, 648 (S.D.NY, 1974); United States v. United Shoe Machinery Co., 247 US 32, 53 (1918). And see United States v. Phillips Petroleum Co., 1973 Trade Cases §74789 (C.D. Ca., 1973); United States v. El Paso Natural Gas Co., 376 US 651 (1964).

6. United States v. Penn-Olin Chemical Co., 378 US 158, 194 (1963).

7. FTC v. Proctor & Gamble Co., 386 US 568 (1967); Kennecott Copper Corp. v. FTC, 467 F.2d 67 (CA10, 1972). Cf. Phillips Petroleum, supra.

8. Timken Roller Bearing Co. v. United States, 341 US 593, 1950-51 Trade Cases §62837 (1951).

9. United States v. Minnesota Mining and Mfg. Co., 92 F. Supp. 947, 1950-51 Trade Cases §62687 (D. Mass 1950).

10. United States v. Griffith, 334 US 100, 92 L ed 1236; United States v. Klearflax Linen Looms, Inc., 1944-7 Trade Cases §57407 (D. Minn., 1945). United States v. Southwestern Greyhound Lines, Inc., 1952-53 Trade Cases §67470 (N.D. Okla. 1953); United States v. Terminal Railroad Association, 224 US 383 (1912). Concerted group action is similarly illegal. Fashion Originators Guild v. Federal Trade Commission, 312 US 457, 85 L ed 949 (1941). Misuse of monopoly power is not legitimated by the existence of a business motive, United States v. Arnold, Schwinn & Co., 388 US 365, 375 (1967); Otter Tail Power Co. v. United States, 410 US 366, 380 (1973).

11. Terminal Railroad, supra; Lorain Journal Co. v. United States, 342 US 143,96 L ed 162 (1951); Associated Press v. United States, 326 US 1, 89 L ed 2013; United States v. Aluminum Company of America, 148 F.2d 416 (CA2, 1945); and Otter Tail Power Co., supra.

12. Standard Oil Co. v. United States, 337 US 293 (1949); United States v. Arnold, Schwinn & Co., 388 US 365 (1967); and Anderson v. American Automobile Ass'n., 1972 Trade Cases 73793 (CA 9, 1972).

13. Siegel v. Chicken Delight, Inc., 448 F.2d 43, 51 (CA 9, 1971), 1971 Trade Cases §73,703, cert. denied, 405 US 955 (1972); Copper Liquor, Inc. v. Adolph Coors Co., 506 F.2d 934, 942-43 (CA 5, 1975), 1975-1 Trade Cases §60128.

Anderson, supra, teaches that: "To sustain the restraint, it must be found to be reasonable both with respect to the public and to the parties and that it is limited to what is fairly necessary, in the circumstances of the particular case. . . ." Dr. Miles Medical Co. v. Park & Sons Co., 220 US 373, 406 (1911).

"The promotion of self-interest alone does not invoke the rule of reason to immunize otherwise illegal conduct. It is only if the conduct is not unlawful in its impact in the market place or if the self-interest coincides with the statutory concern with the preservation and promotion of competition that protection is achieved." United States v. Arnold, Schwinn & Co., supra, p. 375.

14. See, A.B.A., Antitrust Law Developments (1975), p. 43.

15. Standard Oil Co. of California v. United States, 337 US 293, 314 (1949).

16. Cf., Tampa Electric Co. v. Nashville Coal Co., 365 US 320 (1961); and Antitrust Law Developments, supra, pp. 44-45.

17. See United States v. Loew's, Inc., 371 US 38, 55 (1962); and Northern Pacific Ry. v. United States, 356 US 1, 6 (1958).

18. Standard Oil Co. v. United States, 337 US 293, 305-6 (1949).

19. Northern Pacific Ry. v. United States, 356 US 1, 6 (1958).

20. United States v. Jerrold Electronics Corp., 187 F. Supp. 545; affirmed per curiam, 363 US 567 (1961) (new product).

21. F. M. Scherer, Industrial Market Structure and Economic Performance (Chicago: Rand McNally, 1971), pp. 505-6.

22. Accommodation of regulatory and antitrust legislation is sought in cases such as Ricci v. Chicago Mercantile Exchange, 409 US 289 (1973) (primary jurisdiction); Cantor v. The Detroit Edison Company, supra (state regulation); Northern Natural Gas Co. v. PFC, 399 F.2d 953 (CADC, 1968); and FPC v. Conway Corporation, 426 US 271 (1976).

23. Silver v. New York Stock Exchange, 373 US 341 (1963); Gordon v. New York Stock Exchange, 422 US 659 (1975); and U.S. v. National Association of Securities Dealers, 422 US 694 (1975).

24. These duties are either implicit in a public interest standard, or explicit. Northern Natural Gas Company v. FPC, supra; Gulf States Utilities v. FPC, 411 US 747 (1973); FPC v. Conway Corporation, 426 US 271 (1976), 1976-1 Trade Cases 60912 (public interest standard); Mineral Leasing Act of 1920, 30 USC 181, 184 (h), (k), 185, 202, 43 USC 970 (forfeit pipeline rights-of-way if antitrust laws violated). Cf., Denver Petroleum Corp. v. Shell Oil Co., 306 F. Supp. 289 (D.C. Colo. 1969); 43 USC 1334 (c) (oil pipelines); 43 USC 31 (The director and members of the Geological Survey shall have no personal or private interests in the lands or mineral wealth of the region under survey, and shall execute no surveys or examinations for private parties or corporations); 43 USC 485 (h), 522 (preference for public agencies and REA cooperatives); 43 USC 617 (d) (transmission line use from Boulder Canyon [Hoover Dam] Project); Deepwater Ports, 33 USC 1501, 1503 (c), 1504 (c), 1505 (i), 1506-07, 1511, 1513.

25. Cf., Otter Tail, supra; Ricci v. Chicago Mercantile Exchange, 409 US 289 (1973); Jaffe, Primary Jurisdiction, 77 Harv. L. Rev. 1037, 1043-47 (1964).

26. California v. FPC, 369 US 482 (1962); United States v. Philadelphia National Bank, 374 US 321, 350-51 (1963); Otter Tail, supra.

27. See, for example, Gulf States Utilities Co. v. F.P.C., 411 US 747 (1973); and, F.M.C. v. Svenska Amerika Linien, 390 US 238 (1968).

28. Jaffe, Primary Jurisdiction Reconsidered, 102 U. Pa. L. Rev. 577 (1954).

29. 376 US 398 (1964).

30. See 22 U.S.C. 2370 (c) (2).

31. See Litton Systems, Inc. v. Southwestern Bell Telephone Co., 1976-2 Trade Cases para. 61084 (CA 5, 1976) wherein the court suggests this approach. It is not clear how far the courts may do so under the present Federal judicial code. Cf. FPC v. Transcontinental Gas Pipe Line Corp., US, 46 L. Ed. 2d 533 (1976). Legislation could easily remedy any deficiencies in that respect.

32. This concern is reflected in provisions of the Hepburn Act making oil pipelines common carriers, the Federal Water Power legislation which culminated in the Federal Water Power Act of 1920 (now Part I of the Federal Power Act), the 1911 antitrust proceeding against the Standard Oil Trust, the provisions of the Mineral Leasing Act of 1920, and the reports of government agencies and Senator LaFollette's Committee on Manufacturers; S. Report No. 1263 67th Cong., 4 Sess. (1923), cited in Report on the Petroleum Industry Competition Act of 1976, U.S. Congress, Senate, Judiciary Committee, Part 1, Rept. 94-1005, 94th Cong., 2 Sess. (1976).

33. See, for example, the Connally Hot-Oil Act permitting state prorationing of production, the Natural Gas Act requiring certification of interstate gas pipelines, oil import restrictions imposed in the late 1950s, and provision of the Outer Continental Shelf Act not requiring the secretary to use other than cash bonus bidding when leases are let, and not requiring offshore pipelines to be common carriers. 43 USC 133, 1334(c) and 1337.

34. V. Oppenheim, "Why Oil Prices Go Up: The Past: We Pushed Them," Foreign Policy 25 (Winter 1976-77): 24. The safety-net proposal is discussed in U.S. Congress, Senate, Foreign Relations Committee, Multinational Oil Corporations and U.S. Foreign Policy, 93rd Cong., 2 Sess. (1975), p. 4; and U.S. Congress, Joint Economic Committee, The State Department's Oil Floor Price Proposal: Should Congress Endorse It?, 94th Cong. 1 Sess. (1975); Testimony of M. A. Adelman before Joint Economic Committee, January 12, 1976, reprinted at pages S.956-59, Congressional Record, Vol. 122 (January 18, 1976). Under this concept high oil

prices are sought to encourage conservation and to encourage investment in synthetic fuels.

35. The International Energy Agreement (IEA), justified as a burden-sharing program in the event of another oil embargo or general shortage, operates through a voluntary agreement among major petroleum companies to develop a worldwide emergency contingency plan and, to that end, to exchange information so as to permit firm coordination of production, transportation, and refining. Actions taken pursuant to the voluntary agreement (41 FR 13998, April 1, 1976; 41 FR 24772, June 18, 1976) have antitrust immunity. Energy Policy and Conservation Act of 1976, Sec. 252. These actions, such as meetings of company representatives, have received sparse supervision from the Justice Department and from the FTC. During a recent trial run of the contingency plan, the companies were permitted to communicate with each other and to exchange data directly without government supervision. Section 5, Voluntary Agreement and Plan of Action, as amended, and see 41 FR 41459 (September 22, 1976). While the need to prepare for emergencies certainly may justify waiver of antitrust law provisions in appropriate cases, the general approach of such matters is that waivers should never go beyond the scope essential to meeting the emergency, nor should antitrust supervision of side effects be neglected.

36. Energy Policy and Conservation Act, Public Law 94-163, 89 Stat. 871 et seq.

37. Letter of March 2, 1976 to Senator John Tunney from Assistant Attorney General Kauper.

38. United States v. General Dynamics, 415 US 486, 39 L. ed 2d 530 (1974) (Acquisition of United Electric Coal Companies in 1959 by firm owning Freemont Coal Mining Corp, which firm in turn was acquired by General Dynamics which thereby became the nation's fifth largest coal producer).

39. Kennecott Copper Co. v. FTC, 467 F.2d 67 (CA10, 1972) cert. denied, 416 US 909 (1974).

40. The paucity of antitrust activities of the FTC and of the Justice Department which sought to go beyond local price-fixing schemes has been criticized by congressional committees. This history has been recounted in the U.S. Congress, Senate, Judiciary Committee, Petroleum Industry Competition Act of 1976, Report No. 94-1005, 94th Congress, 2 Sess. (1976), p. 94 et seq.; the report of the Subcommittee on Multinational Corporations, Senate Foreign Relations Committee, Multinational Oil Corporations and U.S. Foreign Policy, 93rd Cong., 2 Sess. (1975), p. 33 et seq.; the reports of the Special Subcommittee on Integrated Oil Operations, Senate Interior Committee, The Burmah-Signal Merger, 93rd Cong., 2 Sess. (1974), pp. 5-7; and An Analysis of the Proposed Standard-

Occidental Merger, 94th Cong., 1 Sess. (1975); the report of the Subcommittee on Consumer Economics, Joint Economic Committee, International Economics and Priorities and Economy in Government, 93rd Cong., 2 Sess. (1974); and the report of the Subcommittee on Special Small Business Problems, House Select Committee on Small Business, Anticompetitive Impact of Oil Company Ownership of Petroleum Products Pipelines, 92d Cong., 2 Sess. (1972), pp. 29-31.

41. The Public Utility Holding Company Act of 1935 (15 USC 79 et seq.) was aimed at rationalizing the industry structure. It struck at the acquisition and retention of scattered, nonintegrated properties through the "single integrated system" standard of §11 (15 USC 79k). Thus it tended to promote vertical integration by restricting the activities of a firm to an area where such integration was feasible. In recent years, the merger and acquisition sections of the act have been found to require SEC scrutiny of joint generation projects that might adversely affect distribution competitors precluded from participation in a joint generation enterprise, and to limit horizontal acquisitions having anticompetitive consequences. For example, the SEC has barred a merger of several New England systems. New England Electric Systems, ___ SEC ___ (1976). In Municipal Electric Assoc. of Massachusetts v. SEC, 413 F.2d 1052 (C.A.D.C., 1969), the SEC was instructed to consider the anticompetitive effect of creating a jointly owned nuclear power generating company on a municipal power system excluded from participation therein.

42. See Cottonwood Mall, supra; Gelmar v. PSC, 67 Utah 222, 247 P.2d 284 (regulated monopoly, not competition is Utah policy); San Miguel Power Assn. v. PSC, 292 P.2d 511 (Utah, 1956) (Co-op is not a public utility and so can not object before PUC to extension by Utah Power and Light Company into its service area); Public Service Co. of Colo. v. PUC, 350 P.2d 543 (Colo., 1960) (excluding co-op from cities and from company's unserved but certified area); and Western Colorado Power Company v. PUC, 411 P.2d 785 (Colo., 1966) where Colorado-Ute, an entrant generation and transmission cooperative, sought to build a 150mw unit. This unit was opposed by Western Colorado Power Company (then a subsidiary of Utah Power and Light) and by Public Service Company of Colorado on grounds of duplication, asserting that the co-ops should get power via wholesale purchases from them. The court held that the protestants were regulated monopolies, that new markets should be protectively secured for existing suppliers, and that there was no need for the new entrants' facility as protestants could provide alternative sources of power. For a discussion of state franchises, and territorial restrictions see Meeks, Concentration in the Electric

Power Industry: The Impact of Antitrust Policy, 72 Col. L. Rev. 64 (1972).

43. Section 10 (2) (a) of the Federal Power Act, 15 USC 803 (a).

44. See, for example, 43 CFR 2851.1-1(a)(j)(71). In Utah Power and Light Company v. Morton, 504 F.2d 728 (CA9, 1974), the company unsuccessfully contested the secretary of the interior's right to require a private utility seeking a right-of-way across public lands for construction and operation of electric transmission lines, to wheel energy from a federal hydroelectric generating facility over the proposed lines' excess capacity. The wheeling provision was found by the court to be within the secretary's authority (under 43 USC 961). For purposes of conservation and power marketing, the Interior Department also reserves the right to increase, at government expense, line capacity. However, a permittee need not wheel to its own nonpreference customers (for example, industries) and it can utilize all of the line capacity. This authority to require wheeling has not been used by the Bureau of Reclamation in California.

45. See, for example, Flood Control Act of 1944, Section 5, 16 USC 825(s); the Reclamation Laws, 43 USC 485 (h) (c); and the Bonneville Act, Sec. 4, 50 Stat. 731 et seq., 16 USC 831-33.

46. 122 Cong. Rec. §10615, 10620 (June 25, 1976).

47. Northern Natural Gas Co. v. Federal Power Commission, 399 F.2d 953, 959 (CADC, 1968).

48. Cf. Pennsylvania Water & Power Co. v. FPC, 343 US 419-20 (1952).

49. Pennsylvania Water & Power Co. v. Consolidated Gas, Electric Light & Power Co., 184 F.2d 552 (C.A. 4, 1950), supplemented, 186 F.2d 934, cert. den., 340 US 906, and, Consolidated Gas, Electric Light & Power Co. v. Pennsylvania Water & Power Co., 194 F.2d 89 (C.A. 4, 1952), cert. den. 347 US 690.

50. 184 F.2d 560.

51. 194 F2d 89. Affirmed Pennsylvania Water & Power Co. v. Federal Power Commission, 353 US 414, 96 L Ed 1042 (1952) (the Court noted that the control over Penn Water by Consolidated had been terminated, and upheld the FPC in requiring reduced rates under a continued policy arrangement, over a dissent arguing that FPC approved arrangements perpetuated the antitrust problems). Cf. Northern Natural Gas Company v. FPC, 399 F.2d 953, 959-61 (C.A.D.C., 1968).

52. Monopoly power is the power to control prices or to exclude competition from the market. U.S. v. E. I. DuPont de Nemours and Co., 351 US 377, 391 (1956).

53. 410 US 366 (1973), affirming, in part, 331 F. Supp. 54 (D.C. Minn., 1971).

54. For instance, the FPC cannot compel wheeling or construction of new capacity (load growth coordination). See Otter Tail, supra; and Otter Tail Power Company v. Federal Power Commission, 473 F.2d 1253 (CA8, 1973).

55. In granting certificates for facilities, federal and California state authorities must consider antitrust allegations. See, for example, Northern California Power Agency v. P.U.C., 5 C.3d 370, 96 Cal. Reptr. 18, 486 P.2d 218 (1971) (PUC must consider and make findings on antitrust allegations made in Geysers units certification proceeding); Northern Natural Gas Co. v. FPC, 130 US App. DC 220, 399 F.2d 953 (1968); California v. FPC, 369 US 482 (1962); and McLean Trucking Co. v. United States, 321 US 67 (1944).

56. In the absence of state legislative authority, a territorial allocation agreement among two utilities filed and approved by the Florida Public Service Commission was successfully attacked by the Justice Department in a Section 1 Sherman Act suit. The suit, which asserted that sales for resale were beyond the state commission's jurisdiction, was settled by consent decree, United States v. Florida Power Corp., (M.D. Fla., Tampa Division, Div. No. 68-297-T); 1971 Trade Cases, 73,637 (August 19, 1971).

57. Under the doctrine of Parker v. Brown, 317 US 341 (1943), state mandated action can result in an exemption from antitrust laws. The action must be mandated by the state, not just permitted. Goldfarb v. Virginia State Bar, 421 US 773, 44 L.ed 2d 572 (June 16, 1975) (must be action of state as sovereign); and see Kinter, The State Action Antitrust Immunity Defense, 23 Am U. L. Rev. 527 (1974). Also, the decision must be effectively that of state officers, not of private business advisors whose decision is adopted by the state. See ABA Section of Antitrust Law, Antitrust Developments 1955-1968, 211; Cantor v. Detroit Edison Co., 428 US 579 (1976).

58. 15 USC 717 et seq. and 16 USC 793 et seq., respectively.

59. See, for example, the Bonneville Act, Section 6 and 7; the Fort Peck Act of 1944, Section 5, 16 USC 833d.

60. See Section 311 of the Federal Power Act, 16 USC 825 (j). Section 30 of the Public Utility Holding Company Act of 1935, 15 USC 79z-4 authorizes and directs the SEC to make studies and investigations of public utility companies, the territories they serve or can serve, and the manner of service; these investigations are to concentrate on structural features of the industry.

61. Gulf States Utilities v. FPC, 411 US 747 (1973); FPC v. Conway Corp., 426 US 271 (1976); City of Huntington v. FPC, 498 F.2d 778 (CADC 1972).

62. California v. FPC, 369 US 486; cf., Pacific Gas and Electric Company, Project No. 2735 (Order issued April 1, 1976).

63. See, for example, Georgia Power Company v. FPC, 373 F.2d 485 (CA 5, 1967). Current case law seeks to require that agencies consider effects on competition as part of the public interest. FMC v. Aktiebolaget Svenska Amerika Linien, 390 US 238 (1968); McLean Trucking Co. v. United States, 321 US 67 (1944); Denver & R.Gr.RR v. United States, 387 US 488 (1967); Northern Natural Gas Co. v. FPC, 399 F.2d 953 (CADC, 1968); City of Pittsburgh v. FPC, 237 F.2d 741 (CADC, 1956); and, Marine Space Enclosures, Inc. v. FMC, 420 F.2d 577 (CADC, 1969) (Licensing cases).

64. Section 203 of the Federal Power Act, 16 USC 824 (b). This section has been construed to not encompass local distribution facilities, Duke Power Co. v. FPC, 401 F.2d 930 (CA 4, 1968), and to not require a showing of positive benefit, Pacific Power & Light Co. v. FPC, 111 F.2d 1014 (1940). FPC approval of a merger, with SEC approval, does not confer immunity from attack under the Clayton Act, §11, 15 USC 21 California v. FPC, 369 US 482 (1964); California v. FPC, 367 US 482 (1962). In the recent case of Kansas Power and Light Co. v. FPC, CADC #75-2080, decided April 7, 1977, the court held that while the FPC need not defer its proceeding pending district court action, it could exercise its discretion so as to defer its proceedings pending court proceedings.

65. Section 10(a), Part I, Federal Power Act, 16 USC 803(a). Licensed projects should not be a part of a program by pool members of preemption of generating alternatives, or other enhancement of their market position, 16 USC 803 (h).

66. Pacific Gas and Electric Company, FPC Docket E-7435 (Order Accepting Rate Schedule for Filing and Denying Request for Suspension and Hearing, issued November 6, 1968).

67. In Pacific Gas and Electric Company, Project No. 2735 (Order issued April 1, 1976), the FPC said it would grant a license for a new pumped storage facility, without deciding antitrust claims advanced. It did this even though resolution of such claims against licensee might necessitate changes in project operations and electric equipment, and even though the FPC's ability retroactively to require new license conditions, without licensee consent, is open to some doubt. Antitrust claims were merely set for later hearings.

68. Along the lines of Section 105 of the Atomic Energy Act.

69. Cf. Report on S.2028, supra.

70. Council of Economic Advisors, 1970 Economic Report of the President, pp. 107-08.

71. More extensive judicial scrutiny appears to be available in environmental than in economic regulatory cases. Compare Citizens to Preserve Overton Park v. Volpe, 401 US 402, 415, 28 L. Ed 2d 136, 152 (1971) with FPC v. Transcontinental Gas Pipeline Corp., 423 US 326, 46 L. Ed. 2d 533 (1976).

72. The Congressional Research Service has estimated that of the $9.2 billion of increased electric rates in 1975, $5.9 billion were derived from fuel adjustment clauses. "Electric and Gas Utility Rate and Fuel Adjustment Clause Increases," 1975. The FPC has recently (Federal Register, May 5, 1977) terminated a proposed Federal Advisory Committee (FAC) rulemaking instituted in 1975 when fuel adjustment clauses were under congressional scrutiny. In ending the rulemaking the FPC noted that those seeking more disclosure of fuel purchasing data lacked specific cases of FAC abuse, that staff audits, restricted to checking vouchers against fuel bills and not reviewing fuel purchasing practices, had turned up only minor discrepancies, and that disclosure of fuel purchasing contracts might harm purchasers in noncompetitive fuel markets.

73. FPC termination of rulemaking proposal, filing of rate schedules, fuel adjustment provisions (April 26, 1977 at 42 F.R. 22897, 22899, May 5, 1977). Contrary to the FPC in this notice, the audit reports for fuel adjustment provisions indicate no review of the adequacy of procurement policies.

74. 42 USC 2135.

75. The NRC's predecessor, the Atomic Energy Commission (AEC), held that there must be a nexus between the complained-of practices and the activities under the license that are alleged to create or maintain a situation inconsistent with the antitrust law. Louisiana Power & Light Co., Dkt. 50-282 A, Memorandum and Order RA I-33-9.619 (28 September 1973). In the Wolf Creek proceeding the NRC appears to have concurred in the view that once it is established that activities under the license would create or maintain a situation inconsistent with the antitrust laws, the commission is required to seek to remedy the situation, and relief need not be limited to the activities under the license that create or maintain a situation inconsistent with the antitrust laws. Initial decisions have also been rendered by hearing boards in cases involving the power pool in Northern Ohio, Toledo Edison Co., Docket 50. (January 6, 1977), and Alabama Power Co., Docket 50. (April 8, 1977).

76. This topic was discussed by the Working Group on Joint Action Among Utilities of the Conference on Research Relating to Small Energy Utilities. See Proceedings, NSF Conference on Research Relating to Small Energy Utilities (Washington, D.C.: National Planning Association, October 1975), Proposal 15, p. 11. Implementation of agreed upon license terms often will require filings with the FPC, an agency not noted for its concern for antitrust matters. Cf. Richmond Lt. & Pr. v. FPC, 481 F.2d 490 (C.A.D.C., 1973).

77. The Board has found against the applicant in Toledo Edison Company, supra, and the Alabama Power Company, supra, cases,

while holding for the applicant in a case involving Consumers Power Company.

78. Arizona Rev'd Stat. Anno., Sec. 40-281 and 40-360. (Plant of 100 mw or more, having 115 kv or more); Colo. Rev'd Stat., 40-5-101; Ore. Rev'd Stat. 758-015 (transmission lines for which land is condemned); 6A Utah Code Anno. 59-4-25; New Mexico Stat. Anno. (1953) 68-7-1. In California, a heating company is a utility, Sec. 244, Public Utility Code. The Public Resources Code, Sec. 25000 et seq., provides for licensing of power plants in California by the California State Energy Resources Conservation and Development Commission. PUC certificates of convenience and necessity are required for new plants, Sec. 1001, California Public Utilities Code. In California, for instance, a permit is required from the State for construction of any plant over 50 mw. California Electric Power Co., 61 Ca. PUC 799 (1963) (Public utility if in power interchange agreement for purchase, sale or exchange of capacity and energy on short-term basis).

79. See, for example, Tucson Gas, Electric Light and Power Company v. Trico Electric Cooperative, 406 P. 2d 740 (Az App., 1965); Western Colo. Power Company v. Public Utility Commission, 411 P.2d 785 (Colo. 1966) (Reversing PUC order permitting cooperative to build plant where other utility claimed it was ready to provide wholesale service); Western Colo. Power Company v. Public Utility Commission, 428 P.2d 922 (Colo., 1967); Western Power and Gas Company v. Southeast Colo. Power Association, 435 P.2d 219 (Colo., 1968); Public Service Company of Colo. v. Public Utility Commission, 485 P.2d 123 (Colo. 1971), (restricting, inter alia, size of load cooperative could service), and cf. Cottonwood Mall Shopping Center v. Utah Power and Light Company, 440 F.2d 36 (denying antitrust claim based on refusal of power company to provide supplemental service to shopping center providing power to tenants because shopping center is acting as utility in contravention of power company service area monopoly under Utah law).

80. Telephone conversation with Frank Morse, Southern California Gas Company, March 4, 1976.

81. Ibid. Morse stated that utility electric power for back-up carries charges in excess of normal power costs.

82. See Securities Act of 1933 and Regulations thereunder.

83. GAO Report, "Receipt and Coordination of Natural Gas Reserve Data: Federal Power Commission, Securities and Exchange Commission," B-178912 (April 30, 1974).

84. Energy Policy and Conservation Act of 1975, Section 503, 42 USC 6383.

85. SEC decisions regarding utility holding company accounts govern the FPC. Sec. 305, Federal Power Act.

86. GAO, Improvements Still Needed in Federal Energy Data Collections, Analysis, and Reporting (June 15, 1976).

87. G.A.O., "Acreage Limitations on Mineral Leases not Effective," RED-76-117 (June 24, 1976). This report states that limits are widely avoidable through three exclusions: inclusion in a unit plan of development, holdings in the name of individual family members, or holdings of fractional interests which are chargeable only if they exceed 10 percent of the lease and then only in a share proportionate to charged parties' ownership share in the lessee. "The Department stated that, with the complexities of vertical, horizontal, and conglomerate ownerships--which are common in the mineral industry--establishing ownership for the accounting of acreage holdings could be a difficult task. . . ." In Utah and Wyoming, 38,748,593 acres have been leased by the federal government (June 30, 1975). Nationwide, over 88 million acres had been leased for oil and gas. Interior suggested that acreage limits on options to lease should be eliminated. Ibid.

88. Richard Levy, The Regulation of Offshore Crude Oil Pipelines and the Consequences for Competition (Washington: National Science Foundation, 1975).

89. Interviews with H. Banta, W. Measday, Senate Antitrust Subcommittee, August 1976, and J. Galloway, House Commerce Committee, August 1976.

USGS files on OCS operation (at Metairie, Louisiana), are not kept in order. Information developed and maintained by the government on the economic ability to produce federal OSC leases is quite limited and derived from unverified industry data; this information is essential if the United States is to collect maximum royalties and prevent speculative withholding of production. Interview with John Galloway, House Commerce Committee Staff, April, 1976.

90. 43 USC 31.

91. The Coal Leasing Amendments Act of 1975, P.L. 94-377, 90 Stat. 1083, directs the secretary to conduct a comprehensive exploratory program on federal coal lands. The secretary is further directed to prepare and publish detailed maps of federal coal lands. 30 USC 208-1, 90 Stat. 1087-88. No such clear and explicit mandate has been imposed with respect to geothermal energy resources.

92. The Geothermal Steam Act of 1970, 30 USC 1001, 84 Stat. 1556, requires that lands within any known geothermal resources area be leased to the highest responsible qualified bidder by competitive bidding. Lease royalties are, for energy, restricted to between 10 and 15 percent of the value of the energy. These provisions limit but do not eliminate the use of alternatives to cash bonus bidding.

93. Office of Technology Assessment, "An Analysis of the Feasibility of Separating Exploration from Production of Oil and Gas on the Outer Continental Shelf" (May 1975).

94. The comparative proceedings could take into account matters such as the scope of development, royalties, bonuses, potentials for adding competition to the market, and the like, if appropriate preference to public bodies could be allowed. Cf. Section 7(a), Federal Power Act, providing a licensing preference to state or municipal bodies, and Section 7(b) providing that where in the FPC's judgment development should be by the United States, licenses shall not issue and the FPC shall submit its findings to Congress. 16 USC 800(a), (b).

95. See, for example, Federal Non-nuclear Energy Research and Development Act of 1974, Public Law 93-577, 42 USC 5901 et seq.

96. ERDA, "Definition Report: Geothermal Energy Research, Development and Demonstration Program, October 1975," ERDA-86, Washington, D.C.

97. Appropriations are authorized for two federal geothermal demonstration plants. Public Law 94-187, 89 Stat. 1063, et seq.

98. Taken from a tape recording of the proceedings of the panel reviewing the LBL project. The tape was obtained from ERDA under Freedom of Information Act requests by the Environmental Action Foundation and Joseph Lerner, economic consultant.

99. Ibid.

100. Ibid.

101. See Malcolm S. Mason, "Current Trends in Federal Grant Law"--Fiscal Year 1976, 35 Federal Bar Journal 163 (1976) for a resume of federal grant law.

102. Joint efforts to influence public officials do not violate the antitrust laws, even though intended to eliminate competition. United Mine Workers v. Pennington, 381 US 657, 670 (1965); and see, Eastern Railroad Presidents' Conference v. Noerr Motor Freight, Inc., 365 US 127 (1961). But a Sherman Act violation may be found if what appeared to be joint political activity directed toward influencing governmental activity is a mere sham to cover what is actually nothing more than an attempt to interfere directly with the business relationship of a competitor. Noerr, supra, 144. And see United States v. Otter Tail Power Co., 410 US 366, 380 (1973); California Motor Transport Co. v. Trucking Unlimited, 404 US 508 (1972); Woods Exploration & Producing Co. v. Aluminum Co. of America, 438 F.2d 1286, 1296-98 (CA 5, 1971), cert. denied, 404 US 1047 (1972) (holding Noerr-Pennington doctrine inapplicable to alleged filing of false gas production forecasts); and George R. Whitten, Jr., Inc. v. Paddock Pool Bldrs., Inc., 424 F.2d 25, 33 (CA 1), cert. denied, 400 US 850 (1970). Disclosure is also

consonant with enforcement of the provisions and policy of the Federal Advisory Committee Act, 5 USC App. I, 2,4,9 and 10.

103. See, for example, United States v. Georgia Public Service Comm., 371 US 285 (1963).

104. With limited exceptions "no executive agency shall dispose of any plant, plants or other property to any private interest until such agency has received the advice of the Attorney General on the question whether such disposal would tend to create or maintain a situation inconsistent with the antitrust laws . . ." 40 USC 488 and see United States v. Aluminum Co. of America, 91 F.Supp. 333 (S.D. N.Y., 1950) (construing the Surplus Property Act of 1944, 58 Stat. 765 which is the precession to current surplus property legislation).

105. See Section 9, Federal Non-nuclear Energy Research and Development Act of 1974, 42 USC 5908.

106. The Energy Policy and Conservation Act, Public Law 94-163, Sec. 503, 42 USC 6383; 89 Stat. 871, 958-59 (1975) mandated that the SEC assure the development of accounting standards to be used by petroleum companies from the prospecting through production stages. The Uniform System of Accounts for Public Utilities is a system for financial accounting and does not give variable and fixed cost breakdowns needed for cost accounting.

APPENDIX A
PHYSICAL TYPES OF GEOTHERMAL RESOURCES

The several types of geothermal resources vary widely in availability and recoverable energy per unit extracted, as well as in the progress made toward their commercial use.

Vapor-dominated (dry steam) systems produce steam which (with some clearing) can directly drive a turbine. It is a rare resource, having been discovered domestically only in the Geysers area and Yellowstone National Park. Commercial development at the Geysers shows that, despite the low pressure of the steam and environmental and regulatory problems, power can be produced there for about half the cost of power generated from imported residual fuel oil.

Very hot water convection systems (temperatures over 150^{c}C) are more plentiful than dry steam. They would be exploited by extracting the water, whose pressure drop as it rises through the well by natural convection would cause part of it (usually about 20 percent) to flash into steam. The remaining water--often contaminated with minerals--must be disposed of in some economical and environmentally acceptable way. Low steam temperatures and low pressures sharply limit the amount of heat recoverable from a hot water pool. Still, hot water and electricity are being produced at operating plants in Wairakei, New Zealand; Cerro Prieto, Mexico; Japan (four plants operating in 1975), and Iceland. Large but unexploited hot water resources are available in the Imperial Valley of California.

Lower temperature (under 150^{c}C) geothermal water resources will not flash directly into steam. To drive a turbine, they must be used in a binary cycle to heat a second (working) fluid that has a lower boiling point, such as isobutane (used experimentally in a Russian 0.75mw plant) or xylene (employed in an experimental 3.8mw plant in Japan). Lower temperature resources are widely distributed, however, and thus some research into their commercial potential has been conducted.

Water at either high or lower temperatures can be used for secondary applications (for example, space heating and cooling), in combination with the primary power-generating application. Of course, the utility of hot water for space heating depends upon there being a demand for that service in the neighborhood of the wells.

A common problem in geothermal hot water systems is the presence of heavy solutions of mineral salts (hot brines). These brines can corrode metal equipment in wells or generating equipment, and cause scale and fouling. Development of heat exchangers to handle them is a principal research objective.

Dry hot rock is an impermeable, unfractured magma body which, at least in theory, could be used to heat water injected from the surface and subsequently extracted. It is very widely distributed, but its exploitation remains strictly a research topic.

Sealed or trapped systems of hot water under great pressure are believed to exist, at depths of 5,000 to 20,000 feet, in parts of the Gulf Coast. These geopressured zones offer a theoretical source of mechanical and thermal energy, as well as small amounts of dissolved natural gas. The economics of exploitation of geopressured zones are uncertain, but some estimates have been made by oil company investigators. The economic promise held out by these estimates turns on the price of natural gas, a joint product, and the well spacing estimated as necessary.

APPENDIX B
UTILITY FUEL CONTRACTS

Appendix B provides descriptive and background material regarding fuel oil procurement by the major private utilities in California.

THE ATLANTIC RICHFIELD COMPANY (ARCO)--PACIFIC GAS AND ELECTRIC (PG&E) CONTRACT

The price adjustment provisions in the ARCO-PG&E resid procurement contract originally provided that the original base price would

> increase or decrease concurrent with and by the same amount per barrel as the combined average of the posted prices for 35.0-35.9 Gravity Crude Oil as posted by Atlantic Richfield Company at Cook Inlet Pipeline Company and Kenai Pipeline Company, plus the average of the ICC common carrier rates from Cook Inlet Pipeline Company and Kenai Pipeline Company to ships rail.

It further provided that

> in the event that the average posted price of said Gravity Crude Oil as posted in Cook Inlet Pipeline Company and Kenai Pipeline Company by all producers of greater than 10% of the total monthly production of such crudes at those points, except Atlantic Richfield Company, is $1.10 or more per barrel below the average price posted by Atlantic Richfield Company, the average of the posted prices of these companies, other than Atlantic Richfield Company, shall be the basis for price changes under this provision.[1]

Commencing April 1, 1974, prices began to increase or decrease by the same percentage as the percentage increase or decrease in the average cutter stock shipped by various concerns; ARCO supplied oil to PG&E in the first quarter of 1975 for $14.71. The contracts between PG&E and ARCO and Union Oil were described in testimony given to the PUC in 1974. Contracts with ARCO and Union Oil were entered into in April and November of 1972, respectively. The PG&E contract with ARCO was to terminate in 1976, but in June 1973, a new contract was negotiated extending the contract term through 1981 and increasing the volume of oil to be supplied to 8.4 million barrels for the calendar year 1974 and to 9 million barrels per year thereafter through 1981.

> At a meeting on November 9, 1973, Arco advised PG&E that the posted price of crude oils was being raised. This price is computed under a formula such average being weighted on the basis of twenty percent (20%) for the average of the Alaskan Crudes as identified in the next preceding paragraph, and eighty percent (80%) for the average of Canadian marketable crude oil (U.S. Dollar Price) having a quality of 42° or higher API Gravity and containing less than 5% sulfur by weight, as posted at Edmonton Terminal, Canada by all producers of ten percent (10%) or more of the total monthly production available at that point.[2]

Also, beginning April 1, 1974, the applicable prices were further adjusted by applying 80 percent of any change in any duty, tariff, or other charge imposed by any governmental agency on subject Canadian crude oil.

The result of the above-mentioned adjustments and other factors is a weighted formula based 20 percent on Alaskan and 80 percent on Canadian crudes. The formula price is given as 20 percent of the new crude price in Alaska plus 80 percent of the price in Canada divided by the base crude price multiplied by the initial price for low-sulfur fuel oil.

It should be noted that the Edmonton price is set by the provincial government after it negotiates with the federal government of Canada; the current price ($8.00) pertains to 82 percent of Alberta production with the remaining 18 percent of production being set by a few private companies, notably Imperial (Exxon) and Shell Oil. These private postings follow the Canadian provincial government postings. The government posting is based on a net-back price from Ontario markets intended to keep the price in Ontario comparable to the price in the United States across the border from Ontario. That price has been gradually heading toward the price that pertains in Canada east of what is known as the Ottawa line, where foreign oil is used.

The contract further provides that prices must track any duty or tariff imposed by the Canadian or U.S. governments. Renegotiation clauses are included regarding technical breakthrough changes in refinery technology, extreme variation in the market price of low-sulfur fuel oil with the same specifications as covered herein, equivalent volume term changes in import regulations, and changes in product specifications. In the event that any of the above occur and mutual agreement is not reached within 60 days, the contract may terminate two years and 60 days after the initial discussion date. The fuel is a low-ash, low-sulfur fuel with a 6 million Btu per barrel heat content.

The contract contains in article 11 an exculpatory clause excusing buyer from performance when California or federal or other law prevents it from burning the fuel oil supplied. The force majeure clause includes the usual range of activities plus acts of public administrators. The clause also includes interference by foreign nations; abnormal increases in cost of transportation caused by war or hostilities between any nations; shortage of fuel oil deliverable due to shortages of refined products, crude oil, natural gas, or raw products comprising the fuel oils agreed to be delivered; lack of capacity in equipment used, or insufficient equipment for producing, manufacturing, refining, or transporting oil and its products.

Seller, in the event that it is unable to make delivery to all of its customers, is authorized to prorate its deliveries among its customers; and buyer acknowledges in the agreement that seller is currently unable to meet all contractual demands for certain hydrocarbons and, as a consequence, is allocating products in accordance with the voluntary allocation policy.

The remaining provisions are typical for such a contract: harmless clauses, imposition of sales tax on the buyer, and the like, except that specific reference is made to two of ARCO's tankers, specifying the amount of time they can be tied up before buyer shall pay seller demurrage, and setting forth the rate for demurrage for each of the two ships. The demurrage rates are subject to adjustment as established by the seller's marine department.

The price of oil when the ARCO agreement took effect was \$5.07 a barrel, but in April of 1974, it had jumped to \$12.88 a barrel. The \$12.88 price was maintained at least until October of 1974.

Correspondence filed with the California PUC indicates that the first announcement of a price rise, to \$5.42, was in a letter dated October 5, 1973, and it paralleled a rise of 35 percent in the posted prices of Alaskan oil. The commencement of Canadian tax burdens is mentioned in a November 20, 1973 letter that refers to an export fee of up to \$1.50 a barrel to become effective December 1, 1973. Another letter of November 20 from ARCO to PG&E states that ARCO is raising its posted price in Alaska by \$1.30 a barrel for new and released oil.[3]

By telegram of March 8, 1974, ARCO sought to explain the basis of its price escalation provisions in the June 1973 contract. The telegram states that the crude oil escalation provision is the basic index mechanism intended to maintain prices reasonably in line with the finished product marketplace and at the same time to recover some measure of its increased crude and other costs. While the index is tied to specific designated crude oils and percentages, it does not necessarily reflect the actual mix of feedstocks, since use is made of a mix of feedstocks from Alaskan, Canadian and other foreign crude oil. The telegram goes on to state that ARCO is purchasing foreign low-sulfur oil at a delivered cost of $13.32 a barrel; as much as $15.45 a barrel is being paid for other foreign crude oil, but no quantities are stated. The PG&E company acquiesced by telefax on March 12 changing the starting date from April 1, 1974 to January 1, 1975.

> At a meeting on November 9, 1973 ARCO advised that its costs for crude were increasing so greatly that it would be necessary to renegotiate the price provision in our June 1, 1973 contract. Early this year, ARCO pointed out that the oil it was importing from Canada was subjected to a $6.40 per barrel export tax and, furthermore, that there was an extreme variation between the market price for low-sulfur fuel oil of the same specifications as covered by the contract and the effective price under the contract price provision.[4]

ARCO was then seeking an immediate pass-through of the increases. As negotiations progressed, this was extended to February, then finally to April 1, 1974. In these negotiations PG&E feared that it was paying more than the price to other utilities. This fear is reflected in an ARCO communication of March 26, 1974 to PG&E stating that ARCO was making every effort "to renegotiate increases in current contract prices for low sulfur fuel oil to its other utility customers"[5] in keeping with increased costs. On March 27, 1974, ARCO refused a request by PG&E to include a most-favored-nation clause to the effect that the price charged by seller, excluding spot sales to other purchasers in District V for low-sulfur oil in quantities over 50,000 barrels per month, would not be exceeded by the price charged to PG&E. ARCO's rejection is stated to be in part based on its legal department's opinion that the provisions requested were unlawful under the antitrust laws. The company merely stated it was attempting to renegotiate its other contract prices.

The contract price rose on April 1, 1974 to $12.88, based on a Canadian crude price of $6.81 with an export duty of $4.00 a barrel and no change in the posted price in Alaska. The June price increase to $13.85 appeared to be based on an increased Canadian export duty and on a $1.20 increase in the duty on crude, and in July, prices rose to $14.69. However, ARCO stated in a July 16 letter[6] that it was granting an allowance of $1.14 per barrel off the contract price, thus temporarily maintaining the $12.88 price per barrel. On August 12, the discount was removed increasing prices to $14.02, with the provision that 348,000 barrels of July volume owed PG&E would be billed at July prices, and that deferral might be made for payment of prices in excess of $12.88 a barrel until such time as ARCO's negotiations to increase SCE's current contract price were concluded.

R. P. Benton of PG&E in his testimony before the California Public Utilities Commission in application No. 55541 stated that the price of $14.02 became effective from ARCO January 1, 1975.[7] He described the previous discounting in the following manner. In mid-1974 PG&E reduced fuel oil requirements, combined with a widening gap between the price being paid by PG&E to ARCO compared with the prices charged by ARCO to SCE and LADWP, caused PG&E to negotiate the postponement of any increase in price beyond the $2.88 per barrel which had become effective on April 1, 1974. The negotiations resulted in ARCO's agreement to waive price increases in June and July, and ARCO withdrew the temporary allowance in August but agreed to permit deferral of any payments in excess of $12.88 per barrel until it had negotiated a price increase with SCE. By letters dated November 14, 1974 and December 2, 1974, ARCO served notice that the deferred payment arrangement was terminated as of November 1, 1974 and that termination of the contract would be recommended if PG&E failed to pay the price demanded. PG&E agreed to make this payment on or before December 30, 1974, feeling it was obligated by contract to pay the deferred price effective August 1, 1974.*

PG&E has, besides its contract with ARCO, low-sulfur fuel oil contracts with Union Oil, Phillips Petroleum, and Perta Trading Company.

---

*ARCO, it should be noted, never entered into a contract with SCE and their then current agreement expired by its own terms on December 31, 1974. Subsequent deliveries to SCE would be under FEA allocations and ceiling prices which are in excess of the price PG&E pays ARCO under the contract. Apparently, LADWP renegotiated its contract with ARCO and is now paying up to 280 cents per Btu ($14.80 per barrel).

The basic agreement with Union Oil is dated November 1, 1972, and runs from January 1, 1973 to December 31, 1980. The original terms of the contract, which has been substantially amended with regard to price, provide for delivery of 480,000 barrels of low-sulfur fuel oil in 1973 and between 2 and 2.2 million barrels for each year thereafter.* Delivery is generally to be in 100,000 barrel cargo lots,† or in 19,000 barrel lots to PG&E barges.

The initial price was set at $4.47 for barge lots and $4.61 for tank ship deliveries, with escalation based on the price of crude oil from Alaska. The escalator clause tracked, penny for penny, the price of Alaskan crude oil, which is determined by averaging four posted prices plus applicable pipeline tariffs of the Cook Inlet and the Kenai Pipelines. The posted prices are those of Union Oil at the entrance to the Cook Inlet Pipeline, plus the tariff to the Drift River Terminal; the average of prices posted by ARCO at two entrances plus the pipeline tariff to the Drift River Terminal, and the price at the entrance to facilities of the Kenai Pipeline, plus the pipeline tariff to the Nikiski Terminal; the price posted by Mobil Oil Corporation at the entrance to the facilities of the Cook Inlet Pipeline, plus the tariff to the Drift River Terminal; and the price posted by SOCAL at the entrance to the facilities of the Kenai Pipeline, plus the tariff to the Nikiski Terminal.

Price renegotiation may be requested by either party six months before the end of any contract year. If agreement is not reached within 90 days, either party may cancel the agreement upon 90 days' notice.

Performance is excused if Union is prevented from or delayed in producing, manufacturing, transporting, or delivering in its normal manner any product or products covered by the contract or the materials from which those products are manufactured.

Exhibit No. 9, dated March 24, 1975, in the proceeding upon Application No. 55511 of PG&E before the California PUC, was submitted by a PG&E witness, Mr. R. P. Benton. An exhibit submitted in proceedings before the California PUC[8] purports to set out price negotiation memoranda and correspondence between PG&E and Union Oil pertaining to the agreement described above, for the purchase and sale of residual fuel oil.[9]

In June 1974, Union began renegotiating in an attempt to raise prices to a level of $16.32 per barrel, effective January 1, 1975,

---

*The fuel oil is required to have 10° API minimum gravity, maximum sulfur content of 0.5 percent by weight, and contain 6 million Btu per barrel.

†Demurrage is specified at $350 per running hour.

with a temporary discount of $3.74 per barrel. Union also sought to change the price escalation provision so that, instead of tracking Alaskan, it would follow, penny for penny, the established Eastern Hemisphere export price for Sumatran light crude oil as established by Caltex Indonesia Oil Company for purchasers utilizing such crude oil in the Eastern Hemisphere or reselling it for such utilization.

When PG&E personnel, at a meeting on September 26, 1974 with Union Oil personnel, inquired, "Did Union Oil not secure sufficient quantities of Alaskan crude oil to meet this contractual obligation prior to execution of the contract in November 1972?" Union is reported to have replied by stating that Alaskan crude production has diminished and has adversely affected their source of supply.[10]

Union, according to this memorandum, expressed interest in modifying its renegotiation request so as to link price escalation to its average cost of crude, and indicated its intention to supply quantities of resid in excess of contract volumes, perhaps doubling contract volumes in 1975 and 1976.

Subsequent to this meeting, Union sent a letter in which it sought a price ranging between $12.26 and $12.58 for the first half of 1975, and rising by $2 per barrel on July 1, 1975. Union is said to have noted, at a November 22, 1974 meeting, that its price for low-sulfur fuel oil to customers assigned under the FEA allocation program was $13.76 per barrel.[11]

On December 16, 1974, Union Oil asserted in a letter to PG&E that it could cancel the agreement as of December 31, 1974; that letter states that PG&E's initial basic problem with Union's proposal was that the discount could fluctuate upon 30 days' notice; and that subsequently, Union might charge PG&E a price greater than its average price to other utilities.

On December 20, 1974, PG&E acquiesced in Union's proposal. In testimony to the California PUC, Mr. Benton of PG&E states that the primary consideration of PG&E's decision to agree to this price change was

> continuation of the contract, which yields a price significantly lower than that which Union would charge under Federal Energy Administration allocation and price regulations in the absence of a contract. We also considered the certainty of paying a price higher than $12.26 per barrel to replace Union flow combined with the additional certainty of being unable to negotiate a contract term longer than two years with some other supplier.[12]

The Union contract provides for up to 2 million barrels of oil per year.

PG&E purchases fuel oil from Phillips Petroleum on the basis of Phillips' monthly postings for such oil.* An agreement for the purchase and sale of low-sulfur fuel oil was entered into between these parties for the eight-month period April 1 to December 31, 1974.

Phillips Petroleum had a contract with PG&E under which 2.7 million barrels of oil were delivered during 1973 and 450,000 barrels in the first quarter of 1974. This contract expired and the renegotiated price terms provided for use of a posted market price for high-sulfur oil with a premium for low-sulfur content.

On January 4, 1974, PG&E contracted with Perta Oil Marketing Corporation to acquire resid purchased by a Bahamian firm--Puerto Oil Marketing Corporation Limited--from the Indonesian state company "Pertamina." Under the contract, which expires October 31, 1976, PG&E acquires a rather light (30.2 to 20.5^{o} API gravity) oil with a very low sulfur content (0.2 percent by weight) in quantities of 200,000 barrels per month (plus or minus 10 percent), at an initial price of $13.70 per barrel subject to Pertamina postings.† PG&E takes delivery in Indonesia.[13]

PG&E also has a contract, dated May 28, 1974, to purchase low-sulfur crude oil from Perta which, in turn, has acquired the oil from its Bahamian affiliate which again in turn, purchases from Pertamina and from Tesoro Petroleum Corporation. This low-sulfur oil is purchased in quantities of 180,000 barrels per month (plus or minus 10 percent) at a price which begins at $13.25 per barrel and tracks Pertamina's price. Delivery is in Indonesia and this oil, at least in part, appears to be used in an exchange or processing agreement with a refinery in Hawaii, Pacific Resources, under which PG&E receives resid.[14] The contract with Pacific Resources commenced October 1974, and runs through March 1977, with minimum total delivery of approximately 3 million barrels of Indonesian oil. Because this is essentially an exchange agreement, the delivery quantity is tied to production in certain Indonesian fields. Other purchases by PG&E consisted of foreign source fuel oil at generally prevailing world market spot prices.

Because of delays in scheduled nuclear units, long-term agreements were entered into with foreign sources in 1973, including two involving Indonesian crude and fuel oil. Indonesian fuel oil carried an average delivery price of $14.75, and the crude sold for $17.50 per barrel, with prices escalating in response to Indonesian posted prices.

---

*The residual fuel oil is to have a maximum 25° API and a maximum sulfur content of 0.5 percent by weight.

†The price declined in 1974.

SAN DIEGO GAS & ELECTRIC COMPANY (SDG&E)

SDG&E, lacking hydroelectric resources, is more dependent upon fossil fuels than are other CALPP companies. In early 1975, SDG&E was purchasing resid from Union Oil at $12.24 per barrel, from Tesoro Alaskan at slightly less, and from HIRI at $11 per barrel.[15] (In 1974, Indonesian oil was purchased from Edgington Oil Company.) SDG&E has followed a policy of using long-term contracts for fuel purchasing.[16]

SDG&E apparently had an oversupply of petroleum in 1974, which allowed it to sell 730,936 barrels of resid to SCE, and to sell 2,271,876 barrels to major refineries for further processing into other petroleum products.

SDG&E purchases most of the crude refined at the Tesoro Alaska refinery. Its contract with the Tesoro Alaskan Petroleum Corporation provides for the purchase of 250,000 to 300,000 barrels of resid made from Alaskan crude. The resid price rises with Tesoro's costs and with tariff rates on the Kenai pipeline. Tesoro sold SDG&E 3,815,000 barrels of resid in the year ending September 30, 1975.

Under its contract with SDG&E, HIRI has been required to expand its Hawaiian refinery, increasing crude runs from 45,000 barrels per calendar day in 1974 in a series of steps to 75,000 barrels per calendar day in 1977. Production of fuel oil is to rise from 12,500 barrels per calendar day in 1974 in a series of steps to 28,000 barrels per calendar day in 1977. SDG&E is to purchase fuel from HIRI at a mark-up of 87 cents over HIRI's weighted average monthly crude prices (for the period August 1, 1974 through July 31, 1984). This mark-up was scheduled to be renegotiated in 1976.

A contract between Union Oil and SDG&E provided for the purchase and sale of 1 million barrels of fuel oil in the second half of 1974 and a like quantity in the first half of 1975. Prices, initially set at $12 per barrel, before taxes, subject to a $3.26 discount revocable on 30 days' notice, were to be renegotiated in 1975 or the contract would end.

Union, on October 15, 1974, proposed a price of $15 per barrel with a discount of $2.76 revocable on 30 days' notice; the discount was reduced to 76 cents effective April 1, 1975. The $2 increase was deferred following a March 19, 1975 meeting between Union Oil and SDG&E personnel. SDG&E assigned or resold a portion of this resid to LADWP.

SOUTHERN CALIFORNIA EDISON COMPANY (SCE)

Following are some observations on contracts made by SCE for the purchase of residual fuel oil. These contracts were filed with the Oversight and Investigations Subcommittee of the House Committee on Interstate and Foreign Commerce in 1975.

SCE has contracts with Standard Oil of California for 22.2 million barrels per year, with Texaco for 5 million barrels per year, with ARCO for 4 million barrels (1974), with Exxon for 6.3 million barrels per year, with Coastal States for 16,560,000 barrels per year, and with MacMillan Ring Free Oil Co. for 3,500 barrels per day.

The SOCAL and Texaco agreements began respectively on April 1, 1971 and December 12, 1970. Both expired December 31, 1975, forcing SCE to enter a seller's market for a tremendous quantity of fuel. The partners in Caltex, SOCAL, and Texaco, appear to supply so high a portion of SCE's oil as to be inconsistent with normal practices of seeking a diversity of sources. In this regard, it should be noted that the force majeure clauses in these contracts excuse performance by a seller in the event that either normal supply or normal transportation sources are in short supply.

The quality specifications for resid, in these contracts, vary. There are no specifications for specific gravity in the contracts with ARCO and SOCAL.

Fuel reports filed with the FPC (Form 423) indicate that SCE gets residual fuel oil having a lower heat content than do PG&E or SDG&E. Very importantly, since its 1967 contract with SOCAL, SCE has not included a clause in its fuel purchase agreements providing for price adjustments tied to the heat content of delivered fuel.

PRICE ADJUSTMENTS

A contract normally entails mutuality of obligation among the parties. The price escalator, force majeure, and price renegotiation provisions of SCE fuel contracts place all risk and obligation on the buyer.

Prices uniformly rise, penny for penny, with increases in prices of crude oil, even though only a small portion of a barrel of crude is refined to resid in U.S. refineries.* Crude oil escalator

---

*Use of Indonesian tax-reference prices as a cent-for-cent escalation by SOCAL for crude oil is interesting because these prices do not represent direct increases in costs to sellers, being at most

clauses are frequently pegged to an arrangement in which SCE guarantees the transportation expenses, indeed paying phantom freight charges* and tax expenses, and SOCAL some ambiguous portion of refinery equipment costs. The arrangement is subject to cut-off by marginal supply interruptions, to supply allocations by its vendor, and to numerous price escalator clauses which do not descend below minimum provisions, while SCE, with a utility's duty to serve, purchases resid in contract quantities so large as to prevent substitution of vendors.

The escalator clauses for taxes and crude prices are too vague to be audited; they attribute the entire costs of crude and its transport to resid. Price escalators based on either New York prices (Texaco) or on tankerage rates from the Caribbean to north of Hatteras use indexes that are under the joint control of major oil companies. This is the case because the tanker rates are set so as to compete with rates on the Colonial pipeline, a joint venture of the majors; and the New York harbor price is set by these transportation rates and by production from a very highly concentrated Caribbean refining industry that is dominated by major companies.[17]

Use of phantom shipping, taxes, and pricing (for example, Exxon contract provision tied to posted prices in Boston harbor), loading of oil costs onto crude oil, the use in contracts with SOCAL of escalators for the price of natural gas and California crude (for example, apparent commodity value pricing)--coupled with the direct and indirect board interlocks and mutual relations among SCE, banks, and vendor oil companies--do not bespeak competitive field prices posted by sellers, while crude sources are unspecified and their prices unverifiable. The same applies to tax escalation clauses.

Just as crude price escalations are phantom, so are tanker rate escalations, but with an additional twist, which is the tendency to use the rates from the Caribbean to the United States north of Hatteras (New York Harbor rates) as do Texaco and Exxon, or from

---

only partially reflected in sellers' costs. Similarly, use in the SOCAL contract of an escalation pegged to prices of Signal Hill California crude is interesting as SOCAL posts the price and controls the marketing of this crude (through a private carrier pipeline) which is only a small segment of California low-sulfur production.

*Under this contractual phantom freight provision, SCE guaranteed the peak of SOCAL's tanker rates. These peaks could be further hiked by cross-chartering arrangements.

Dumai, Indonesia, to El Segundo, irrespective of crude source (SOCAL contract).

Reductions in resid prices are unlikely because tanker rate clauses only go up from high rates; because minimum prices occur; because no correction is made for the use of larger, cheaper tankers, or more efficient refining; and because there are most-favored-buyer clauses (SOCAL and Texaco contracts). The combination of price-reopener provisions and huge contracts also works to that end.* Resid prices paid by SCE exceeded those paid by neighboring utilities. Use of crude oil prices as the basis of penny-for-penny price escalation indexes is peculiar since resid is only one product.† The long-term contracts for the purchases of resid by SCE are quite similar to each other with the exception of one contract with a small refiner.

Essentially the contracts with SOCAL, Texaco, Exxon, and to a lesser degree, ARCO and MacMillan Ring Free Oil reflect one-way-street bargaining. So does the fact that SCE has acquired an option to purchase all of the resid produced by an independent refiner, MacMillan. The imposition of arms-length bargaining and competitive fuels procurement cannot readily go forward so long as fuel adjustment clauses make cozier arrangements attractive.

---

*The use of phantom costs indexes is reflected in the Exxon contract provisions for renegotiating prices on the basis of those Exxon posts in New York Harbor (#6 oil) or Boston (#4 oil). The tanker indexes used refer to rates for single voyages, not the lower-term charters that would be actually used in whole or part.

†In the first half of 1975, the Mineral Industry Surveys report runs to stills in California of 372,085,000 barrels (247.41 million domestic, and 124.675 million foreign crude oil), and PAD District V refinery inputs of 466,359,000 barrels (263,982,000 domestic and 202,377,000 foreign). Residual fuel oil production was 91,292,000 barrels (and foreign imports were 7,713,000 barrels). Thus, resid output was 19.6 percent by volume of runs to stills.

NOTES

1. ARCO-PG&E, Low Sulfur Fuel Oil Agreement, June 1, 1973, filed in California Public Utilities Commission, Application 55222, Exh. No. 7.

2. Additional prepared testimony of John F. Roberts, Jr., PG&E, before California Public Utilities Commission (March 1974), p. 3.

3. Letters and memoranda between ARCO and PG&E have been filed in various PG&E fuel adjustment proceedings before the California PUC, beginning with the application in case number 55222. See Exhibit 12, App. No. 55222.

4. Additional prepared testimony of John F. Roberts, Jr., pp. 3-4.

5. Telex of March 27, 1977 to PG&E from D. R. Diendonne, ARCO; which is part of Exhibit 12, California PUC Application 55222.

6. Letter of July 16, 1974 to PG&E from D. R. Diendonne, ARCO, found in Exhibit 12, ibid.

7. Testimony of R. B. Benton, PG&E, before California Public Utilities Commission, Application 55541 (March 24, 1975).

8. Proceedings on Application No. 55541 of PG&E before the California PUC, Exhibit No. 9, dated March 24, 1975, submitted by R. P. Benton.

9. The price of residual fuel oil from Union rose to $8.54 by the fall of 1974, with deliveries being less than the contract called for because of FEA allocations. Exhibits 2 and 10, Application No. 55541, California PUC.

10. Memorandum of October 3, 1974, PG&E Materials Department, Exhibit 9, App. 55541.

11. Ibid.

12. Exhibit 2, App. 55541.

13. Exhibit 10, App. 55222, filed with the California PUC, October 16, 1974.

14. The contract is reproduced as Exhibit 11, ibid.

15. Exhibit 2B, Application No. 55506, Ca. P.U.C. (April 1975).

16. See Exhibit 6, p. 3-1, ibid.

17. Use of posted prices where these are set in whole or part by vendor is also peculiar. SOCAL sets many field prices in California and California crude prices are one escalator in SOCAL contractors. See, California Crude Oil Market Control, Report of State Committee on the Public Domain, California State Legislature.

APPENDIX C
FINANCE

The large-scale involvement of the petroleum industry with the financial community of the United States is both necessary and unavoidable in some respects, and, at least in some degree, is a potential barrier to major competitive enterprises.

A major portion of petroleum company-finance company relationships can be briefly categorized as follows:

Long-term debt accounted for about 28 percent of capital employed by huge petroleum companies in 1970. Such debt is often intimately tied to specific projects and associated income expectations.

Banks and trust departments have substantial equity interests in most major petroleum companies--on occasion over 10 percent of stockholdings.

Petroleum companies hold large quantities of marketable securities, at any given time, and these accounts are growing.

There is an extensive, far-ranging set of interlocks between bank and petroleum company board-of-director personnel.

The course of dealings involved in the above-sketched relationships involves extensive information exchanges and a community of viewpoint among financial intermediaries and the petroleum companies.

The institutions involved in these relationships reach from the country bank to the peaks of the financial pyramids. The very largest financial institutions are included, as is to be expected. Because of this, the relationships between leading banks and petroleum companies will affect the set of correspondent bank relationships through which leading institutions mobilize capital throughout the commercial banking system. These close relationships are a mixed blessing. They help mobilize capital--certainly they are intended to do so. But in an industry with competitive restraints, they

---

Much of the following discussion is taken from an unpublished paper, Bierman and Stover, "A Preliminary Review of the Relationship Among Banks and Petroleum Companies" (Spring 1976). Work on that paper was supported, in part, by the National Science Foundation, RANN Program Grant GI41470. The material, opinions, and conclusions in the paper are those of the authors and do not necessary reflect the view of the National Science Foundation.

also may tend to add inertia, keeping the capital flow in existing channels. When there is a need for alternatives to existing energy sources, the alternatives may face greater difficulties in mobilizing capital than would prevail if the dominant industry were more competitive, if extensive director interlocks did not exist, and if financial institutions had a lesser stake in the success of existing enterprises. This is especially the case for new entrants lacking established financial connections.

THE RELATIONSHIPS AMONG BANKS AND PETROLEUM COMPANIES

Historically, the petroleum industry has turned to banks for a substantial portion of its external financing. This tendency appears to be getting stronger. For the Chase Manhattan group of large petroleum companies, in the period 1956-70, outside funds provided 16 percent ($24 billion) of the $150 billion of cash available. Sources of outside funds were the public (50 percent), banks (33 percent), insurance companies (5 percent), and others (12 percent).[1]

As shown in Table A.C.1, external financing of large oil companies has included little equity money. Large oil companies do not raise capital by issuance and sale of additional common stock. As shown in Table A.C.1, the number of shares of common stock outstanding of larger oil companies was fairly level between 1970 and 1974. Annual reports of larger oil companies indicate that such rises in the number of outstanding shares as occurred were frequently made up of shares issued for acquisitions, or pursuant to stock option plans, or upon the conversion of a senior security.

While long-term debt provided 14 percent of the funds available in the period 1956-70, its share rose from 3.3 percent of new financing in 1959 to a high of 23.7 percent in 1968. Long-term debt first exceeded 20 percent of total capital employed in 1970. At the same time, use of off-balance-sheet financing, provided largely through leasing services of financial institutions (principally banks), expanded greatly.[2] If long-term leases are treated as long-term debt, total debt rose from 20 to 28 percent of the total capital employed by large petroleum companies in the 1958-70 period.

The rising dependence of large petroleum companies on bank financing is shown by the general tendency for lease rentals to be an increasing part of capital costs, in many cases exceeding debt interest expenses.[3] See Table A.C.2, comparing interest costs of debt securities and loans with net lease rentals.[4]

TABLE A.C.1

Common Stock Outstanding
(thousands of shares)

Company	1973	1970
Amerada Hess	21,661[a]	12,606
American Petrofina	9,157[b]	7,657
APCO Oil Corp.[c]	2,837[b]	2,810
Ashland Oil	22,703[a]	21,195
Atlantic Richfield	46,640[a]	45,186
Belco	7,500[b]	7,500
British Petroleum	386,100	358,844
Cities Service	26,084[a]	27,993
Clark Oil	7,108[a]	7,118
Clinton Oil	50,426[b]	51,556
Commonwealth Oil	13,296	12,455
Continental Oil	50,405[a]	49,796
Crown Central Petroleum	1,530	891
Diamond Shamrock	14,855[b]	14,592
El Paso N. Gas	27,869[b]	26,390
Exchange Oil & Gas	5,253[b]	5,252
Exxon	224,089[b]	221,704
Forest Oil	6,857[b]	4,855
General American	6,122[a]	5,609
Getty	18,669[a]	19,038
Gulf Oil	197,250[b]	207,593
Hamilton Brothers	3,381[b]	3,362
Kerr-McGee	24,989[a]	22,147
Kewanee Oil	8,402[b]	6,691
Louisiana Land & Exploration	36,275[a]	18,126
Marathon Oil	30,472[d]	30,293
McCulloch Oil Corp.	16,607	6,892
Mission Corp.	3,425	3,425
Mesa Petroleum Co.[e]	10,975[a]	2,535
Mobil Oil	101,856[a]	101,313
Murphy Oil	5,859[a]	5,348
Pennzoil Company	23,969[a]	19,476
Phillips Petroleum	75,709	74,062
Rock Island Refining	80[f]	--
Shell Oil Company	67,365[b]	67,385
Standard Oil (Indiana)	69,801[a]	68,848
Standard Oil (California)	169,839[a]	169,674
Standard Oil (Ohio)	27,539[b]	18,008[g]
Sun Oil Company	36,834[a]	30,973
Tenneco, Inc.	67,698[b]	58,404
Texaco, Inc.	271,904[b]	272,344
Union Oil Co. of California[a]	28,407[b]	28,318

[a]Number of shares issued at December 31.
[b]Average number shares outstanding in year.
[c]Adjusted for stock dividends.
[d]Not stated if average or year end.
[e]Mesa sold 2 million shares in 1973.
[f]Voting shares; also 720,000 non-voting shares.
[g]Included special stock held by BP Oil Corporation.

TABLE A.C.2

Interest and Lease Expenses of Selected Companies

Company	Year	Company Interest Expenses (millions of dollars)	Company Lease Rents (Net) (millions of dollars)
Amerada Hess	1973	38	83
	1972	35	70
Ashland	1973	25	--
	1972	20	--
Atlantic Richfield	1973	65	37
	1972	62	31
Clark	1973	4	18
	1972	4	12
Cities Service	1973	45	25
	1972	43	26
Conoco	1973	65	93
	1972	56	68
Exxon	1973	122	503
	1972	81	422
Gulf	1973	135	93
	1972	147	103
Marathon	1973	29	--
	1972	24	--
Mobil	1973	97	239[a]
	1972	86	230
Phillips	1973	62	51
	1972	59	50
Shell Oil	1973	61	34
	1972	59	32
Standard Oil (Indiana)	1973	79	109
	1972	67	79
Standard Oil (California)	1973	69	186[b]
	1972	73	144
Standard Oil (Ohio)	1973	12[b]	31
	1972	21	37
Sun Oil Company	1973	16	31[c]
	1972	5	24
Tenneco	1973	179	36[d]
	1972	151	32
Texaco	1973	363	143
	1972	288	108
Union Oil Co. of California	1973	41	32
	1972	41	27

[a]Rent on financing leases of $50 million and $48 million in 1973 and 1972, respectively.

[b]Does not include interest expense on Trans-Alaska pipeline and interest on advance sale of Prudhoe Bay crude oil ($13 million in 1973, $10 million in 1972).

[c]Approximation from annual report notes to financial sheets.

[d]Total rental expense.

Source: Compiled from annual reports.

The continuing importance of banks is seen in recent oil industry financings such as Ashland Oil's financing of the $2 million cost of Alaska lands by loans from institutional investors for 80 percent of lease costs in return for a 90 percent interest in net revenues to pay-out and then a 50 percent net revenue interest, with the proviso that if not paid out in seven years, the investors may exchange unrecovered notes for Ashland's common stock at set prices. Other examples include BP's loan from a bank consortium for the Forties field with payments of about 60 percent of oil flows; BP's domestically obtained $200 million production loan payable with 50 percent of production from Columbia Gas for the North Slope; bridge financing of $450 million obtained by American Natural Gas Production Co., Hamilton Bros., and Placid through a bank line of credit (including demand provisions); a $300 million credit arrangement between SOHIO and a group led by Chase Manhattan Bank for Alaska work; and a $150 million loan by a group of banks to Occidental Petroleum's British subsidiary to develop the Piper field in the North Sea.

Petroleum companies make extensive use of private placements. Because banks (and life insurance companies) are among the biggest buyers at private placements,[5] their frequent ownership of petroleum company securities is therefore to be anticipated. The use of private placements is also paralleled, as shown below, by a lack of equity sales. Only 12.4 percent of all private placements were equity in 1972 (up from an average of 5.9 percent in the 1965-72 period).[6]

The reliance on debt for external financing is traditional for smaller drilling contractors.[7] The importance of bank loans to small- and medium-size oil companies has been increased with the drying up of an alternative financing mechanism which once appeared to have great promise--the drilling program.[8]

Commencing in the mid-1960s, at least several billion dollars of capital have been raised for drilling exploratory and development wells through drilling programs.* These programs generally consist of an oil company general partner and a group of limited-partner investors, although some such programs are set up as incorporated investment funds. These drilling programs have provided significant funding for smaller oil companies, many of which have been basically management companies for annual drilling programs. At the same time drilling programs have offered a source of financing to the general-partner oil company and a tax shelter, primarily through the allocation of intangible drilling costs, to the limited-partner

---

*A few funds also operate to purchase producing properties.

investors. The general partner frequently obtains a nonrecourse loan for this share of expenses.

A review of oil company annual reports and SEC Form 10-K annual reports indicates that the number of offerings, and the amount being raised by drilling funds, have been falling off.

The amounts raised through such programs are not great when compared with the total sums spent for exploration and development in North America (the area where programs have funded work). However, these funds have been an important source of financing for the exploration and development work done by smaller companies such as Petro Lewis, Calvert Exploration Company, and Beacon Resources. The funds further provided a type of financing that partially shielded these companies from downside risk; they were an alternate source of funding to the sale of mineral interests to other oil companies[9] or to private placements with companies that are substantial users of energy.

Such private placements do not appear to be conducive to corporate independence and, in fact, may harm the public interest by replacing arm's-length buying by a situation in which the management of the energy-using firm can justify its oil investment only by pointing to higher oil prices. Corporate independence will not be furthered when the inevitable equity sweetener is included in the price of money.*

For large companies such as Pennzoil, a drying up of the market for drilling programs means the impairment of a source of leveraged financing that did not dilute equity, and which provided the industry, in nine POGO-type offerings, $584 million between 1970 and 1972.† The series of amendments of the plans for one of the last such offerings, which progressively shielded investors, may further indicate how the market for such ventures has soured.‡

---

*See, for example, terms of $12 million loan to Ashland Oil for Alaskan exploration work, or the reported carried interest given the insurance company lender by Tenneco in 1972 for offshore leases.

†POGO is Pennzoil Offshore Gas Operators, Inc.; Pennzoil, subsequent to POGO, in 1972, organized PLATO--Pennzoil Louisiana and Texas, Inc.

‡Increased rates of capital outlay by larger oil companies, and successful private placements by smaller firms, show that profits are adequate to bring forth investment in oil exploration. The use of internal financing from depreciation and retained earnings, in lieu of a more balanced capital structure for financing capital expenditures with sales of securities, may be expected to drive up the price of oil.

While the petroleum industry is reliant upon bank financing, as shown in Table A.C.3, many large petroleum companies have been picking up substantial amounts of marketable securities. These quick assets, as specifically reflected in annual reports of some companies, include large amounts of certificates of deposits[10] sold by banks. For the large firms listed in Table A.C.4, holdings of marketable securities rose $2,713,904,000 between 1972 and 1973. These holdings of marketable securities would appear to create a continuing active relationship between petroleum companies and banks.

TABLE A.C.3

Registration Activity in Tax Shelter Securities, 1971-73
(in thousands of dollars)

	Oil and Gas Exploration and Extraction		
Year	Number of Registrations Filed	Value of Registration	Amount of Cash Sales
1971	107[a]	908,823	855,594
1972	110[b]	978,336	853,153
1973	95[c]	707,365	621,030
Percent change, 1971-72	+2.8	+7.7	-0.2
Percent change, 1972-73	-13.6	-27.7	-27.2

[a]Six registrations were withdrawn, with value of $34.9 million and proposed cash sales of $29 million.

[b]Five registrations were withdrawn, with value and proposed cash sales of $48.15 million.

[c]Four registrations were withdrawn, with value and proposed cash proceeds of $7.65 million.

Source: Securities and Exchange Commission, Office of Economics, 1975.

TABLE A.C.4

Company Holdings of Marketable Securities
(thousands of dollars)

Company	1973	1972	Net Change	Percent Change
Amerada Hess	45,000	9,956	35,044	4.52
Atlantic Richfield	201,555	20,104	181,451	10.03
BP[a]	261	88	173	2.97
Cities Service	144,480	89,568	54,912	1.61
Clark Oil	12,904	11,900	1,004	1.08
Continental Oil	154,889	114,327	40,562	1.35
Exxon	2,525,429	1,210,379	1,315,050	2.09
Getty	129,209	199,699	(70,490)	0.65
Gulf Oil	921,000	562,000	359,000	1.64
Louisiana Land & Exploration[b]	41,466	17,345	24,121	2.39
Marathon Oil	53,168	90,960	(37,792)	0.58
Mobil Oil	624,409	449,846	174,563	1.39
Murphy Oil[c]	102,394	62,750	39,644	1.63
Pennzoil Co.[d]	55,643	169,401	(113,758)	0.33
Phillips Petroleum	68,162[e]	50,764	17,398	1.34
Shell Oil Co.	383,542	356,384	27,158	1.08
Standard Oil (Indiana)	452,932	293,774	159,158	1.54
Standard Oil (California)	430,972	205,222	225,750	2.10
Standard Oil (Ohio)	101,941	--	101,941	--
Sun Oil	61,070[e]	26,052	35,018	2.34
Tenneco, Inc.	13,428	44,482	(31,054)	0.30
Texaco, Inc.	324,852	283,619	41,233	1.15
Union Oil Co. of California	78,141	14,813	63,328	5.28
Total Increase			2,713,904	

Note: Parentheses show a decrease in 1973.

[a]Treasury bills and government securities.

[b]Certificates of deposit (CD).

[c]Includes CD's of $83,538,000 in 1973 and $52,324,000 in 1972.

[d]Temporary cash investments.

[e]Short-term investments.

Source: Compiled from annual reports.

Petroleum companies have other major quick assets. The Chase Manhattan Bank's group of 37 petroleum companies increased their total current asset holdings by some 9.9 billion dollars between 1972 and 1973.* Current assets consist of cash, marketable securities, and notes and accounts receivable.

INTERLOCKING RELATIONSHIPS BETWEEN BANKS AND PETROLEUM COMPANIES

Interlocks between boards of directors of banks and petroleum companies are frequent, and as shown in Table A.C.5, involve every substantial petroleum company. Interlocks exist both with major banks and with smaller country banks. The directorate interlocks raise questions about anticompetitive consequences of board loan and investment policy decisions, and anticompetitive effects of special information going to banks.

Directors of banks who are also affiliated with petroleum companies are in a position to influence bank loan policies and company investment policies in ways disadvantageous to potentially conflicting investments of bank customers, for example, competitive energy supply projects. Harmonizing of investment projects to advance the interest of both the bank and the petroleum company is an inviting path.

As shown in Table A.C.5, banks are frequently numbered among the largest holders of equity and debt securities of petroleum companies. This also obviously creates some community of interest. Because of the use made by petroleum companies of bank financing, banks receive, as creditors and potential creditors, financial and operating information from loan-seeking petroleum companies. This information is not generally available to the public, and to the extent that it is received by bank directors affiliated with oil companies, it could give them special knowledge regarding present or potential competitors.†

According to the testimony of an officer of a very large bank heavily engaged in such financing, petroleum-company-affiliated directors of banks are on the bank board because of their expertise.

---

*Some of these 37 firms do not separately disclose marketable security holdings in their annual reports.

†Loan volume data, alone, could at times be highly informative, for example, the amount of standby lines of credit obtained before federal leasing sales.

TABLE A.C.5

Financial House Directorate Interlocks and Securities Holdings in Petroleum Companies

Petroleum Company	Financial Houses with Whom Petroleum Company Director is Affiliated	Financial Houses among Largest Corporate Equity Holders	Financial Houses among Largest Corporate Debt Holders
Allied Chemical	Bowery Savings Bank (N.Y.)	Bankers Trust Co.	Chase Manhattan
	Chase Manhattan	Chase Manhattan	Liberty Mutual
	Chemical Bank	Lazard Freres & Co.	U.S. Trust Co.
	Donaldson, Lufkin, & Jenrette, Inc.	Loeb, Rhoades & Co.	
	Fidelity Union Bancorp.	Merrill Lynch	
	First National Bank (Southaven, Ms.)	Morgan Guaranty Trust	
	First National City Corp.	National Shawmut Bk. of Boston	
	Loeb, Rhoades & Co.	State Str. Bank & Trust (Boston)	
	N.Y. Life Ins. Co.	Swiss Bank Corp.	
Amerada Hess	Com'l Union Assur. Co. Ltd.	Bank of California, N.A.	Chase Manhattan
	Chemical Bank	Bank of Delaware	Chemical Bank
	First Nat'l Bank of Jackson, Ms.	Bank of New York	Equitable Life Assur. Soc.
	First Nat'l State Bank of N.J.	Chase Manhattan	First Nat'l Bank of Chicago
	Lamar Life Ins.	Equit. Life Assur. Soc.	First Nat'l City Bank
	Mutual Benefit Life Ins. Co.	Manufacturers Hanover	Metropolitan Life Insurance
	N.Y. Life Ins. Co.	Morgan Guaranty Trust	Morgan Guaranty Trust
	Thos. Jefferson Life Ins. Co.	Prudential Ins. Co.	
		State Str. Bank & Trust (Boston)	
American Petrofina[a]	None		American Nat'l Insurance Co.
			Chase Manhattan[b]
			Dry Dock Savings Bank (N.Y.)
			European-American Banking Corp.
			First Nat'l Bank--Dallas[c]
			Mass. Mutual Life Ins. Co.
			National Bank of Detroit
			Southwestern Life Insurance Co.

(continued)

TABLE A.C.5 (continued)

Petroleum Company	Financial Houses with Whom Petroleum Company Director is Affiliated	Financial Houses among Largest Corporate Equity Holders	Financial Houses among Largest Corporate Debt Holders
APCO Oil	American Nat'l Fire Ins. Co.	Bache & Co.	Bank of Montreal
	Central Nat'l Corp. (N.Y.)	Bank of Montreal	Chase Manhattan
	Continental Reinsur. Co.	Brown Bros. Harriman & Co.	First Nat'l Bank of Chicago
	Federated Capital Corp.	Central Nat'l Corp. (N.Y.)	Wilmington Tr. Co. (Del.)
	Great American Ins. Co.	Chase Manhattan	Loeb, Rhodes & Co.
	Jefferson Ins. Co. of N.Y.	First Nat'l City Bank	Madison Fund, Inc.
	Lehman Bros.	E. F. Hutton & Co., Inc.	Merrill Lynch
	Loeb, Rhoades & Co.	Loeb, Rhoades & Co.	Metropolitan Life Ins. Co.
		Madison Fund, Inc.	
		Merrill Lynch	
		Morgan Guaranty Trust	
		Pershing & Co., Inc.	
		Wilmington Trust Co. (Del.)	
Ashland Oil	Bank of Bluegrass (Lexington, Ky.)	Bank of N.Y.	Bank of Delaware
	Criterion Ins. Co.	Cede & Co.	Bankers Trust Co.
	Drovers State Bank (Minn.)	Continental Bank (Ill.)	Chase Manhattan
	First Security Nat'l Bank & Trust Co. (Lexington, N.Y.)	Morgan Guaranty Trust	Dresdner Bank of Germany
	First Union Nat'l Bank of N.C. (Charlotte)	Nat'l Shawmut Bank (Boston)	
	GEICO		
	GE Financial Corp. & Gov't Employees Life Ins. Co.		
	May Ave. Bank & Trust (Ok. City)		
	Second Nat'l Bank (Ashland, Ky.)		
	Second New Haven Bank (Ct.)		
	Security Ins. Group (Textron)		
	Security-Conn. Life Ins. Co.		
	U.S. Trust Co. of N.Y.		

Atlantic Richfield	Not Reported	Bankers Trust Co.	Aetna Life Ins. Co.
		Cede & Co.	Amsterdam-Rotterdam Bank N.V.
		Chase Manhattan	Chase Manhattan
		Manufacturers Hanover	Metropolitan Life Ins. Co.
			Morgan Guaranty Trust
			N.Y. Life Ins. Co.
Beacon Oil	None	P. H. Greer Co., Inc.	
Belco Petr.	None	U.S. Trust Co. of N.Y.	Aetna Life Ins. Co.
			Chase Manhattan
			El Paso Nat'l Bank
			First Nat'l City Bank
			Morgan Guaranty Trust
Champlin Petr. (Union Pacific Corp. Subsidiary)	Amer'n Bkrs. Ins. Co. of Fla.	Bank of N.Y.	
	Brown Bros. Harriman & Co.	Brown Bros. Harriman & Co.	
	Bus. Men's Assur. Corp.	Chase Manhattan	
	Chemical N.Y. Co.	Equit. Life Assur. Soc. of U.S.	
	Colo. Nat'l Bank (Denver)	Manufacturers Hanover	
	First Nat'l Bank of Oregon	Merrill Lynch	
	First Nat'l City Corp.	Nat'l Shawmut Bank of Boston	
	First Sec. Bank of Idaho, N.A.	State Str. Bank & Trust (Boston)	
	First Sec. Bank of Utah, N.A.		
	First Security Corp.		
	First United Bancorporation		
	Guarantee Mutual Ins. Co.		
	Irving Trust Co.		
	Charter N. Y. Corp.		
	Metropolitan Life Ins. Co.		
	Mutual Life Ins. Co.		
	Omaha Nat'l Bank		
	Real Banc, Inc.		
	The Seaman's Bank for Savings		
	Security Pacific Corp.		
	United Mo. Bancshares (Kansas City)		
	W. Omaha Nat'l Bank		
	White Weld & Co., Inc.		
	World Service Life Ins.		

(continued)

TABLE A.C.5 (continued)

Petroleum Company	Financial Houses with Whom Petroleum Company Director is Affiliated	Financial Houses among Largest Corporate Equity Holders	Financial Houses among Largest Corporate Debt Holders
Cities Service Oil Co.	Canadian Imperial Bank of Commerce First Nat'l Bank of Tulsa Harlem Savings Bank Kuhn Loeb Loeb, Rhoades & Co. Manufacturers Hanover Trust Morgan Guaranty Trust J. P. Morgan & Co. New Amsterdam Casualty Co. State Nat'l Bank--Greenwich, Ct.	Bank of New York Brown Bros. Harriman Cede & Co. Manufacturers Hanover Morgan Guaranty Trust	Chase Manhattan First City Nat'l Bank (Houston) Metropolitan Life Ins. U.S. Trust Co. of N. Y.
Clark Oil	Association Life Ins. Co. Loewi & Co.	Cede & Co. Marine Nat'l Exchange Bank (Milwaukee)	Aetna Casualty--Aetna Life Nat'l Bank of Chicago Jefferson Std. Life Ins. Lutheran Mutual Life Ins. Mass. Mutual Life Ins. Nat'l Life Ins. New England Mutual Life Ins. Northwestern Nat'l Life Ins. Prudential Ins. Teachers Ins. & Annuity
Clinton Oil	Central State Bank (Wichita) City Bank & Trust Co. (Jackson, Mi.) First Nat'l Bank of Wichita Gude, Winmill & Co. Northern States Bancorp (Detroit)	Cede & Co. Merrill Lynch	First City Nat'l Bank (Houston) Reserve Life Ins. Co. Rothschild Intercontinental Bk. Ltd.
Commonwealth Oil	Banco Credito y Ahorro Ponceno Banco de Economia de P. R. Banco Popular de P. R. Carib. Fed. S & L First Penn Corp. Fireman's Fund Amer'n Ins. Co.'s First Boston Corp.	Bache & Co. Blyth, Eastman Dillon Union Sec. & Co. First Nat'l Bank of Boston Loeb, Rhoades & Co. Merrill Lynch Pitcairn Co. (Wilmington, Del.)	First Nat'l City Bank[d] New York Life Ins. Co. Riggs National Bank (D.C.)

	First Nat'l Bank of Boston		
	First Nat'l City Bank		
	Puerto Rico Inv. Funds Inc.		
	Putnam Tr. Co. (Greenwich, Ct.)		
Continental Oil	Bankers Trust Co.	Bank of New York[e]	Bankers Trust Co.
	Canada Life Assurance	Bankers Trust Co.	Chase Manhattan
	Canadian Imperial Bank of Commerce	Cede & Co.	First Nat'l City Bank
	Cont. Ill. Corp.	Chase Manhattan[f]	Mellon Bank
	Federated Capital Corp.	Mellon Nat'l Bk & Trust (Pittsburgh)	Morgan Guaranty Trust
	Johnson City Bank (Tx)		
	Morgan Guaranty Trust		
	J. P. Morgan & Co., Inc.		
	Putnam Trust Co.		
	Royal Globe Ins. Co.		
	Trust Co. of Ga.		
	U.S. Trust Co. of N.Y.		
Crown Central Petr. Co.	Chase Manhattan, N.A.	Barclay's Bank Co. (N.Y.)	Aetna Life Ins. Co.
	Hallgarten and Co.	W. E. Hutton & Co.	Conn. General Life Ins.
	Union Trust Co. of Md.	Loeb, Rhoades & Co.	Equitable Life Assur.
		Merrill Lynch	Interstate Life & Accident Ins.
		Tom and Barut, Ltd.	Jefferson Std. Life Ins. Co.
		Union Trust of Md.	Mutual Life Ins. Co.
			New England Mutual Life Ins. Co.
			Penn. Mutual Life Ins. Co.
			Union Trust Co. of Md.
Diamond Shamrock Oil & Gas	Cleveland Trust Co.	American Bk. & Trust Co. of Pa.	Bankers Trust Co.
	First Nat'l Bank of Amarillo, Tx.	Bank of N.Y.	Chase Manhattan
	First Nat'l Bank of St. Louis	Cede & Co.	Morgan Guaranty Trust
	First Union Corp.	Cleveland Trust Co.	N. J. Nat'l Bank
	Mellon Nat'l Corp.	First Nat'l Bank of Chicago	Savings Bank Tr. Co.
	St. Louis Union Trust Co.	Irving Tr. Co.	Wilmington Tr. Co.
	Society Nat'l Bank of Cleveland	Mellon Nat'l Bk. & Tr. (Pittsburgh)	
		Pittsburgh Nat'l Bk.	
		State Str. Bk. & Tr. Co. (Boston)	
		Swiss Bank Corp.	
		Wilmington Trust Co. (Del.)	

(continued)

TABLE A.C.5 (continued)

Petroleum Company	Financial Houses with Whom Petroleum Company Director is Affiliated	Financial Houses among Largest Corporate Equity Holders	Financial Houses among Largest Corporate Debt Holders
El Paso Natural Gas Co.	American Bank (Odessa, Tx.)	First Nat'l City Bank	Aetna Life Ins. Co.
	Bank of New York	Merrill Lynch	Chase Manhattan
	Desert Ins. Co. Ltd.	Reynolds Securities	Conn. Gen'l Life Ins. Co.
	El Paso Nat'l Bank (Tx.)		Equitable Life Assurance
	First City Bancorporation of Texas		John Hancock Mutual Life Ins.
	First State Bank		Metropolitan Life Ins. Co.
	Home Savings & Loan (Odessa)		Mutual Life Ins. Co.
	Int'l Ins. Agency, Inc.		N.Y. Life Ins. Co.
	Mellon Bank & Trust Co.		Northwestern Mutual Life Ins.
	Pacific Mutual Life Ins. Co.		Sun Life Assurance Co. of Canada
	Permanent Bank & Trust Co. (Odessa, Tx.)		Travelers Ins. Co.
	San Angelo Nat'l Bank		
	TX Comm. Bank N.A.		
	TX Comm. of Houston		
	TX Comm. of Lubbock		
	Western Bancorp (CA.)		
Exchange Oil & Gas	American Tidelands Life Ins. Co.		
	Hibernia Nat'l Bank in N.O.		
	Wertheim & Co.		
EXXON	American Gen'l Ins. Co.	Bank of New York	Bankers Trust Co.
	Chase Manhattan	Bankers Trust Co.	First Nat'l City Bank
	Chemical Bank	Cede & Co.	John Hancock Mutual
	Dry Dock Savings Bank (N.Y.)	Chase Manhattan Bank	Metropolitan Life Ins.
	Equit. Life Assur. Soc.	Chemical Bank	Mutual Benefit Life Ins.
	First City Bankcorporation of Texas	First Nat'l Bank of Boston	Travelers Ins. Co.
	First Nat'l City Bank	Manufacturers Hanover	
	Metropolitan Life Ins.	Morgan Guaranty Trust	
	Morgan Guaranty Trust	State Str. Bank & Trust (Boston)	
	J. P. Morgan & Co.	United States Trust Co.	
	Prudential Ins. Co.		
	Texas Commerce Bancshares, Inc.		

Farmland Industries	Farmers Elev. Mutual Ins. Co.		Bank of the Southwest
	Farmers Life Co.		Bankers Trust Co.
	Farmland Life Ins. Co.		Chase Manhattan Bank
			Commerce Bank (Kansas City)
			First Nat'l Bank of Chicago
			First Nat'l City Bank
			Omaha Bank for Co-ops
			St. Louis Bank for Co-ops
			Wichita Bank for Co-ops
Felmont Oil Corp.	The Franklin Corp.	Cede & Co.	
Fletcher Oil & Ref'g Co.	Idaho Fidelity Corp.		Northwestern Nat'l Life Ins. Co.
Forest Oil	Nat'l Bank of Commerce (San Antonio)	Bank of CA	Chase Manhattan, N.A.[g]
		Bank of NY	Federal Life & Cas.
		Cede & Co.	John Hancock Mutual
		First Nat'l Bank of Nev.	Lehman Bros., Inc.
		Metropolitan Life Ins. Co.	Metropolitan Life Ins. Co.
		State Str. Bank & Trust (Boston)	Morgan Guaranty Trust
			Mutual Life Ins. Co.
			Northwestern Mutual Life
			People's Home Life Ins.
			State Str. Bank & Trust (Boston)
General American Oil Co. of Texas	American Nat'l Bank of Jacksonville	Cede & Co.	
	Blyth Eastman Dillon & Co.	Chase Manhattan	
	Deposit Guaranty Nat'l Bank (Miss.)	New England Merchants Nat'l Bank (Boston)	
	The Excelsior Life Ins. Co.	Republic Nat'l Bank of Dallas	
	Matthews & Co., Ltd. (CAN)		
	The People's Nat'l Bank Tyler, Tx.		
	Republic Nat'l Bank of Dallas		
	Standard Life Ins. Co.		
	State Bank of Jacksonville (Fla.)		
General Crude Oil	American General Ins. Co.	Bank of New York	None
	Glenmede Trust Co. (Pa.)	Glenmede Trust (Pa.)	
	Meyerland State Bank	Manufacturers Hanover	
	Manufacturer Hanover Corp.		

(continued

TABLE A.C.5 (continued)

Petroleum Company	Financial Houses with Whom Petroleum Company Director is Affiliated	Financial Houses among Largest Corporate Equity Holders	Financial Houses among Largest Corporate Debt Holders
Hamilton Bros. Petr. Corp.	Chase Int'l Inv. Corp. First Nat'l Bancorp. First Nat'l Bank in Dallas Lamar Life Ins. Life & Casualty Co. of TN Nat'l Shawmut Bank of Bos. Trust Co. of N.Y.	Not reported	Bank of Montreal
Hunt Oil Co.	Exchange Bank & Trust Co. (Dallas)		Federal Savings & Loan Assoc. (Thibodeau, La.) Mutual Life Ins. Co. N. American Life & Casualty Co.
Husky Oil	Cont. Ill. Nat'l Bank & Trust First Security Corp. (Utah)	Bank of Montreal Can. Imp. Bank of Comm. Chase Manhattan Cont. Ill. Nat'l Bank & Trust Royal Bank of Can. Montreal Trust Co. Canada Permanent Trust Co.	Bank of New York Cont. Illinois Nat'l Bank (Chicago) Federal Life & Casualty Co. First Sec. Bank of Utah Morgan Guaranty Trust New York Life Ins.
Kerr-McGee Corp.	American Bank of Edmond (Ok.) American Fidelity Assur. Co. American Fidelity Ins. Co. Capital Hill St. Bank & Trust Co. (Ok. City) Citizens Bank of Ok. City Fidelity Bank, N.A. (Ok. City) First Nat'l Bank, Alex. Ok. First Nat'l Bank & Trust, Muskegee Ok. First State Bank, Blanchard, Ok. First State Bank & Trust Co. of Oklahoma City F & M Bank of Tulsa	Cede & Co. Chase Manhattan Fidelity Bank, N.A. First Nat'l Bank of Boston Lehman Corp. State St. Bk. & Trust (Boston) Metropolitan Life Ins. Co. Wilmington Trust Co. (Del.)	Bank of Southwest First Nat'l Bank of Chicago First Nat'l City Bank John Hancock Mutual Prudential Ins. Co. of America N.Y. Life Ins. Co.

	Lehman Bros., Inc. Liberty Nat'l Corp. (Ok. City) Reserve Nat'l Ins. Co. Soc. First Nat'l Bank of L.A. S.W. Title & Trust Co. (Ok. City)		
Kewanee Oil	Blyth Eastman Dillon & Co. Philadelphia Nat'l Bank Tanney, Montgomery, Scott, Inc.		
Lion (Tosco Subsidiary)	Exchange Bank & Trust Co. (Dallas)		The Equitable Life Ins. Soc. The First Nat'l Bank of St. Paul
Lone Star Gas Co.	Bank of Commerce Employers Casualty Co. Republic Nat'l Bank of Dallas San Angelo National Bank Texas Employers Ins. Assoc.	Chemical Bank (trustee) Equitable Life Assur. Soc. First Nat'l Bank in Dallas First Nat'l City Bank Mellon Nat'l Bank and Trust (Pittsburgh) Merrill Lynch Nat'l Shawmut Bank Northwestern Pa. Bank & Trust Republic Nat'l Bank of Dallas	Mellon Nat'l Bank & Trust (Pittsburgh)
La. Land & Exploration	American Life Ins. Co. of N.Y. Bank of N.Y. Citizens Nat'l Bank of Hammond, La. Depository Trust Co. Dry Dock Savings Bank First City Nat'l Bank of Houston Hibernia Nat'l Bank of New Orleans La. & Southern Life Ins. J. P. Morgan & Co. N.Y. Life Ins. Northern Ins. Co. of N.Y. Whitney Nat'l Bank (New Orleans)	Bank of New York Bankers Trust Co. Brown Bros. Harrison & Co. Cede & Co. Chase Manhattan Delaware Trust Co. Goldman Sachs & Co. Harris Trust & Savings Bank (Chicago) Lehman Bros. Lewco Securities Corp. Manufacturers Hanover Trust Md. Nat'l Bank Mellon Nat'l Bank & Trust (Pittsburgh)	

(continued)

TABLE A.C.5 (continued

Petroleum Company	Financial Houses with Whom Petroleum Company Director is Affiliated	Financial Houses among Largest Corporate Equity Holders	Financial Houses among Largest Corporate Debt Holders
La. Land & Exploration (continued)		Merrill Lynch Morgan Guaranty Trust Nat'l Bank of Tulsa Nat'l Bank of Rutherford (N.J.) Northwestern Nat'l Bank of Minn. Savings Bank & Trust Co. Spencer Trask & Co., Inc. State Str. Bank & Trust Co. (Boston) U.S. Trust of N.Y. Nat'l City Bank of Cleveland	Bank of New York First Nat'l Bank of Commerce (New Orleans, La.)
Marathon Oil[h]	The First Boston Corp. First Nat'l Bank of Findlay (Ohio) Nat'l City Bank of Cleveland N.Y. Life Ins. Co. Philadelphia Nat'l Bank Sears Bank & Trust Co. (Chicago) Toledo Trust Co.		
McCulloch Oil[i]	Bache & Co.	Cede & Co. Merrill Lynch Mutual Life Ins. Co.	Bank of America Chase Manhattan City Nat'l Bank & Trust (Kansas City) First Nat'l Bank in Dallas Mellon Bank & Trust Co. (Pittsburgh) Merrill Lynch Savings Bank & Trust Co. State Str. Bank & Trust (Boston)
Mobil Oil	American Sec. & Trust Co. (D.C.) Bankers Trust Co. Brooklyn Savings Bank Chemical N.Y. Corp. Federal Ins. Co. First Nat'l City Corp. Schroder Trust Co. Vigilant Ins. Co.	Not Reported	Not Reported

Monsanto	Bank of America, N.Y. Boatmen's Nat'l Bank of St. Louis Charter Nat'l Life Ins. Co. Equitable Life Assur. Soc. First Nat'l Bank (St. Louis) First Nat'l City Bank Great American Ins. Co. Liberty Mutual Fire Ins. Liberty Mutual Life Ins. Mercantile Bancorp Inc. (St. Louis) Merchants Nat'l Bank (Cedar Rapids, Ia.) Metropolitan Life Ins. N.Y. Life Ins. Co. St. Louis County Nat'l Bank St. Louis Union Trust Co.	Bankers Trust Co. Brown Bros. Harriman Chase Manhattan Merrill Lynch Nat'l Bank of Detroit	Chase Manhattan Commerzbank, A.G. Metropolitan Life Ins. N.Y. Life Ins. Northwestern Mutual Life Ins.
Murphy Oil	Allied Bank Int'l First Nat'l Holding Co. First Nat'l Bank of El Dorado First Tennessee National Corporation Louisiana & Southern Life Ins. Co.	Bankers Trust Co. Cede & Co. Ft. Worth Nat'l Bank First Trust Co. of St. Paul Merrill Lynch Morgan Guaranty Trust United Mo. Bank (Kansas City)	Equitable Life Assur. First Nat'l Bank of Chicago Mercantile Bank of Canada Mitsubishi Bank Ltd. Nat'l Westminster Bank Ltd. Royal Bank of Canada Union Bank of Switzerland
Occidental Petr. Corp.	Cleveland Trust Co. Florida Nat'l Bank of Jacksonville Home Life Equity Fund National Liberty Corp.	Cede & Company[j] Chase Manhattan Merrill Lynch	Aetna Life Ins. Co. Chase Manhattan Conn. Gen'l Life Ins. Equitable Life Assur. Soc. First Nat'l City Bank John Hancock Mutual Mutual Life Ins. Co. N.Y. Life Ins. Co. Teachers Inst. & Annuity

(continued)

TABLE A.C.5 (continued)

Petroleum Company	Financial Houses with Whom Petroleum Company Director is Affiliated	Financial Houses among Largest Corporate Equity Holders	Financial Houses among Largest Corporate Debt Holders
Panhandle[i]	Commerce Bancshares	Bankers Trust Co.	Aetna Life Ins. Co.
Eastern Pipe Line	Eur-American Banking Corp.	Cede & Co.	First City Nat'l Bank
	Morgan Guaranty Trust	Chase Manhattan	Life Ins. Co. of Va.
		Chemical Bank	Metropolitan Life Ins. Co.
		Fidelity-Philadelphia Trust Co.	N.Y. Life Ins. Co.
		First City Nat'l Bank	Savings Bank Trust Co. (N.Y.)
		Irving Trust Co.	U.S. Trust Co.
		Manufacturers Hanover	
		Mellon Nat'l Bank & Trust Co. (Pittsburgh)	
		Morgan Guaranty Trust	
		Nat'l Shawmut Bank of Boston	
		Wilmington Trust Co. (Del.)	
Pennzoil	Brit. Assur. Trust Ltd.	Brown Bros. Harriman	Chemical Bank
	C.A. Casualty Ins. Co.	E. F. Hutton & Co.	First Nat'l Bank of Chicago
	Commonwealth Assur. Co.	First Nat'l City Bank	Lincoln Nat'l Life Ins.
	Federal Capital Corp.	First Nat'l Bank of Shreveport (trustee)	Mass Mutual Life Ins. Co.
	First Bancorp (Tulsa)	State Str. Bank & Trust (Boston)	Merrill Lynch
	First Nat'l Bank (Midland)		Mutual Benefit Life Ins.
	Mellon Nat'l Bank & Trust (Pittsburgh)		New England Merchants Nat'l Bank
	Valley Forge Insurance Co.		Savings Bank Trust
			State Str. Bank & Trust
Phillips Petr.	American Reinsurance Co.	Bankers Trust Co.	Chase Manhattan
	First Bancshares Inc. (Bartlesville)	Chase Manhattan	Cont. Illinois Nat'l Bank
	First Nat'l Bank in Bartlesville	First Nat'l Bank, Bartlesville	First Nat'l City Bank
	First Nat'l Bank & Trust Co. of Tulsa	Fidelity Bank (Phila., Pa.)	Metropolitan Life Ins.
	First Nat'l City Bank & Citicorp	Merrill Lynch	N.Y. Life Ins. Co.
	Franklin N.Y. Corp.	U.S. Trust Co. of N.Y.	Prudential Ins. Co.
	Nat'l Bank of Tulsa		
	Teachers Inst. Annuity Assoc.		
	Union Nat'l Bank (Bartlesville, OK)		
	Zions Utah Bancorp		

Rock Island Refinery	American Fletcher Nat'l Bank & Trust Co.	First Nat'l Bank in Wichita Mercantile Trust Co.	None
Standard Oil (Ind.)	American Nat'l Bank & Trust Co. (Chicago) Bank & Trust Co. of Arlington Heights, Ill. Chase Manhattan Chemical N.Y. Corp. Chicago Bank of Commerce Continental Ill. Corp. First Nat'l Bank & Trust Co. of Tulsa Harris Trust & Savings Bank (Chicago) Nat'l Blvd. Bank of Chicago Union Cent. Life Ins. The Wilmette Bank (Ill.)	Cede & Co. Chase Manhattan First Nat'l Bank of Chicago Prudential Ins. Co.	Chemical Bank Metropolitan Life Ins. Mitsubishi Bank Ltd. S. Nationale de Credit a l'Indus. Bank The Sumito Bank, Ltd. U.S. Trust Co. of N.Y.
Standard Oil (Ohio)[k]	Central Bancshares Corp. (Cleveland) Cleveland Trust Co. Lincoln National Corp. National City Bank of Cleveland	Bankers Trust British Petr. (Oversea) N.Y. Chase Manhattan Nat'l City Bank Wilmington Trust Co. (Del.)	Chase Manhattan Chemical Bank First Nat'l Bank--Chicago First Nat'l City Bank Morgan Guaranty Trust U.S. Trust Co. of N.Y.
Shell Oil Company[l]	Bank of California NA Bank of the Southwest Capital National Bank (Austin, Tx.) Charter N.Y. Corp. Chase Manhattan Chemical Bank Conn. Mutual Life Ins. Co. Dean Witter & Co. Inc. Lehman Brothers Ranger Ins. Co. Seaboard Surety Co.	Manufacturers Hanover Trust Co.	Bankers Trust Co. First Nat'l City Bank Irving Trust Co. N.Y. Life Ins. Co. U.S. Trust Co.

(continued)

TABLE A.C.5 (continued)

Petroleum Company	Financial Houses with Whom Petroleum Company Director is Affiliated	Financial Houses among Largest Corporate Equity Holders	Financial Houses among Largest Corporate Debt Holders
Southland Oil Company	First Nat'l Bank of Jackson (Miss.)	Lamar Life Ins.	Chemical Bank
	Lamar Life Ins. Co.		First Nat'l Bank-Memphis
			Whitney National Bank
Standard Oil (CA)[i]	Banca D'America e d'Italia	Crocker Nat'l Corp.	Bank of Int'l. Settlements (Switzerland)
	Bank of America NT & SA		Caisse Gen. d'Epargne et de Retraite (Brussels)
	Bank of California		Mediobanca Sp. A.
	Crocker Citizen's Nat'l Bank		Soc. Nationale de Credit L'Industrie (Brussels)
	Crocker Nat'l Corp.		U.S. Trust Co. of N.Y.
	Equitable Life Ins.		
	Fireman's Fund Amer'n. Ins.		
	Fireman Fund Ins. Co.		
	First Nat'l City Bank		
	Seattle First Nat'l Bank		
	United CA. Bank		
Sun Oil Company	Bay State Corp.	Not reported	Chase Manhattan
	Fidelity Union Life Ins.		Fidelity Union Trust Co.
	First Nat'l Bank & Trust Co. of Tulsa		Metropolitan Life Ins. Co.
	Girard Bank (Pa.)		U.S. Trust Co. of N.Y.
	Glenmede Trust Co. (Pa.)		
	Mutual Life Ins. Co.		
	Northeastern Bank of Pa.		
Tenneco Inc.	American Gen'l Life Ins.	Cede & Co.	Aetna Life Ins.
	Employers Ins.-Wisconsin	Houston Nat'l Bank	Bankers Trust Co.
	Houston Nat'l Bank	Merrill Lynch	Conn. Gen'l Life Ins. Co.
	Harris, Upham & Co.	Stone & Webster	Equitable Life Assur.
	First Wisconsin Mortgage Trust		John Hancock
	N. W. Bancorp.		First Nat'l City Bank
	TX Commerce Bank		Metropolitan Life Ins.
			Prudential Life Ins.
			Teachers Ins. & Annuity

TESORO Petr.	E. F. Hutton & Co., Inc.	Not reported	The First Jersey Nat'l Bank First Nat'l Bank, Ft. Lauderdale First Nat'l Bank of Topeka Harris Trust & Savings Bank E. F. Hutton & Co. Janney Montgomery Scott, Inc. Merrill Lynch
Texaco	Bessemer Securities Corp. Brown, Harriman & Int'l Banks, Ltd. Continental Ill. Corp. Equitable Life Assur. Soc. Financial Gen'l Bancshares First Nat'l Bank of Washington Gen'l Reins Corp. Mutual Life Ins. of N.Y. Nat'l Blvd. Bank of Chicago Nat'l City Corp. Sun Life Assur. of Canada State Bank of St. Charles, Ill. United Services Life Ins. Co.	Not reported	Not reported
Texas City Ref'g.[i]			American Nat'l Bank of Mobile First Nat'l Bank of Memphis Mutual Life Ins. Co.
Texas Oil & Gas Co.	Spencer Trask & Co.	Bankers Trust Co. Bank of N.Y. Cede & Co. Chemical Bank Ford Foundation First Nat'l Bank (Denver) Manufacturers Hanover Morgan Guaranty Trust Phicar & Co. (Newark, N.J.) Wall St. Trust Co. U.S. Fidelity & Guaranty Trust Co.	American Nat'l Ins. Co. Bank of Delaware Cleveland Trust Co. Jefferson Std. Life Ins. First Nat'l City Bank N.J. Nat'l Bank U.S. Trust Co. of N.Y.
Total-Leonard (Subsidiary of Total Petr. Ltd.)	Bank of Alma, Mich. Isabella Co. State Bank		Aetna Life Ins. Co. Conn. Mutual Life Ins. Fr. Amer'n Bkg. Corp.

(continued)

TABLE A.C.5 (continued)

Petroleum Company	Financial Houses with Whom Petroleum Company Director is Affiliated	Financial Houses among Largest Corporate Equity Holders	Financial Houses among Largest Corporate Debt Holders
Total-Leonard			Lincoln Nat'l Life Ins.
(continued)			Mich. Nat'l Bank
Union Oil Co. of	Bank of America	Bank of N.Y.	Chase Manhattan
California	Benefit Trust Life Ins. Co.	Cede & Co.	John Hancock
	Canadian Imperial Bank of Comm.	First Nat'l City Bank	Metropolitan Life Ins.
	Korea Exch. Bank of CA.	Merrill Lynch	N.Y. Life Ins. Co.
	Pacific Mutual Life Ins.	Prudential Ins.	U.S. Trust Co.
	Palatine Nat'l Bank	Sec. Pacific Nat'l Bank	
	Suburban Nat'l Bank of Woodfield		
	Union Bancorp, Inc.		
	Western Bancorp.		

[a]American Petrofina Holding Co. is listed as largest equity holder.

[b]For Prudential Ins., N.Y. Life Ins. Co., Penn. Mutual Life Ins., and Conn. Gen'l Life Ins. Co.

[c]Lease trustee for eight insurance companies: Pacific Mutual Life Ins. Co., Penn. Mutual Life Ins. Co., Commonwealth Ins. Co., State Farm Ins. Co., Acacia Mutual Life Ins. Co., Country Life Ins. Co., Lutheran Mutual Life Ins. Co., United Benefit Life Ins. Co.

[d]In group for credit agreement with Chase Manhattan, Chemical Bank, 1st Nat'l Bank of Boston, B. Credito y Ahorro Ponceno, Banco de Ponce, Banco Pop. de P.F.

[e]Nominee for: General Investors Company, Inc.; The Johnston Mutual Fund, Inc. Petroleum Corporation of America; United Funds, Inc.; and The United States Fund.

[f]Nominee for: American Mutual Fund, Inc.; Chemical Fund, Inc.; Eberstadt Fund, Inc.; Equity Fund, Inc., International Resources Fund, Inc.; and Investment Company of America.

[g]Nominee for: Manufacturer's Hanover Trust Company; Marine Midland Bank (Western); First National Bank and Trust Company of Tulsa; National Bank of Commerce (San Antonio).

[h]Only partial public response.

[i]Only partial response.

[j]Cede & Company holds common stock for a number of companies, including Bache & Co., E. F. Hutton, Merrill Lynch and Reynolds Securities, Inc. who each have 800,000 or more shares of common stock.

[k]British Petr. (Oversea) N.Y. is largest holder of equity shares.

[l]Shell Petr. N.V. which owns 69.6 percent of Shell Oil, is in turn owned 60 percent by Royal Dutch Petr. Co. and 40 percent by the "Shell" Transport and Trading Co., Ltd. Shell Petr. N.V. has pledged 43.2 percent of Shell Oil common stock to Morgan Guaranty Trust under a trust indenture of Shell Funding Corp.

Source: Information submitted to U.S. Senate, Interior Committee, Subcommittee on Integrated Oil Company Operations, 1974; bank filings with Comptroller General, 1974.

In the witness's bank, the board routinely summarizes for the petroleum company directors smaller loans involving petroleum, and presents larger loans in detail, for petroleum company director review.[11] This is entirely understandable. It also might present conflicts of interest, and provide the petroleum people on the bank board with unusual market information opportunities.

Indeed, the presence of petroleum-bank interlocks may readily be expected to have a discouraging effect on persons considering seeking bank loans for petroleum operations, but not wanting to show their books to competitors. This discouragement could occur irrespective of actual bank practice regarding loan information provided board members.

The local would-be driller-producer, if he goes, as is usually the case, to a large country bank for financing (for example, for production loans), might be affected in his proposal by the fact that on the bank's board is a representative of an oil company which is a competitor in production or an owner of area gathering and transportation facilities.* Similarly, persons considering larger projects must turn to major regional banks and New York banks where, also, oil companies are represented on boards of directors. The psychological impact on project planners of such bank interlocks, whether justified or not, could readily affect the competitive nature of plans; and board members will have both a legal obligation and a psychological tendency to harmonize their responsibilities to all of the firms on whose boards they sit.

CORRESPONDENT BANKING

Just as there are extensive joint venture relationships between large and small petroleum companies (for example, farm-outs of drilling prospects, or oil exchanges), there are extensive relationships between banks, and among banks and utilities. See Table A.C.6.

The relationship between small and large banks is a product of the practice of correspondent banking.[12] Correspondent banking is an interbank practice whereby city correspondent banks provide a cluster of services to smaller country banks in exchange for interbank deposits. Correspondent banking is not, however, clearly dichotomized into country and city bank categories; instead, the system exhibits a pyramidal structure wherein moderate-sized

---

*Persons planning local projects upstream of production will frequently find the representatives of established local refiners on bank boards.

TABLE A.C.6

Interlocks among Boards of Directors

Bank	Utility	Energy Company
Ban Cal Tri-State Corp.	San Diego Gas & Elec. Co.	Natomas Co.
(Bank of California)		Baker Oil Tools, Inc.[a]
Bank America Corporation	Southern California Edison Co.	Dillingham Oil Company
		Getty Oil Co.
		Kaiser Industries
		Standard Oil (CA)
		Union Oil Company of California
Bankers Trust New York Corp.	Baltimore Gas & Elec. Co.	Continental Oil Co.
	Consumers Power Co.	Hudson's Bay Oil & Gas Co., Ltd.
		International Paper Co.
		Mobil Oil Corp.
		Universal Oil Products Co.
Bank of New York Co.	--	El Paso Natural Gas Co.[b]
Boatmen's Bancshares Inc.	Union Electric Co.	Capital Coal and Coke Co.
Central Bancshares of the South, Inc.	--	--
The Chase Manhattan Corporation	--	Allied Chemical Corp.
		Exxon Corp.
		R. J. Reynolds Industries, Inc.
		Shell Oil Co.
		Standard Oil Co. (IN)
Chemical New York Corporation	Consolidated Edison	Amerada Hess
		Aramco
		Exxon
		Mobil Oil Corp.
		Texas Gas Transmission Corp.
		The Hillman Company
		Union Pacific Corp.
Clevetrust Corp.	--	Diamond Shamrock Corp.
		Standard Oil Co. (Ohio)

First Chicago Corp.	Commonwealth Edison Co.	Northwest Industries, Inc.
		Burlington Northern, Inc.
		Atlantic Richfield Co.
First City Bancorporation of Texas	--	Exxon Co. (USA)
		United Gas, Inc.
		Texas Eastern Transmission Corp.
		Panhandle Eastern Pipeline Co.
		Quintana Petroleum Co.
		Robertson Coal, Inc.
		PetroLewis Corporation
		The Superior Oil Company
		Halliburton Company[c]
		Highland Oil Company
		Northwest Exploration
		Coquina Oil Corporation
		Seven Oil, Ltd.
		El Paso Natural Gas Company
		Transco Companies, Inc.
First International Bancshares, Inc.	Dallas Power & L. Co.	SEDCO, Inc.
	Duke Power Co.	Blue Crown, Inc.
		Blue Crown Petroleum, Ltd.
		Various Delhi Oil Concerns
		Gas Producers Corp.[d]
		Hamilton Bros. Petroleum Corp.
		Murchison Oil Co.
		Thermal Energy Co.
		Baptist Foundation & Affiliates of Texas
		Vaughn Petroleum, Inc.
		American Liberty Oil Co.
First National Bancorporation, Inc. (Denver)	Public Serv. Co. of Colorado	Denver & R.G.W. RR. Co.
		Hamilton Bros. Oil Co.
First National Boston Corp.[e]	Stone & Webster, Inc.[f]	International Paper Co.
	Boston Edison Co.	Cabot Corp.
	New England Elec. System	Mississippi River Transmission Corp.

(continued)

TABLE A.C.6 (continued)

Bank	Utility	Energy Company
First National Cincinnati Corp.	Cincinnati Gas & Electric Co.	Roberta Coal Co.
First National City Corp.	Consolidated Edison Co. of N.Y.	Union Pacific Corp.
	Stone & Webster, Inc.[f]	Monsanto Co.
		Exxon Corp.
		W. R. Grace & Co.
		Standard Oil Co. of Calif.
		Phillips Petroleum Co.
First Oklahoma Bancorporation, Inc.	Oklahoma Gas & Elec. Co.	Eason Oil Co.
		Mustang Fuel Corp.
		Katy Industries[c]
		Oceanography International
First Pennsylvania Corp.	Philadelphia Electric Co.	Berwind Corp.
First Tulsa Bancorporation	Central & South West Utilities	Reading & Bates Offshore Drilling Co.
	Corp. and its subsidiary,	Apache Exploration Co.
	Public Service Company of	Pennzoil Co.
	Oklahoma	Sun Oil Co.
		Helmerich & Payne Inc.
		Phillips Petroleum Co.
		Mabee Petroleum Corp.
		Skelly Oil Co.
		Cities Service Co.
		Williams Companies
		Warren-American Oil Co.
		Oriole Oil Co.
		Standard Oil Co. (IN)
		Bigheart Pipe Line Corp. & Affiliates
J. P. Morgan & Co.	Niagara Mohawk Power Corp.	Continental Oil Co.
		Louisiana Land & Expl. Co.
		Airforce Pipeline, Inc.
		Exxon Corp.
		Cities Service Co.
		Burlington Northern, Inc.
		Texas Gulf, Inc.
		Panhandle Eastern Pipe Line Co.
		Canadian Pacific, Ltd.

Manufacturers Hanover Corp.	Consolidated Edison Co. of N.Y.	Cities Service Co.
	Public Serv. Elec. & Gas Co.	General Crude Oil Co.
		Southern Pacific Co.
Mellon National Corp.	Duquesne Light Company	Continental Oil Company
		Diamond Shamrock Corp.
		El Paso Natural Gas. Co.
		Gulf Oil Corp.
		Lone Star Gas Co.
National City Corp. (Cleveland)	Cleveland Electric Illum. Co.	Beatrice-Pocahontas Co.
		Marathon Oil Co.
		Dow Chemical Co.
		Texaco, Inc.
		Ariel Petroleum Co., Ltd.
		Hanna Petr. Co. & Affiliates
		Standard Oil Co. (Ohio) & Affiliates
Republic National Bank of Dallas[g]	Dallas Power & Light Co.	Lone Star Gas Co.
	Southwestern Public Service Co.	Developers Oil Co.
	Texas Power & Light Co.	Rainbow Oil Producing Co.
		Halliburton Co.
		Gen. Amer. Oil Co. of Tx. and Affiliates
		Exchange Oil & Gas Co. Funds MNOP & Q
		Petro Oil & Gas Co.--Fund A & B
		Excalibar Oil Corp.
		Neuhoff Oil & Gas Corp.
		Concho Petroleum Co.
		Ranchers Exploration & Development Corp.
		Kirby Petroleum

Note: List, not necessarily complete, of interlocks, direct and indirect, existing in 1973, with some updating to 1974. Authors' investigation of such interlocks is still in progress.

[a]Equipment supplier.

[b]Gas pipeline.

[c]Oil field equipment and services.

[d]Ownership interest.

[e]Trust accounts (market value, 12/31/74): Exxon Corp. $123 million, Texaco $35 million.

[f]Utility design and engineering (service co.).

[g]Owns United Petroleum Corp., Boyce Oil Co., and McBean Oil Co.

Source: Annual reports to the Federal Reserve Board; annual reports of electric companies to the SEC.

institutions normally play a double role by supplying services while themselves relying on large banks for correspondent aid.[13] A standard part of the service package of correspondent banks is participation by the larger banks in the smaller bank's loans, and occasional farming out of participation in the other direction.[14]

By and large, the country bank is required to hold demand deposit balances with the city bank as a means of payment for correspondent services. These services include collection services, providing credit information, purchase of currency, trust facilities, real estate mortgage funds, purchase and sale of government bonds and securities, safe-keeping facilities, demand deposits, financial counsel, office facilities, loan participations, deposit referrals, repossession facilities, and forms and audit systems information.[15]

The average number of banks with which correspondent relations are maintained increases with the size of the bank.[16]

The relationship between smaller petroleum companies and smaller banks may buttress the relationship between smaller and larger petroleum companies arising out of joint operational endeavors. This buttressing could arise through the smaller petroleum company and smaller banks being mutually aware of their reliance on larger banks and petroleum companies with which larger banks have interlocks.

Just as the smaller bank looks to larger banks for services, smaller petroleum companies must rely upon larger companies for direct or indirect (exchange) transportation services, farm-outs of drilling opportunities, and some geophysical prospecting, among other things. Also, larger banks must be looked to for major financing. Thus the interests of both are such as to be conducive to their adopting a cooperative approach to relations with bigger petroleum firms. The linking of inter-petroleum-firm relations to interbank relations, through petroleum firm financing practices, securities holdings, and ownership and board interlocks with banks, coupled with correspondent banking practices, creates a danger of an anticompetitive community of thought.[17]

Development drilling is a fairly sure proposition, and explorationists have traditionally obtained outside funding for development and production through the sale of production loans to banks. A production loan is a nonoperating interest in production, which is limited in total amount (in terms of money, time, or physical limits of production). The production payment is expressed either as a certain amount of money (with or without explicit interest) or a certain number of units of product; a production-payment holder is free of the burden of development costs and production expenses and is entitled to a fraction or percentage of gross production.[18] Production payments put lenders into the petroleum business, and give them a stake in preventing supply gluts driving down the price of petroleum and thereby reducing the lenders' security and loan profits.

There has been some concern in several quarters with tendencies toward channeling capital flows to a relatively few large companies. This sort of concern was recently addressed by the Senate Finance Committee, as follows, in dealing with the institutional investor issue.

> Institutional investors--trust departments of large U.S. banks, insurance companies, mutual funds, pension funds, large endowment funds, foundations--today dominate market transactions, accounting for over 70 percent of the dollar value of New York Stock Exchange trading, compared with 35 percent in 1963. . . .
>
> . . . "In the name of playing safe with their clients' money, large institutional investors have been concentrating their activity in an ever-narrowing circle of investment choices," says James Needham, Chairman of the New York Stock Exchange. . . .
>
> . . . According to an article in Business Week . . . the 10 leading institutional investors are as follows:
>
> The Leading Institutional Investors:
> Most of the top 10 are banks

Institution	Investment Portfolios (billions of dollars)*
Morgan Guaranty Trust	$27.2
Bankers Trust	19.9
Prudential Insurance	18.3
First National City Bank	17.2
U.S. Trust of New York	17.0
Metropolitan Life Insurance	16.5
Manufacturers Hanover Trust	10.9
Mellon National Bank & Trust	10.5
Investors Diversified Services	9.7
Chase Manhattan Bank	9.2

*Excludes real estate investments.

> These 10 institutional investors hold $156.4 billion in their portfolios. Chairman Paul Kolton of the American Stock Exchange estimates that total equity holdings of financial institutions today

> are $310 billion, with banks holding $170 billion, mutual funds $45 billion, insurance companies $42 billion, and with foundations investment counsellors and smaller institutions holding the rest. This $310 billion--36 percent of the total amount outstanding ($1,160 billion) is disproportionately concentrated in the small companies. Thus, there has been created a "two-tier" market. . . .[19]

A two-tier market in energy supply ventures is more than a mere possibility. The well-established positions of petroleum companies, and the large capital requirements and risks associated with entrants into alternative energy sources, with outsiders dependent on bank loans (with equity markets limited) or loans from major companies,[20] seem to point toward such a situation.[21]

Entrant firms are faced with having to go to banks having an interest in the well-being of their competitors, and, perhaps more importantly, through interlocks and the course of business over time, a shared viewpoint as to what is and is not feasible.

NOTES

1. Cossey, Financing Oil and Gas Exploration--Past, Present, and Future, Southwest Institute (1973).

2. Lease financing is generally obtained from commercial bank-affiliates. Samuel L. Shapiro, Equipment Leasing (New York: Proctrung Law Institute, 1973), p. 184.

3. That is, rental paid under an arrangement where a bank or banks acquire and hold title to a property and lease it to a petroleum company. A concise description of personal-property lease financing from the banker's point of view appears in Bank Administration Institute, Bank Administration Manual (1970), pp. 516-19.

4. Joe Zeppa, "The Drilling Contracting Business," in Oil & Gas Operations: Legal Considerations in the Tidelands and on Land, ed. Ralph Slovenka (Baton Rouge, La.: Claitor's Book Store, 1973), reports the importance of bank loans to small exploration companies.

5. Institutional Investor Study, H.R. Doc. No. 92-64, 92d Cong., 1 Sess. (1969).

6. S.E.C. Statistical Bulletin 31, no. 3, pp. 18, 19; and see, regarding sales to institutions, Irwin Friend, Investment Banking and the New Issue Market (Cleveland: World, 1967), p. 337; Note on Private Placements, 59 U. Va. L. Rev. 886 (1973).

7. Traditionally, the financing of small drilling contractors who drill incremental wells has been based on internal cash flow, bank loans, or farm-outs. See, for example, Zeppa, op. cit. These firms look to both banks and large oil companies for financing. Zeppa also relates how farm-outs are needed by independent oil operators who can drill but lack the resources to bid for leases or to do seismic or other exploratory work. Majors benefit by farming out properties from vast land inventories, enabling evaluation of these domestic holdings while they drill overseas. The ability to finance drilling through farm-outs is hindered when independents have difficulties getting drilling equipment.

8. Tax Code provisions for depletion allowances and deduction as expenses of intangible drilling costs do not provide initial funding for a project or firm. The allowance for depletion (26 USC 3613) refers to a depletion deduction from "gross income from property," which means income from the sale of oil and gas actually extracted therefrom. Big Four Oil & Gas Co. v. United States, 118 F. Supp. 958 (W.D. Pa., 1954); I.R.S., Regs. §1.613.3 (a). Similarly, the option to expense intangible drilling costs (26 USC §263 [c]), may be exercised as to ". . . expenditures made by an operator for wages, fuel, repairs, hauling, supplies, etc., incident to and necessary for the drilling of wells and the preparation of wells for the production of oil or gas." [I.R.S., Regs. §1.612-4 (a)].

9. See, for example, Calvert Exploration Company Annual Report for 1973.

10. See, for example, James C. Van Horn, Financial Management and Policy, 334 et seq. (Englewood Cliffs: Prentice-Hall, 1968).

11. Testimony of Wallace Wilson, U.S. Congress, Senate, Judiciary Committee, Hearings on the Industrial Reorganization Act: Part 9--The Energy Industry, 94th Cong. 1 Sess. (1975).

12. Austin, A New Antitrust Problem: Vertical Integration in Correspondent Banking, 122 U. Pa. L. Rev. 366 (1973).

13. Ibid., p. 376.

14. Ibid., pp. 368-69.

15. U.S. Congress, House, Committee on Banking and Currency, Subcommittee on Domestic Finance, Correspondent Relations: A Survey of Banker Opinion, 88th Cong., 2 Sess. (1964).

16. A Report on the Correspondent Banking System, U.S. House of Representatives, Banking Committee, 88th Cong., 2 Sess., (1964).

17. Cf. Austin, op. cit., pp. 387-88, citing Phillips, "Competition, Confusion and Commercial Banking," Journal of Finance, March 1964, p. 32; and Solomon, "Bank Merger Policy and Problems: A Linkage Theory of Oligopoly," Journal of Money, Credit & Banking, August 1970, pp. 323, 331.

18. Stanley P. Porter, Petroleum Accounting Practices (New York: McGraw-Hill, 1965), pp. 73-74, 188-95.

19. U.S. Congress, Senate, Finance Committee, Report on the Role of Institutional Investors in the Stock Market, 93d Cong., 1 Sess.

20. For example, advance payments or drilling partnerships; see "New Drilling-Money Spout Opening for Independents," Oil and Gas Journal (September 2, 1974), p. 15.

21. Cf. The Role of Institutional Investors in the Stock Market, pp. 4-5.

APPENDIX D
PRIMARY AND EXCLUSIVE JURISDICTION, STATE ACTION, AND FOREIGN STATE ACTION DEFENSES

PRIMARY AND EXCLUSIVE JURISDICTION

Judicial enforcement of antitrust laws in regulated sectors of the economy involves some particular problems. In regulated sectors of the economy, including the electric utility industry, attempts to enforce antitrust policy in the courts raise questions as to whether a matter should be tried there or before a regulatory agency. The courts in these cases are concerned about the proper court-agency division of responsibility for making decisions.

The allocation of decisional responsibility in cases where like matters come, or could come, before both courts and agencies has spawned a considerable volume of law. This law, under the respective headings of primary and exclusive jurisdiction, seeks to allocate decisional functions by determining either which tribunal may hear a matter, or which tribunal must hear all or part of a matter first before it may be taken up in a second forum. Because of the costs involved, and the legal restraints and proclivities of different forums, decisions determining jurisdiction to hear a matter may determine its eventual outcome--or even whether it will be heard at all.

Like most legal doctrines, the doctrine of primary and exclusive jurisdiction has been developed through a series of judicial decisions seeking to balance conflicting needs and interests so as to reconcile several statutory schemes. A priori rules are sought through a series of ad hoc and post hoc determinations. The problem of overlapping jurisdiction among federal agencies has usually been treated as a political problem in the absence of conflicting agency actions. However, recently the FTC was held to be collaterally estopped from seeking to investigate data relating to trade-association estimated proved reserves because the FPC, according to the court, had determined that the data was reliable. FTC v. Texaco, F.2d, CADC No. 74-1551 (August 8, 1975). Until recently, the lessons of these decisions for future matters have been less than clear.

Exclusive jurisdiction defenses to antitrust complaints can be maintained in situations in which only a federal administrative agency has authority; and having acted, its actions can be subjected to judicial review only by direct appeal, not by collateral antitrust action. See Gordon v. New York Stock Exchange, 422 US 659 (1975); U.S. v. National Assn. Securities Dealers, 422 US 694 (1975); Hughes Tool

Co. v. T.W.A., 409 US 363 (1973); Pan American Airways v. United States, 371 US 296 (1963); and William Schwarzer, "Regulated Industries and the Antitrust Laws--An Overview," 41 I.C.C. Prac. J. 543 (1974).*

These cases hold that the antitrust laws are repealed by implication but only when and to the extent necessary to permit schemes of regulatory acts to work. See Silver v. New York Stock Exchange, 373 US 341 (1963). Courts look to find either a pervasive regulatory scheme, Gordon, previously cited, p. 689, Otter Tail Power Co. v. U.S., 410 US 366, 373-75 (1973), or a specific statutory provision and regulation thereunder. See Assn. Securities Dealers, previously cited; and cf. FMC v. Seatrain Lines, Inc., 411 US 726 (1973). The statutory provisions must be plainly repugnant to antitrust provisions (U.S. v. Philadelphia National Bank, 374 US 321 (1963); Gordon, previously cited, p. 682; cf. Georgia v. Pennsylvania R. Co., 324 US 439 (1945), or the regulation pervasive, National Assn. Securities Dealers, previously cited.

With the decision in Otter Tail, exclusive jurisdiction problems are unlikely to arise on the basis of assertions of pervasive regulations of electric utilities.† However, the issue could still be raised in regard to wholesale rates filed with the FPC. In that situation, the court's limited retreat in the National Association of Securities Dealers case from its requirement of active regulatory supervision (not just statutory authority) voiced in Gordon and Seatrain, could prove difficult for plaintiffs.

Where a regulatory statute explicitly provides an antitrust immunity flowing from agency action, prior agency approval is usually required for immunization from antitrust prohibitions. See Clayton Act, 15 USC §18 (1970); Coultas, "The Doctrine of Primary

---

*In connection with judicial review, two relevant and much-discussed doctrines are "ripeness" and "exhaustion of remedies." The doctrine of ripeness means simply that to obtain judicial review of administrative action, the effect of that action on the party seeking review must be sufficiently concrete, and the question sufficiently clear for judicial review. Abbott Laboratories, Inc. v. Gardner 387 US 136; Toilet Goods Ass'n v. Gardner, 387 US 158; and Gardner v. Toilet Goods Ass'n, 387 US 167; Bantam Books v. Sullivan, 372 US 58 (1963). The exhaustion-of-remedies doctrine performs a similar "traffic cop" function for judicial caseloads, and by requiring a full use of existing administrative remedies before judicial appeal, reflects a priori congressional allocations of workload. McKart v. United States, 395 US 185 (1969).

†A district court has rejected this defense in an oil pricing case in which pervasive FEA regulations were asserted.

Jurisdiction: Determination of Express and Implied Immunity from the Antitrust Laws," J. Air L. 39 (1973), p. 559; Gordon, previously cited; and Keogh v. Chicago & Northwestern Ry., 260 US 156 (1922).

Antitrust and other issues before courts sometimes involve issues of fact also relevant to matter before federal administrative agencies. The allocation of decisional functions between courts and agencies is treated generally under the doctrine of primary jurisdiction.

> "Primary jurisdiction . . . applies where a claim is originally cognizable in the courts, and comes into play whenever enforcement of the claim requires the resolution of issues which, under a regulatory scheme, have been placed within the special competence of an administrative body; in such a case the judicial process is suspended pending referral of such issues to the administrative body for its views." United States v. Western Pacific Railroad Co., 352 US 59, 64 (1956).

The doctrine of primary jurisdiction has been stated in United States v. Radio Corporation of America, 358 US 334 (1959), to have originated in the case of Texas & Pacific R. Co. v. Abilene Cotton Oil Co., 204 US 426, and to be grounded on the necessity for administrative uniformity, and according to Justice Brandeis in Great Northern R. Co. v. Merchants Elevator Co., 259 US 285, 291, in the need for administrative skill "commonly to be found only in a body of experts" in handling intricate facts. Also see, Far East Conference v. United States, 342 US 570, 574-75 (1952).*

No fixed formula exists for application of the doctrine of primary jurisdiction. In every case the question is whether the reasons for the existence of the doctrine are present and whether the purposes it serves will be aided by its application in the particular litigation. United States v. Western Pacific Railroad Co., 352 US 59 (1956).

The basis of the doctrine in the argument for uniformity overlooks the fact that administrative agencies are, unlike courts, not bound by their prior decisional law. As to the argument of agency

---

*A good historical discussion of the doctrine of primary jurisdiction is found in Convisser, Primary Jurisdiction: The Rule and its Rationalizations, 65 Yale L.J. 315 (1956). This article generally casts doubt on the validity of the doctrine as well as its practical value.

expertise, it should be noted that the Justice Department and the FTC are the expert agencies on antitrust matters (and are primarily prosecutorial rather than regulatory bodies), while the courts are the governmental bodies experienced in deciding such cases. California v. F.P.C., 369 US 482 (1962).

Davis, in his Administrative Law Treatise, states:

> At the heart of the problem of primary jurisdiction in antitrust cases lies the choice of substantive policy in favor of either (1) enforced competition, (2) regulation, or (3) some mixture of enforced competition and regulation. In the fields for which Congress has provided a considerable amount of regulation, making this choice in particular contexts is often difficult.[1]

However, the literature appears to use (or seek) a priori approaches to the doctrine, which is considered to be a confused one. See, for example, Kestenbaum, Primary Jurisdiction to Decide Antitrust Jurisdiction: A Practical Approach to the Allocation of Functions, 55 Geo. L.J. 812 (1967).

While noting the notion of Federal Maritime Board v. Isbrandtsen Co., 350 US 481, 498 (1958), that practical considerations may dictate a division of functions between courts and agencies, under which the latter make a preliminary, comprehensive investigation of all the facts, analyze them, and apply them to the statutory scheme as it is construed, the literature does not discuss policy considerations of divisions of labor but turns largely on whether administrative jurisdiction exists. See, for example, Coultas, previously cited, and matter therein cited.

> What is at stake in terms of the primary jurisdiction doctrine is that the resolution of the question of which law applies be properly allocated between the two tribunals, so that the location of the interface between the antitrust and regulatory statutes be decided in the most efficient, orderly, and equitable manner. Less a matter of substantive law than of judicial administration, procedure, and esthetics, it is yet of compelling importance to court, agency, and litigants.[2]

Also see Kestenbaum, previously cited, pp. 812, 814.

As noted by Kestenbaum, much of the writing in this area concerns the location of the boundary between antitrust jurisdiction and antitrust immunity in specific cases under specific statutes. Confused merging of the question of jurisdiction with repeal of antitrust laws is noted--for example, as arising in cases where the court, after finding jurisdiction in the courts, proceeds to discuss primary jurisdiction, even though recourse to an agency lacking jurisdiction would be futile. An agency should not be asked to decide issues unless the issues are material to the question of antitrust immunity and unless the agency can contribute something substantial to their resolution. See Otter Tail, previously cited (lack of FPC jurisdiction); Jaffe, Primary Jurisdiction, 77 Harv. L. Rev. 1037, 1043-47 (1964); cf. Ricci v. Chicago Mercantile Exchange, 409 US 289 (1973). In Ricci, a proceeding under Section 1 of the Sherman Act, 15 USC §1, was stayed pending administrative proceedings deemed available under the Commodity Exchange Act, 7 USC §1 et seq. Administrative proceedings were pending, and their institution was a matter of administrative discretion. Upon a proceeding, the agency could order that the complained-of action be ceased.

The complaint alleged that plaintiff was deprived of his membership on a commodity exchange contrary to the rules of the exchange, the Commodity Exchange Act, and pursuant to an anticompetitive conspiracy. "The problem to which the Court of Appeals addressed itself is recurring. It arises when conduct seemingly within the reach of the antitrust laws is also at least agreeably protected by another regulatory statute . . . (footnote omitted)."[3]

The Ricci court held that a stay was appropriate when three premises are found to pertain: namely, the regulatory scheme is somewhat incompatible with the maintenance of an antitrust action; some facet of the dispute in the case is within the agencies' jurisdiction; and adjudication of the dispute by the agency promises to be of material aid in resolving the immunity question.*

It was held that the issue of whether the action complained-of was done pursuant to a valid rule must be first addressed before the question of implied immunity can be reached (409 US 303-04). Furthermore, the court sought to obtain the assistance of the agency's superiority in gathering the relevant facts and in marshalling them

---

*The Ricci Court noted that there was no congressional intent to confer general antitrust immunity in the relevant area; the act contains no categorical exemption, and the administrative authority is not particularly focused on competitive considerations. Cf., California v. FPC, previously cited, p. 302, fn. 13.

into a meaningful pattern--citing Fed. Maritime Bd. v. Isbrandtsen Co., 356 US 481 (1956).

After Ricci, the law appears to remain that where the activity challenged is not "arguably lawful" under a regulatory scheme, reference of issues to an agency is not appropriate. Carnation Co. v. Pacific Westbound Conf., 383 US 213 (1966); Kestenbaum article, previously cited, pp. 823-25; Coultas article, previously cited, fn. 54.

In spite of the delays occasioned by occasional references to agencies, the case law does reflect an intention to require antitrust aspects to be considered, either by refusing to allow courts to be ousted of jurisdiction (California v. F.P.C., 369 US 482 [1962]; United States v. Philadelphia National Bank, 374 US 321, 350-351 [1963]; Seatrain Lines, Inc. v. F.M.C., 411 US 726 [1973]), or by requiring agencies to consider antitrust aspects. See, for example, Gulf States Utilities Co. v. F.P.C., 411 US 747 (1973); F.M.C. v. Svenska Amerika Linien, 390 US 238, 245 (1968); Conway, previously cited; Northern Natural Gas Co. v. F.P.C., 399 F.2d 953 (CADC, 1968);* Marine Space Enclosures, Inc. v. F.M.C., 420 F.2d 577 (CADC, 1969), Stickells, Antitrust Laws (1972), at §527-35.

In "Judicial Doctrine of Primary Jurisdiction as Applied to Antitrust Suits," Staff Report to Subcommittee No. 5, House Judiciary Committee (1957), the authors call for an end to the use of the doctrine of primary jurisdiction in cases where the United States sues "in vindication of policies whose enforcement Congress has expressly reposed in the Attorney General through resort to the courts. The sole reliable counterweight to industry-mindedness of specialized agencies is provided by this mandate to the unspecialized law enforcement agency."[4] And cf. Schwartz, Legal Restriction of Competition in the Regulated Industries: An Abdication of Judicial Responsibility, 67 Harv. L. Rev. 436, 464-71 (1954).

Several legislative measures attempt to improve agency antitrust policy enforcement by mandating a role for the Department of Justice in licensing or merger proceedings. See, for example, Atomic Energy Act §105 [42 USC 2135 (c)], and the Bank Merger Act, 12 USC §1828.†

---

*This case also highlights the need for coordination between the SEC and the FPC.

†The Bank Merger Act provides for de novo antitrust review by district courts at the behest of the Justice Department after approval of mergers by the Federal Reserve Board. 12 USC 1879. Cf. U.S. v. Citizens National Bank, 422 US 86, 45 L. Ed. 2d 41 (1975).

Aggrieved parties having access to federal courts are far more likely to achieve redress of antitrust grievances than parties confined to administrative agencies. As noted by the four dissenting justices in Ricci, access to be effectual must be timely. The expense and delay of a prior administrative proceeding are more than many persons and enterprises can bear. The anticompetitive tendencies of administrative agencies are well documented. See U.S. Congress, Senate, Judiciary Committee, Report on the Competition Improvements Act of 1976, S.2028, Report No. 94-1045, 94th Cong., 2 Sess. (1976).

The rule of California v. FPC, that primary responsibility for trying antitrust cases is in the courts, is essential if remedies are to be practically afforded.

STATE ACTION DEFENSE

Antitrust claims are sometimes defended on the grounds that the action complained of was taken pursuant to the mandate of state law. This defense reflects an effort by the courts to reconcile apparent conflicts between federal antitrust law and state regulation. It serves a decision-allocating role, and is parallel in that respect to the doctrine of primary jurisdiction. As an attempt to balance state and federal law it is anomalous to the general supremacy of federal enactments in areas of conflict with state law. This anomaly is derived from the legislative intent of the Sherman Act.

The state action doctrine has been raised against attacks on anticompetitive practices for which tariffs were filed. See, Cantor v. The Detroit Edison Company, 428 US 579 (1976); Gas Light Co. v. Georgia Power Co., 440 F.2d 1135 (CA 5, 1971), cert. denied, 404 US 1062 (1972); and, Washington Gas Light Co. v. Virginia Electric & Power Co., 438 F.2d 248 (CA 4, 1971); and see Mazzola v. Southern New England Telephone Co., 1975 Trade Cases 66926 (Conn. Supr. Ct. August 19, 1975) (and telephone cases cited therein).

The doctrine which traces its origin to the case of Parker v. Brown, 317 US 341 (1942), has been subjected to narrowing in recent cases. It is applicable only insofar as the action taken was mandated by action of a state in its sovereign capacity, and not merely prompted. Goldfarb v. Virginia State Bar, 421 US 773 (1975). Under the recently decided Cantor case, previously cited, the Supreme Court examined the doctrine.*

---

*Cantor, a retail druggist selling light bulbs, alleged that defendant is using its monopoly power in the distribution of electricity to restrain competition in the sale of bulbs. Defendant distributes

In this case, the court was concerned about the fairness of holding a citizen to obeying a federal law in disobeyance of the command of state law, and about whether Congress intended to supervene state regulation. The first concern was rejected as contrary to the basic effect of antitrust law and precedent even in the event of government participation, Continental Ore Co. v. Union Carbide and Carbon Corp., 370 US 691. The Cantor court noted the active role played by utilities in initiating rate practices. "Nevertheless, there can be no doubt that the option to have, or not to have, such a program is primarily respondent's, not the Commission's."[5] This statement corresponds to the requirement of Goldfarb, previously cited, that more than passive state approval or even prompting is required.

State regulation was held not to, a priori, oust the applicability of antitrust law. It was noted that regulation and antitrust may be consistent. (There is no logical inconsistency.) Even with inconsistency, antitrust laws remain applicable in essentially unregulated areas such as the market for light bulbs, and exemptions are not lightly implied elsewhere.

The standards for determining the existence and scope of an implied exemption because of conflict between state regulation and federal antitrust policy "must be at least as severe as those applied to federal regulatory legislation."[6] The Court has consistently refused to find that regulation gave rise to an implied exemption without first determining that exemption was necessary in order to make the regulatory act work "and even then only to the minimum extent necessary."[7]

The Cantor decision was accompanied by two concurrences, and a dissent by three justices. The plurality opinion, in a section concurred in by the majority, would restrict the state action defense to state officials; one concurring justice stressed the ancillary nature of the practice attacked; one justice in concurring urged that a test be applied in which the benefit of state sanctioned anticompetitive activity would be weighed against its potential harm.

The approach taken in Cantor restricting the state action to cases reconcilable with federal antitrust policy where the two are in conflict* was presaged by several court decisions limiting the

---

light bulbs at no separate charge to its customers, thereby foreclosing a market. The bulbs distribution program had long been in tariffs filed with the Michigan PUC.

*In regard to this approach see Note on Parker v. Brown: A Preemptive Analysis, 84 Yale Law Jour. 1164 (1975) which urges that state action be permitted to stand only if it serves to protect

doctrine. In Woods Exploration & Production Co. v. Aluminum Co. of America, 438 F.2d 1286 (CA 5, 1971), cert. denied, 404 US 1047 (1972) it was held that regulatory decisions based on false information supplied by defendants create no defense. In a series of cases it was held that the defense may not stand in the absence of a clear statement by a state of its intention to preclude competition. Traveler's Ins. Co. v. Blue Cross, 298 F. Supp. 1109 (W.D. Pa. 1969); Geo. R. Whitten, Jr., Inc. v. Paddock Pool Builders, Inc., 424 F.2d 25 (CA 1), cert. denied, 400 US 850 (1970); and, cf., Hecht v. Pro-Football, Inc., 444 F.2d 931 (CADC, 1971), cert. denied, 404 US 1047 (1972).* In International Telephone and Telegraph v. G.T. & Elec., 351 F.Supp. 1153, 1202 (D. Hawaii, 1972) prior commission approval of an ownership structures was held not dispositive when antitrust issues were not addressed. The defense of prior state action is available only when the state regulatory program is not contrary to constitutional limits on state action. U.S. Const. Art. VI, Cl. 2.

This analysis consists of first identifying the field in which the challenged state action operates or has its effects and determining whether there is a federal policy in that field.† Rice v. Santa Fe Elev. Corp., 331 US 218 (1947). Second, the court must determine whether the federal policy was intended to be exclusive [see Hines v. Davidowitz, 312 US 52 (1941)]; or whether national conformity is necessary [see, for example, Burbank v. Lockheed Air Terminal, Inc., 411 US 624 (1973); Perez v. Campbell, 402 US 637 (1971); Huron Portland Cement Co. v. Detroit, 362 US 440 (1960); Rice, previously cited; Hines, previously cited]; or whether federal regulation of the area is pervasive.‡

---

the interests forwarded by antitrust policy; low prices, efficient resource allocation, and protection of small firms. Also see Posner, The Proper Relationship Between State Regulation and the Federal Antitrust Laws, 49 NYU L. Rev. 693 (1974).

*This requirement of a clear statement was criticized in a concurring opinion by Justice Blackman in Cantor who suggested that it neglects situations where a statement was deemed unnecessary due to obvious need.

†As explained in Note on Parker v. Brown, previously cited.

‡See Note on Federal Preemption of State Laws; The Effect of Regulatory Agency Attitudes on Judicial Decision Making, 50 Ind. Law Jour. 848 (1975). It is suggested that the attitude of the agency having jurisdiction in the field in question heavily influences the judicial outcome of preemption decisions.

If an area is not totally preempted, the court then determines whether the state program stands as an obstacle to the attainment of a congressional objective. Fla. Lime & Avocado Growers, Inc. v. Paul, 373 US 132, 141 (1962). Thus constitutional preemption encompasses both areas essentially federal, and areas subject to federal regulation.* However, where federal antitrust policy has been construed as not ousting state regulatory jurisdiction (Cantor, cited previously), the two legislative schemes must be conformed or accommodated to one another.

In electric utility matters, federal antitrust and regulatory policy apply. Federal regulatory policy clearly perceives a role for state regulation. The Public Utility Act provided for joint federal-state regulatory proceedings [Section 209, Federal Power Act, 16 USC 824 (h)], and gave the states a major role in determining the structure of utility systems primarily operating within one state. However, this 1935 statute setting out basic federal utility regulatory policy, written when interstate power pooling was far rarer than today, clearly foresees interstate commerce operations governed by federal law,† under a national policy of free and independent competition. See 15 USC §79 (a)(i); SEC v. New England Electric System, 384 US 176 (1966); Gulf States Power Co., previously cited.

The 1935 act does not provide for bulk power supply franchising. Federal power policy, embodied in earlier and later legislation pertaining to federal power marketing and nuclear licensing, recognizes that the aims of competition and regulation can be mutually supportive. It has been specifically held that territorial allocations of bulk power supply trade are contrary to the Sherman Act.‡

Litton Systems, Inc. v. Southwestern Bell Telephone Co., 1976--2 Trade Cases para. 61,084 (CA 5, September 23, 1976) is a recent example of induced efforts to accommodate state regulatory and antitrust policy. There the doctrine of primary jurisdiction

---

*State regulation can also be preempted by federal procurement policy. United States v. Ga. Public Service Comm., 371 US 2851 (1963).

†The Federal Power Act, part of the 1935 enactment, was preceded by the Supreme Court's decision in Public Utilities Commission v. Attleboro Steam & Electric Co., 273 US 83 (1927) that a state attempt to regulate the sale of electricity from a Massachusetts plant to a Rhode Island utility imposed an unconstitutional direct burden on interstate commerce.

‡Pennsylvania Water & Power Co. v. Consolidated Gas, Elec. Light & Power Co., 184 F.2d 552 (Ca 4, 1950).

was held to be inapplicable, and prior reference to state agencies was held to be improper in a case involving allegations of unlawful tying and predatory pricing of a package of telephone branch exchange equipment and telephone service. The court noted that the case did not involve a situation in which the suit ought to have been prosecuted exclusively or initially before an administrative body, and as regards the latter point it noted that the case did not present the need, found in Ricci, to accommodate federal antitrust policy with federal regulatory policy. The court, however, did not end its consideration of the propriety of requiring prior reference to state agencies at this point. Rather, it went on to consider the Ricci tests for prior agency reference in light of the possibility that the doctrine of Parker v. Brown might require deference to state agencies.

To determine whether the purpose served by the doctrine of primary jurisdiction would be aided by its application [see United States v. Western P.R.R., 352 US 59 (1956)], the court inquired as to whether state law was incompatible with the maintenance of an antitrust action, whether some facets of the matters in dispute are within state regulatory jurisdiction, and whether adjudication of such disputes promises to be of material aid in resolving the immunity question.

The court found that no useful purpose would be served by requiring prior reference to state agencies. This is the case because it is Bell's conduct that is being challenged, not the conduct or policy of any state agency or official. The system being attacked was devised by Bell which then garnered nearly automatic approval from the state agencies. Thus a failure to grant immunity from antitrust actions would not be unfair to Bell [Cantor v. The Detroit Edison Company, 428 US 579 (1976)], since its tying activities were not the product of regulatory coercion, and secondly, since ending the tying relationships would not undermine state regulation (and therefore inconsistent state and Sherman Act regulation).

The state regulatory agencies, by their silence with respect to the issue before the court, "have 'spoken' once, and the laudable goal of 'judicial accommodation' would become a nightmare of judicial paralysis were we now to force the agencies to suggest reasons why the tying . . . may be necessary to the regulation of that service."[8]

The court found that the district court would not be incapacitated in the absence of the assistance of state expertise. In regard to the possible need to implement the district court's decree through the filing of tariffs with state agencies, the court found no significant problem:

> The district court's order here could simply direct the defendant to file satisfactory tariffs with the state commissions and retain jurisdiction pending the agencies' consideration of the new tariffs. If the tariffs without the tying provisions are rejected, the district court would then be able to decide anew whether implied antitrust immunity exists. State officers enforcing the regulatory schemes could be joined as parties-defendant at this point in the proceedings.[9]

The goals of the federal antitrust laws are to encourage efficiency and low prices and to hold economic power in check. Where available, competition works better than regulation to achieve these goals because it operates swiftly, because it rewards skills and innovation, and because it returns profits to those who produce what the public wants and buys. While competition is not a spur in the relatively few "natural monopolies," there is still a very large part of the regulated sector in which competition can serve as an effective spur to efficiency and innovation. See Baker, Competition and Regulation: Charles River Bridge Recrossed, 60 Cornell L. Rev. 159 (1975). The state action doctrine, accordingly, should be limited to matters in which the action is complementary to the procompetitive policy of federal antitrust and utility law. Restraints on trade should not be permitted where alternative means exist to achieve a goal. The Cantor decisions, while advancing antitrust law from a state of confusion in which one court indicated that filed tariffs, per se, confer immunity (Washington Gas Light, previously cited), leave important questions unanswered. The outcome of disputes regarding limitations on state-filed tariffs restricting industrial self-generation or permitting anticompetitive rate levels and rate structures is unclear. Disputes may also arise in the context of state certifications to build bulk power facilities.

The defense of prior state action does not seem appropriate in proceedings regarding planning and access to bulk power supply facilities. Utilities operate as part of an interstate flow of electricity and other trade. They function best as part of groups whose plans and operations do not stop at state lines. The economic regulation of a multistate grid is beyond the capability of state agencies.

There is no unfairness due to retroactivity in requiring that these prospective actions be carried out in a competitive manner. There is no reason why costs of related service could not be recovered, protecting utility investment. If such services are priced at their costs, and customers can choose between buying or generating, society is better off. Avoiding or reducing load growth no longer presents a cream-skimming problem:* customers benefit from reduced construction budgets and expedited development of innovative, small-source technology. The growth of plant sizes, if affected, will not be likely to harm customers since economies of scale are questionable beyond current unit sizes which are being replicated, not increased. Furthermore, if unit power sales are offered in competition with geothermal energy, economic growth of unit sizes will not be adversely affected. Customers may expect competition for service to large loads to hold down general operating costs.

State action restricting sales to retail customers, except in certified or franchised service areas, would present a barrier to any putative new power supplier. Restricting bulk sales--for example, to industrial or large commercial users--burdens the interstate commerce in which high voltage lines operate, and does not serve a valid purpose; protecting loads of gas- and oil-fired systems from geothermal competition would not be in the public interest. Geothermal entrants could mount a serious challenge against such trade restriction.

DEFENSE OF FOREIGN STATE ACTION

Antitrust jurisdiction is asserted over the subject matter where the domestic or foreign commerce of the United States is substantially affected. The farthest statement of the point is found in the Alcoa case. "It is settled law that any state may impose liabilities, even upon persons not within its allegiance, for conduct outside its borders that has consequences within its borders which the state apprehends."[10]

Actions attacked may occur outside of the United States [Continental Ore Co. v. Union Carbide and Carbon Corp., 370 US 69, 1962 Trade Cases 70362 (1962)]; may involve foreign as well as domestic firms or associations [United States v. Watchmakers of Switzerland Information Center, 63 Trade Cases 70600 (D.C.N.Y., 1962); OCCF, FTC Docket 6106 (exclusive supply contract between domestic scrap dealers and office for European steel mills)]; and may be entered into here [Timken Roller Bearing Co. v. United States, 341 US 593, 1950-51 Trade Cases 62837 (1951)]; or abroad

---

*As noted by Baker, previously cited, one class of service should not be called upon to subsidize another. Competition increases the availability of opportunities for joint action to reduce costs, that is, open pooling.

[Hazeltine Research, Inc. v. Zenith Corp., 239 F. Supp. 51 (N.D., Ill., 1965), 65 Trade Cases 713 55; rev'd on other grounds, 388 F.2d 25 (CA 7, 1967), 1967 Trade Cases 72310, rev'd 395 US 100, 1969 Trade Cases 72800 (1969); vacated 418 F.2d 212, 1969 Trade Cases 72849 (CA 7, 1969)].

Most cases have involved restrictions on exports (Hazeltine, previously cited) and their marketing [United States v. Minnesota Mining and Manufacturing Co., 92 F. Supp. 942, 1950-51 Trade Cases 62687 (D. Mass., 1950); United States v. Gulf Oil Co., 1960 Trade Case, 69851 (D.C.N.Y., 1950); and, United States v. Anthracite Export Ass'n., 1970 Trade Cases 73348 (D.C. P.A. 1970)]. Others have dealt with restraints on transportation [United States v. Pacific and Arctic Railway and Navigation Co., 228 US 87 (1913); and, Pacific Seafarers, Inc. v. Pacific Far East Line, Inc., 404 F.2d 804 (D.C. Cir., 1968), cert. denied, 393 US 1093 (1969)]. The courts have been more likely to find an effect on U.S. domestic or foreign commerce if a U.S. firm is involved [Wilbur L. Fugate, Foreign Commerce and the Antitrust Laws (Boston: Little, Brown, rev. ed., 1973)].*

A foreign company may be a party to a restraint of trade by a U.S. company by virtue of its contractual relationships with other U.S. firms, where the foreign company knew or should have known that its activities were a substantial contribution to an illegal plan in U.S. markets and that its activities had a direct and substantial effect upon trade [United States v. General Electric Co., 82 F. Supp. 753 (D.C.N.J., 1949) 1948-49 Trade Cases 62353].

Personal Jurisdiction

For a court to have jurisdiction over a person, that person must be amenable to service and service must in fact be made. A foreign firm is amenable to service if it is carrying on business of any substantial character in one of the judicial districts into which the United States is divided [United States v. Scophony Corp., 333 US 795 (1948). Venue lies in any district, 28 USC 1391 (d); Brunette Machine Works, Ltd. v. Kockum Industries, Inc., 406 US 706 (1972)]. If found in this country, a defendant may be served at its home office abroad [International Ford Tractor Sales Co. v. Massey-Ferguson, Ltd., 210 F. Supp. 930, 939 (D. Utah, 1962), aff'd per curiam, 325 F.2d 713 (CA 10, 1963); Fed. Rules of Civil Procedure 4(i)].

---

*A foreign firm needs only a general intent to act so as to affect U.S. commerce, if effects occur.

Special Defenses

In foreign trade matters, special problems arise in regard to participation by governments in business ventures and in regard to conflicting mandates of foreign law.

A foreign sovereign is generally immune from suit, without its consent, in U.S. courts [Banco Nacional de Cuba v. Sabbatino, 376 US 398 (1964) (Act of State doctrine)]. Generally, where a foreign government participates in a business venture on a commercial basis the defense of sovereign immunity does not apply [United States v. Deutshes Kalisyndikat Gesellschaft, 31 F.2d 199 (S.D.N.Y., 1929); In re Grand Jury Investigation of the Shipping Industry, 186 F. Supp. 298 (D.D.C., 1960)]. An exception to this general rule may be found when a foreign government participates in a commercial venture for national security purposes [In re Grand Jury Investigation of World Arrangements with Relation to Production, Transp. Ref., and Distrib. of Petroleum, 13 F.R.D. 280 (D.D.C., 1952)].*

In the event that a complained of act involves the action and motives of a foreign government acting in its sovereign capacity in its country, U.S. courts will not hear the case [Occidental Petroleum Corp. v. Buttes Gas and Oil Co., 1971 Trade Cases 73525, 331 F. Supp. 92 (C.D., Ca. 1971), aff'd per curiam, 461 F.2d 1261 (CA, 9), cert. denied, 409 US 950 (1972); and, Hunt v. Mobil Oil Corp., 1975-2 Trade Cases 60591 (S.D.N.Y., 1975)].

Thus portions of complaints dealing with government actions regarding international boundaries and petroleum concessions have been dismissed (Hunt, previously cited).† The related actions of private firms giving rise to contractual disputes or to other restraints of trade remain actionable.

Compulsion by a foreign government of a locally incorporated subsidiary constitutes a defense [Interamerican Refining Corp. v. Texaco Maracaibo, Inc., 307 F. Supp. 1291 (D. Del., 1970)]. Likewise, a decree will be enforced only as regards foreign matters to the extent permitted in loci forii [United States v. Imperial

---

*Subpoena quashed when Anglo-Iranian Oil Company asserted it had been ordered by British government not to produce documents not located in United States and not related to business transacted in United States. When the successor BP acquired control over SOHIO, the U.S. government resisted the mergers and a settlement requiring partial divestiture was made. United States v. Standard Oil Co., 1970 Trade Cases 72988 (N.D., Ohio, 1970).

†The parts of the complaint pertaining to a sharing and sales agreement among Libyan producer-concessioners was not dismissed.

Chemical Industries, Ltd., 105 F. Supp. 215 (D.C.N.Y., 1952) 1952 Trade Cases 67282; United States v. General Electric, 115 F. Supp. 835 (D.C.N.Y., 1953) 1953 Trade Cases 67576; and United States v. Watchmakers of Switzerland Information Center, Inc., 1965 Trade Cases 71352 (S.D.N.Y., 1965) and 1965 T.C. 80491].

However, agreements made by a U.S. firm with foreign firms to restrict imports to the United States are not protected by the authorization or acquiescence of a foreign government [United States v. R.P. Oldham Co., 152 F. Supp. 818 (N.D., Ca., 1957) 1957 Trade Cases 68790].* Similarly, the delegation of discretionary power by a foreign government is not a defense [Continental Ore Co. v. Union Carbide and Carbon Corp., 370 US 690 (1962)]. Even in the event of actions taken pursuant to foreign government direction, actions taken in the commerce of the United States are not immune [Sabre Shipping Corp. v. American President Lines, Ltd., 285 F. Supp. 949 (S.D.N.Y., 1968), 1968 Trade Cases 72493].†

Joint Ventures and Mergers

Under the U.S. antitrust law, mergers tending to substantially lessen competition are prohibited. These prohibitions apply to acquisitions involving foreign firms as acquiring or acquired parties [United States v. Standard Oil (Ohio), 1970 Trade Cases 72988 (N.D. Ohio, 1969) (consent decree on British Petroleum acquisition of control of Sohio); United States v. Asiatic Petroleum Corp., 1971 Trade Cases 73689 (D. Mass., 1971) (Royal Dutch Shell Co. subsidiaries acquisition of oil distributor: consent decree); United States v. Schlitz Brewing Co., 253 F. Supp. 129, aff'd, 385 US 375 (1966) (acquisition of Canadian brewer); and In re Litton Indus., Inc., FTC Docket 8778 (April 10, 1968)]. They also apply in the case of mergers of U.S. subsidiaries of foreign firms [U.S. v. CIBA Corp., 1970 Trade Cases 73269 (S.D.N.Y., 1970)].

Joint ventures among competitors or potential competitors have been a subject of concern under American antitrust law [United

---

*Conspiracy in Japan, and lawful there, among five U.S. importers of wire nails, an American subsidiary of a Japanese nail exporter, and a number of Japanese firms.

†The recently enacted Foreign Sovereign Immunities Act, Public Law 94-583, 90 Stat. 2891 limits immunity of foreign governments to commerce activities having direct effects on the United States or carried out in the United States. 28 USC 9605. The act does not change the law regarding actions of nongovernment entities.

States v. Penn-Ohio Chemical Co., 378 US 158, 12 L. Ed. 2d 775 (1964); and United States v. Monsanto Co., 1967 Trade Cases 72001 (D. Pa., 1967) (divestiture ordered in joint venture of Monsanto and Bayer)]. Some joint enterprises have been attacked as market division schemes. Swiss Watchmakers, Timken, Minnesota Mining and Manufacturing, and United States v. Imperial Chemical Industries, all previously cited.

Allegations have been made that U.S. antitrust law, particularly as it pertains to joint ventures, weakens the ability of U.S. firms to trade abroad. The Justice Department which, together with the FTC, is charged with enforcing basic antitrust laws, has denied these allegations. The Justice Department has also questioned the usefulness of the Webb-Pomerance Act which provides a limited exception from antitrust law for associations for foreign trade registered with the FTC, operating according to approved bylaws and not attempting to set prices or fix output levels.

CONCLUSION

The increasing importance of international trade and the substantial involvement of governments in such commerce may be expected gradually to lead to a balancing-of-interests test to determine the appropriate choice of laws. At present sovereign actions of a state within its borders are not attackable in U.S. courts, save for uncompensated expropriations of property, while actions of private firms are, if actions are directed to and have a substantial U.S. impact.*

In this regard, American courts will assert jurisdiction over a firm if, as a practical matter, the firm carries on a business--directly, through an agent, or through a closely directed subsidiary --in the United States. The U.S. government has taken an apparently lenient attitude toward overseas joint ventures. However, joint ventures allocating trade and territories may be prosecuted.

---

*The problem of conflicting foreign law is somewhat paralleled by problems arising when state laws conflict with the procompetitive thrust of federal antitrust law. When state laws restrain trade the courts have held that they are not necessarily preempted by federal antitrust law. The lead case in this regard is Parker v. Brown, 317 U. 5341, 87 L.Ed 315. The ability to raise a state law defense to a complaint grounded in the federal antitrust law has been closely limited in recent cases. Cantor v. Detroit Edison Company, 428 US 579 (1976).

NOTES

1. Kenneth Davis, Administrative Law Treatise (St. Paul, Minn.: West Publishing, 1958).

2. William Schwarzer, Regulated Industries and the Antitrust Law: An Overview, 41 I.C.C. prac. J. 543 (1974).

3. Kestenbaum, Primary Jurisdiction to Decide Antitrust Jurisdiction: A Practical Approach to the Allocation of Functions, 55 Geo. L. J. 229-30 (1967).

4. U.S. Congress, House, Judiciary Committee, Subcommittee No. 5, Staff Report, "Judicial Doctrine of Primary Jurisdiction as Applied to Antitrust Suits," 85th Cong., 1 Sess. (1957).

5. Cantor v. The Detroit Edison Company, 428 US 579 (July 6, 1976).

6. Ibid.

7. Ibid.

8. Ibid.

9. Ibid.

10. United States v. Aluminum Co. of America, 148 F.2d 416 (CA 2, 1945).

INDEX

ABOUT THE AUTHORS

SHELDON L. BIERMAN is a lawyer, consulting in energy and utility matters. He currently has his own firm. Previously he has participated in several energy research projects, and was employed by the federal government in utility regulatory and pollution control work. Mr. Bierman holds a B.S. degree in Mechanical Engineering from the University of Missouri-Rolla, and a J.D. degree from Georgetown University.

WILLIAM J. LAMONT is partner in the law firm of Lobel, Novens and Lamont, specializing in antitrust and energy matters. Mr. Lamont was an attorney in the Antitrust Division, Department of Justice from 1954 to 1972. Prior to that, he was an attorney in the Office of Legal Counsel, the Office of the Solicitor General, and the Office of Alien Property. Mr. Lamont received a B.A. degree and a J.D. degree from the University of Iowa.

DAVID F. STOVER is Assistant General Counsel of the Postal Rate Commission. He has practiced law in Washington, D.C., since 1965, specializing in utility, regulatory and antitrust matters; from 1965 to 1971 he served on the legal staff of the Federal Power Commission. Mr. Stover received his A.B. degree from Princeton University and his J.D. from the University of Pennsylvania.

PAUL A. NELSON is Assistant Professor of Management Science and MSBA Program Director at Michigan Technological University. He has served as Management Analyst at the U.S. Army Management School and as Comptroller Staff Officer for the Comptroller of the Army. He received his B.S. from the University of Wisconsin-Milwaukee and his M.S., M.A., and Ph.D. from the University of Wisconsin-Madison. Dr. Nelson has written on benefit-cost analysis, mathematical programming models, and the electric utility industry.

RELATED TITLES
Published by
Praeger Special Studies

ALTERNATIVE ENERGY STRATEGIES: Constraints and Opportunities
John Hagel III

THE ECONOMICS OF NUCLEAR AND COAL POWER
Saunders Miller

ENERGY USE AND CONSERVATION INCENTIVES: A Study of the Southwestern United States
William H. Cunningham
Sally Cook Lopreato

THE HYDROGEN ENERGY ECONOMY: A Realistic Appraisal of Prospects and Impacts
Edward M. Dickson
John W. Ryan
Marilyn H. Smulyan

THE NUCLEAR POWER DEBATE: Major Moral, Economic, Technical, and Political Issues
Desaix Myers III